Selected Titles in This Series

52 **Sanguthevar Rajasekaran, Panos Pardalos, and D. Frank Hsu, Editors,** Mobile Networks and Computing

51 **Pierre Hansen, Patrick Fowler, and Maolin Zheng, Editors,** Discrete Mathematical Chemistry

50 **James M. Abello and Jeffrey Scott Vitter, Editors,** External Memory Algorithms

49 **Ronald L. Graham, Jan Kratochvíl, Jaroslav Nešetřil, and Fred S. Roberts, Editors,** Contemporary Trends in Discrete Mathematics

48 **Harvey Rubin and David Harlan Wood, Editors,** DNA Based Computers III

47 **Martin Farach-Colton, Fred S. Roberts, Martin Vingron, and Michael Waterman, Editors,** Mathematical Support for Molecular Biology

46 **Peng-Jun Wan, Ding-Zhu Du, and Panos M. Pardalos, Editors,** Multichannel Optical Networks: Theory and Practice

45 **Marios Mavronicolas, Michael Merritt, and Nir Shavit, Editors,** Networks in Distributed Computing

44 **Laura F. Landweber and Eric B. Baum, Editors,** DNA Based Computers II

43 **Panos Pardalos, Sanguthevar Rajasekaran, and José Rolim, Editors,** Randomization Methods in Algorithm Design

42 **Ding-Zhu Du and Frank K. Hwang, Editors,** Advances in Switching Networks

41 **David Aldous and James Propp, Editors,** Microsurveys in Discrete Probability

40 **Panos M. Pardalos and Dingzhu Du, Editors,** Network Design: Connectivity and Facilities Location

39 **Paul W. Beame and Samuel R Buss, Editors,** Proof Complexity and Feasible Arithmetics

38 **Rebecca N. Wright and Peter G. Neumann, Editors,** Network Threats

37 **Boris Mirkin, F. R. McMorris, Fred S. Roberts, and Andrey Rzhetsky, Editors,** Mathematical Hierarchies and Biology

36 **Joseph G. Rosenstein, Deborah S. Franzblau, and Fred S. Roberts, Editors,** Discrete Mathematics in the Schools

35 **Dingzhu Du, Jun Gu, and Panos M. Pardalos, Editors,** Satisfiability Problem: Theory and Applications

34 **Nathaniel Dean, Editor,** African Americans in Mathematics

33 **Ravi B. Boppana and James F. Lynch, Editors,** Logic and Random Structures

32 **Jean-Charles Grégoire, Gerard J. Holzmann, and Doron A. Peled, Editors,** The SPIN Verification System

31 **Neil Immerman and Phokion G. Kolaitis, Editors,** Descriptive Complexity and Finite Models

30 **Sandeep N. Bhatt, Editor,** Parallel Algorithms: Third DIMACS Implementation Challenge

29 **Doron A. Peled, Vaughan R. Pratt, and Gerard J. Holzmann, Editors,** Partial Order Methods in Verification

28 **Larry Finkelstein and William M. Kantor, Editors,** Groups and Computation II

27 **Richard J. Lipton and Eric B. Baum, Editors,** DNA Based Computers

26 **David S. Johnson and Michael A. Trick, Editors,** Cliques, Coloring, and Satisfiability: Second DIMACS Implementation Challenge

25 **Gilbert Baumslag, David Epstein, Robert Gilman, Hamish Short, and Charles Sims, Editors,** Geometric and Computational Perspectives on Infinite Groups

For a complete list of titles in this series, visit the AMS Bookstore at **www.ams.org/bookstore/**.

Mobile Networks and Computing

DIMACS
Series in Discrete Mathematics
and Theoretical Computer Science

Volume 52

Mobile Networks and Computing

DIMACS Workshop
Mobile Networks and Computing
March 25–27, 1999
DIMACS Center

Sanguthevar Rajasekaran
Panos Pardalos
D. Frank Hsu
Editors

NSF Science and Technology Center
in Discrete Mathematics and Theoretical Computer Science
A consortium of Rutgers University, Princeton University,
AT&T Labs–Research, Bell Labs (Lucent Technologies),
Telcordia Technologies, and NEC Research Institute

American Mathematical Society

This DIMACS volume contains papers from a DIMACS workshop on Mobile Networks and Computing, held March 25–27, 1999.

2000 *Mathematics Subject Classification*. Primary 68M10, 68M12, 90B18.

Library of Congress Cataloging-in-Publication Data

Mobile networks and computing : DIMACS workshop, mobile networks and computing, March 25–27, 1999, DIMACS Center / Sanguthevar Rajasekaran, Panos Pardalos, D. Frank Hsu, editors.

p. cm. — (DIMACS series in discrete mathematics and theoretical computer science, ISSN 1052-1798 ; v. 52)

"NSF Science and Technology Center in Discrete Mathematics and Theoretical Computer Science, a consortium of Rutgers University, Princeton University, AT&T Labs–Research, Bell Labs (Lucent Technologies), Telcordia Technologies, and NEC Research Institute."

Includes bibliographical references.

ISBN 0-8218-1547-4 (alk. paper)

1. Mobile computing—Congresses. 2. Mobile communication systems—Congresses. I. Rajasekaran, Sanguthevar. II. Pardalos, P. M. (Panos M.), 1954– III. Hsu, D. Frank (Derbiau Frank), 1948– IV. NSF Science and Technology Center in Discrete Mathematics and Theoretical Computer Science. V. Series.

QA76.59 .M68 2000
006.3–dc21 00-023293

---

10 9 8 7 6 5 4 3 2 1 05 04 03 02 01 00

"Learn without flaw what is worthy of learning and lead a life true to what has been learnt"

- Thiruvalluvar (80-20 B.C.) (Thirukkural, Chapter 40: Education, Verse 1)

Contents

Foreword

The DIMACS Workshop on Mobile Networks and Computing was held on March 25–27, 1999 at Rutgers University. We would like to express our appreciation to Panos Pardalos, Sanguthevar Rajasekaran, R. Badrinath, and Frank Hsu for their efforts to organize and plan this successful workshop.

The workshop was part of the Special Year on Networks. We extend our thanks to Stuart Haber, David Johnson, and Mihalis Yannakakis for their work as special year organizers.

The workshop brought together theoreticians and practitioners working on various aspects of mobile computing such as sensor networks, smart spaces, field computing, and channel allocation.

DIMACS gratefully acknowledges the generous support that makes these programs possible. The National Science Foundation, through its Science and Technology Centers program, the New Jersey Commission on Science and Technology, the National Security Agency through its support of the Special Year on Networks, and DIMACS's partners at Rutgers, Princeton, AT&T Labs-Research, Bell Labs, NEC Research Institute, and Telcordia Technologies generously supported the special year.

Fred S. Roberts
Director

Robert Sedgewick
Co-Director for Princeton

Preface

In the context of the 1998-1999 DIMACS special year with focus on **Networks**, a three-day workshop entitled "Mobile Networks and Computing" was held on March 25-27, 1999 at the DIMACS Center, Rutgers University. The workshop was organized by Panos Pardalos, Sanguthevar Rajasekaran, R. Badrinath, and Frank Hsu. More than 50 researchers from universities, institutes, governmental agencies, and industrial companies attended the workshop. Participants from Asia, Australia, and Europe gave the workshop an important international component.

Advances in the technologies of networking, wireless communications, and miniaturization of computers lead to the rapid development in mobile communication infrastructure, and have engendered a new paradigm of computing. Users carrying portable devices can freely move around, while still being connected to the network. This provides flexibility in accessing information anywhere and at any time. On the other hand, the price of this flexibility has introduced new levels of complexity that were not encountered in software and protocol design in wired networking.

The new challenges in designing software systems for mobile networks include location and mobility management, channel allocation, power conservation, among others. This workshop is aimed at bringing together researchers from academia as well as the industry who are working on various aspects of mobile computing. Topics will include sensor networks, smart spaces, field computing, channel allocation, etc.

A total of 24 invited talks were presented. Overall, the workshop achieved a great success. This volume contains a collection of refereed papers from the workshop.

The workshop was sponsored by DIMACS, through grants from the National Science Foundation, the New Jersey Commission on Science and Technology, and other sources. We would like to take this opportunity to thank the sponsors, David Johnson for his support, the DIMACS staff, the participants, the authors, the anonymous referees, and the American Mathematical Society, for making the workshop successful and the publication of this volume possible.

Sanguthevar Rajasekaran, Panos Pardalos, and Frank Hsu
November 1999

DIMACS Series in Discrete Mathematics
and Theoretical Computer Science
Volume **52**, 2000

Stochastic Modeling of a Single TCP/IP Session over a Random Loss Channel

Al-Hussein A. Abou-Zeid, Murat Azizoglu, and Sumit Roy

ABSTRACT. In this paper, we present an analytical framework for modeling the performance of a single TCP session in the presence of random packet loss. This framework may be applicable to communications channels that cause random packet loss modelled by appropriate statistics of the inter-loss duration. It is shown that the analytical model predicts the throughput for LANs/WANs (low and high bandwidth-delay products) with reasonable accuracy, as measured against the throughput obtained by simulation. Random loss is found to severely affect the network throughput, higher speed channels are found to be more vulnerable to random loss than slower channels, especially for moderate to high loss rates.

1. Introduction

TCP/IP has been designed for reliable networks in which most packet losses occur primarily due to network congestion. An important aspect of TCP is its window-based congestion avoidance mechanism [**6**]. In TCP/IP, when a node successfully receives a packet, it sends an acknowledgment (ACK) back to the source. At all times, the source keeps a record of the number of unacknowledged packets that it has released into the network. This number is called the congestion window size, or simply, *the window size*. The source is allowed to increase its window size as long as packets keep being acknowledged. The source detects a packet loss by either the non-arrival of a packet ACK within a certain time (maintained by a *timer*), or by the arrival of multiple ACKs with the same next expected packet number. A packet loss is interpreted by the source as an indication of congestion, and the source responds by reducing its window size so as not to overload the network with packets, thereby indirectly controlling the data rate. Thus modelling the dynamic behavior of congestion window size is key to analyzing TCP/IP throughput performance in a variety of situations. The system (source, network and receiver) is frequently called a 'self-clocked' system since the arrival of ACKs acts as the clock that increases the window size, while the loss of a packet acts as a 'reset' for the system.

1991 *Mathematics Subject Classification.* Primary 54C40, 14E20; Secondary 46E25, 20C20.

An important system parameter that affects TCP throughput is the ratio β of the buffer size available in the bottleneck links of the network to the link bandwidth-delay product. For LANs, the round-trip delay in a connection is small, so that the bandwidth delay product could be much smaller than the buffer size (large β). WANs, on the other hand, have large round-trip delays, so that the buffer size is typically smaller than the bandwidth-delay product (small β). It is easy to see that, for the same bandwidth delay product, increasing the buffer size at a bottleneck link (increasing β) decreases the amount of congestion. It is worth noting that the bandwidth-delay product determines the maximum number of packets that can be in transit between a source-destination pair.

Network congestion is not the only source of packet loss - random packet loss may be significant in many circumstances. For example, it may arise due to intermittent faults in hardware elements in wired networks. In a wireless network, multipath fading that characterizes many terestrial links may be modelled as leading to random packet loss depending on the fade rates relative to data rate on the channel. While random packet loss on the Internet has been reported in [**8**], it was not taken into consideration in the congestion control mechanism in TCP/IP. Previous research [**2, 3, 4, 5**] has shown that random packet loss (which is not due to congestion) may severely decrease the throughput of TCP because TCP interprets random packet loss to be due to congestion and hence lowers the input data rate into the network, and consequently the throughput.

In [**2, 3**], a discrete-time model for random packet loss was used in which any given packet is assumed to be lost with probability q independent of all other packets. This model induces a geometric distribution on the number of packets successfully transmitted between consecutive packet losses, that may or may not be appropriate for specific lossy networks. A different loss model was employed in [**5**] which assumed that packet loss is characterized by an inhomogeneous Poisson process. The steady-state distribution of the window size was obtained in [**5**] under the assumption of infinite buffer size. In contrast, in this paper we assume a continuous-time packet loss model governed by a general renewal process and investigate the effect of finite buffer size on the performance.

A basic system model is shown in Figure 1. An infinite source (i.e. one that always has a packet to send) releases packets into a buffer of size B upon receiving ACKs from the destination. The packets are then sent over a single link with capacity μ packets per second and a net delay of τ (propagation delay through the channel, any other processing delays etc.) is assumed. Define $T = \tau + 1/\mu$ to be the time between the start of transmission of a packet and the reception of an ACK for this packet. Then μT is the bandwidth-delay product and the ratio $\beta = \frac{B}{\mu T}$ is the buffer size normalized by the bandwidth-delay product.

2. Ideal Channels without Random Packet Loss

We first briefly review the operation of TCP for the case of ideal channels, and summarize the key results in [**1, 3**] relevant to our work. There are different versions of TCP - the popular Tahoe version developed by Jacobson [**6**] (TCP-T) as well as the Reno version (TCP-R) that incorporates a fast retransmit option together with a method for reducing the effect of slow start [**7**]. While [**1, 3**] considered both versions, our analysis concentrates on TCP-R only, since our aim is to present

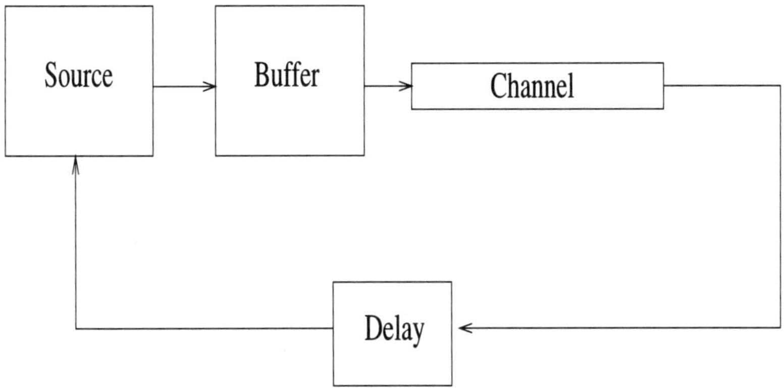

FIGURE 1. Block diagram of the system model

an analytical framework for the analysis of TCP/IP rather than to compare the different versions.

In TCP/IP, there are no explicit ACK or negative ACK signals. When the source sends a packet, it initializes a timer with an expiry time that is set depending on a current estimate of the delay τ. When a destination correctly receives a packet, it sends a signal to the source with the 'next expected' packet number. If this is received by the source prior to timer expiry, it is considered an ACK; otherwise, the packet is considered lost. In some implementations of TCP such as TCP-R, if the source receives multiple ACK signals with the same 'next expected' packet number, it interprets this as a packet loss.

Let $t' = 0$ denote the time of establishment of the TCP session under consideration, and let $W(t')$ denote the congestion window size at time t'. Then the algorithm followed by a TCP session (assuming 'ideal' operation of TCP) can be described as follows:

TCP-Tahoe:

1. Every time a packet ACK is received,
 if $W(t') < W_{th}$, set $W(t') = W(t') + 1$ ... **Slow Start Phase**
 elseif$((ACKcount++) == W(t'))$ $\{$ $ACKcount = 0, W(t') = W(t') + 1$ $\}$ ... **Congestion Avoidance Phase**

2. Every time a packet loss is detected,
 set $W_{th} = W(t')/2$ and set $W(t') = 1$. **(Go back to Slow Start Phase)**

TCP-Reno:

1. Every time a (non-repeated) packet ACK is received,
 if $W(t') < W_{th}$, set $W(t') = W(t') + 1$ ... **Slow Start Phase**
 elseif$((ACKcount++) == W(t'))$ $\{$ $ACKcount = 0, W(t') = W(t') + 1$ $\}$ ... **Congestion Avoidance Phase**

2. Every time a packet loss is detected via duplicate ACK,
 set $W_{th} = W(t')/2$, then set $W(t') = W(t')/2$ (the algorithm doesn't initiate

slow start)

3. Every time a packet loss is detected via timer expiry, the algorithm goes to slow start (as in TCP-T)
set $W_{th} = W(t')/2$ and set $W(t') = 1$.

A TCP-T session typically evolves as follows. A packet is released from the source into the buffer just after the session is established and the transmitter enters 'slow start' phase. Each time an ACK is received, the window size is incremented according to slow start until it reaches W_{th} at which time the algorithm switches to congestion avoidance phase. The source subsequently increases its window size by one only after every window's worth of acknowledgments (congestion avoidance phase). This continues until a packet loss is detected, whence the window size is reset to one and the algorithm re-enters the slow start phase. This cycle repeats itself until all the packets at the source are transmitted and acknowledged at which time the TCP/IP session is terminated. It is worth noting that during slow start, the window size actually increases rapidly; in order to increase the window size by one when an ACK is received, the source releases two packets simultaneously into the buffer, while, to keep the window size constant, the source releases only one packet (to replace the one just acknowledged). In slow start, the source always releases two packets at each ACK reception, while in congestion avoidance, the source releases one packet for each ACK reception (to keep the window size constant) except when a window's worth of acknowledgment is received, in which case it releases two packets.

In a TCP-R session, the first cycle after session establishment is identical to that of TCP-T. However, after the first packet loss, the algorithm does not go back to slow start. Instead, the algorithm reduces the window size to half its value and continues in the congestion avoidance phase. In this description, we have assumed that the algorithm implements Selective Acknowledgments (SACKs) so that loss of multiple packets does not lead to time out phenomena, and hence the algorithm never enters Step 3 in the pseudo-code above.

Finally, we assume that fast retransmit option is used [**11**] in the event of a packet loss detection to avoid lengthy stoppage of transmission.

Denote $w_p = \mu T + B = \mu\tau + B + 1$, and note that when the window size reaches w_p, the bit pipe (the combination of the channel and the transmit buffer) is fully utilized, i.e, the buffer is fully occupied and the maximum number of packets allowable are in transit. A further increase in window size at this stage causes buffer overflow, at which point the window size is reset (for TCP-T) or halved (for TCP-R) and W_{th} is set to $w_p/2$.

Sample functions of the window size obtained from simulations for TCP-Tahoe and TCP-Reno are shown in Figure 2 when the only source of packet loss is buffer overflow.

A note about buffer overflow is in order. As can be noticed in Figure 2, the buffer build up rate in the slow start phase is very high, due to the 'fast' nature of slow start. Hence, one would expect that for small values of β, buffer overflow can take place in slow start. This is actually true, and has been noted in [**3**]. In such a case, for TCP-T, the window size is set back to one and another slow start phase follows with a lower threshold. For TCP-R, the window size is set to half its value

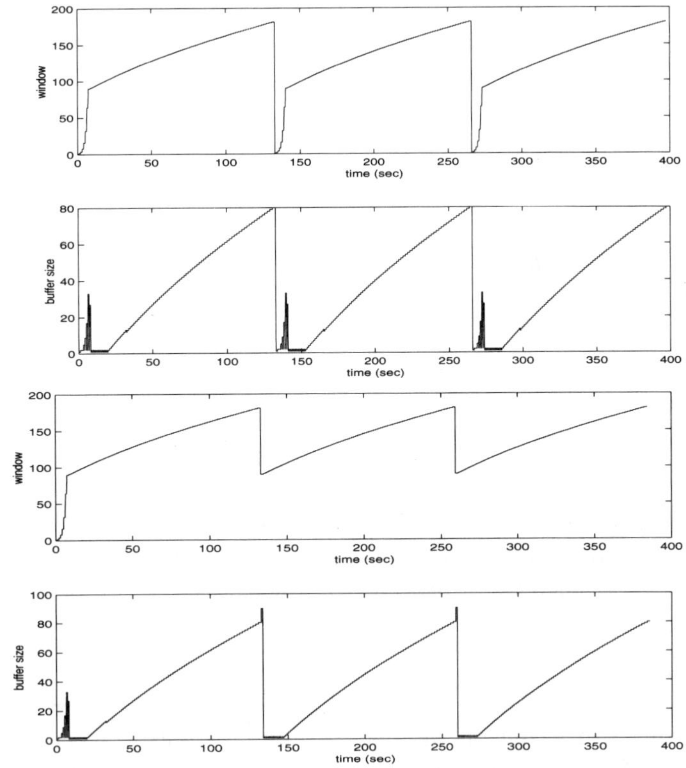

FIGURE 2. Window and buffer size evolution without random loss of packets for $\mu = 100$, $\tau = 1.0$, $\beta = 0.8$. (a)TCP-T, (b) TCP-R.

and the window evolution continues as depicted earlier. The effect of this 'double' slow start is significant for TCP-T but negligible for TCP-R.

The algorithms outlined above, defining the window size evolution during a TCP session have been studied in [**1, 3**] and mathematical expressions have been obtained for the envelope of the window size, denoted, for convenience, also by $W(t')$, for each of the phases (slow start and congestion avoidance). We summarize those expressions in a convenient way (so as to be able to use them in the following section) as follows (refer to Figure 3). Let n denote the number of packets acknowledged during a time interval t. Then the window evolution can be expressed as follows:

1. **Slow Start** $(1 < W(t') < W_{th})$. Consider two instants t_0', $t_0' + t$ in a slow start phase of any of the TCP cycles. Choose t_0' such that $W(t_0') = 1$.Then,

$$W(t_0' + t) = 2^{t/T} \tag{2.1}$$

$$n = W(t_0' + t) - 1 \tag{2.2}$$

2. **Congestion Avoidance - Phase I** $(W_{th} < W(t') < \mu T)$. Consider two instants t_0', $t_0' + t$ in a congestion avoidance phase of any of the TCP cycles.

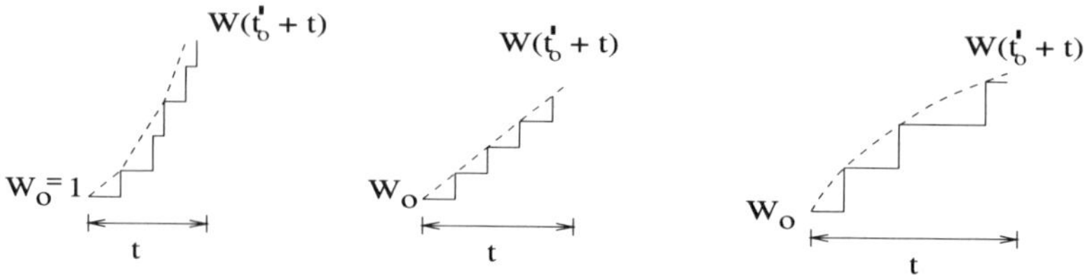

FIGURE 3. Sketch of the exponential, linear and sub-linear $O(\sqrt{t})$ phases for window evolution. Solid lines indicate the actual window size evolution while dotted lines indicate the envelope.

Choose t_0' such that $W(t_0') = W_0$.Then,

$$W(t_0' + t) = W_0 + t/T \tag{2.3}$$

$$n = \frac{1}{T}(W_0 t + t^2/(2T)) \tag{2.4}$$

3. **Congestion Avoidance - Phase II** $(\mu T < W(t') < w_p)$. Consider two instants t_0', $t_0' + t$ in a congestion avoidance phase of any of the TCP cycles. Choose t_0' such that $W(t_0') = W_0$.Then,

$$W(t_0' + t) = \sqrt{W_0^2 + 2\mu t} \tag{2.5}$$

$$n = \mu t \tag{2.6}$$

Note that in the sequel, we focus on time instants t' where $W(t')$ is discrete. Equation (1) follows from considering the time instances at which packets are sent and packet ACKs are received, and noting that two packets are sent for each ACK received. Equation 2 follows from the fact that each time an ACK is received, $W(t')$ is incremented by one; since $W(t')$ is initially equal to one, the total number of packets acknowledged at the end of the slow start phase is one less than equal to the window size at the end of the slow start phase. Equations (3)-(6) follow from a continuous time approximation of the window size. Specifically in the congestion avoidance phase, if we approximate the window size by a continuous time function, then

$$\frac{dW(t')}{dt'} = min\{\frac{1}{T}, \frac{\mu}{W(t')}\},$$

the solution to which yields (3),(5). Note that (3) indicates a linear growth of the window size with time with a slope of $1/T$, while (5) specifies a growth that is $O(\sqrt{t})$. This can be explained as follows - for $W(t') < \mu T$, the channel is able to transmit $W(t')$ packets in T seconds (linear growth with slope $1/T$), until the window size reaches μT, at which time buffer build-up set in. Thus it takes more than T seconds to send $W(t')$ packets subsequently and hence the window increase is slower than linear.

Using (2.1) - (2.6) and taking into consideration the periodic evolution of the window size outlined in the previous discussion, it is straightforward to compute the average packet transmission rate R as the ratio of the number of packets sent in one cycle of the TCP session to the time duration of the cycle. Note that for the case of TCP-R, we neglect the first cycle since it is different from the rest of the cycles (it is the only one that contains slow start). The average transmission rate is given by

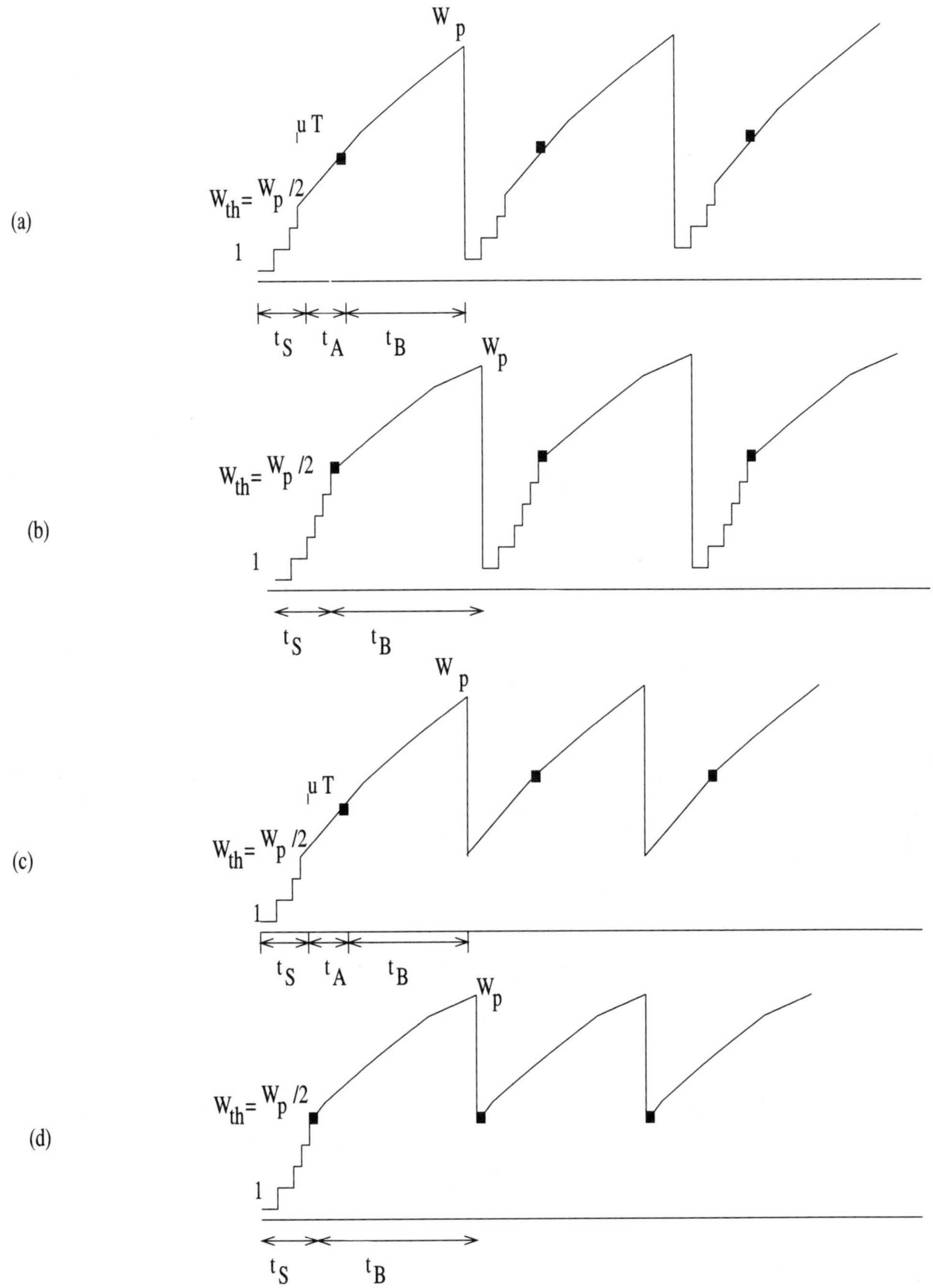

FIGURE 4. Sketch of the TCP window evolution without random loss ; (a) TCP-T with $\beta < 1$ (b) TCP-T with $\beta > 1$, (c) TCP-R with $\beta < 1$, (d) TCP-R with $\beta > 1$.Note that for $\beta < 1$, $w_p/2 < \mu T$, and hence the window evolution has a linear phase, while for $\beta > 1$, $w_p/2 > \mu T$ and hence the window evolution doesn't have a linear phase.

TCP-T:

$$\beta < 1 : R = \frac{n_S + n_A + n_B}{t_S + t_A + t_B} \tag{2.7}$$

$$\beta > 1 : R = \frac{n_S + n_B}{t_S + t_B} \tag{2.8}$$

TCP-R:

$$\beta < 1 : R = \frac{n_A + n_B}{t_A + t_B} \tag{2.9}$$

$$\beta > 1 : R \simeq \mu \tag{2.10}$$

The average throughput is then calculated as

$$\rho = \frac{R}{\mu} \tag{2.11}$$

The values above n_S, n_A, n_B, t_S, t_A and t_B are obtained by substituting for W_0 and $W(t')$ (see Figure 3) in (1)-(6) by the initial and final values of the slow start and congestion avoidance phases as indicated in Figure 4. For example, to compute n_S and t_S for TCP-T, set $W(t_S) = w_p/2$ in (1); to compute n_B and t_B for TCP-R and $\beta < 1$, let $W_0 = \mu T, W(t_B) = w_p$ in (3); to compute n_B and t_B for TCP-R and $\beta > 1$, let $W_0 = w_p/2, W(t_B) = w_p$ in (3) and so on.

Note the difference in the expressions for $\beta < 1$ and $\beta > 1$. Consider TCP-T; for $\beta < 1$, $w_p/2 < \mu T$ and hence the cyclical evolution of the window size consists of an exponential growth (the slow start phase from 1 to $w_p/2$) described by (1), then a linear growth (congestion avoidance phase from $w_p/2$ to μT) expressed by (3) and then $O(\sqrt{t})$ growth (congestion avoidance phase from μT to w_p) described by (5). On the other hand, for $\beta > 1$, $w_p/2 > \mu T$, and hence, the cyclical evolution consists of exponential growth (slow start from 1 to $w_p/2$) followed by $O(\sqrt{t})$ growth (congestion avoidance from $w_p/2$ to w_p) without having a linear phase. Similar observations apply for TCP-R; refer to Figure 4 for a schematic diagram of the window size evolution for each of the two ranges of β.

3. Channels with Random Packet Loss

3.1. Random Loss Model. Let S_i denote the time of the ith packet loss, for $i = 1, 2, \ldots$. Let $X_i = S_i - S_{i-1}$ denote the time between $(i-1)^{\text{th}}$ and i^{th} packet losses with $X_1 = S_1$ by convention. As stated earlier, we will consider $\{X_1, X_2, \ldots\}$ to be a set of IID random variables with probability density function $f(x)$ and distribution function $F(x)$. Thus, the process (pdf) defined by the loss occurrence times $\{S_1, S_2, \ldots\}$ is a renewal process with interrenewal pdf $f(x)$.

Now, suppose that at a certain time instant X_1 $(=S_1)$, the first *random* packet loss event occurs. Denote the window size at that instant by W_1. When the source detects this loss (by the arrival of duplicate ACKs for the case of TCP-R), the window size is halved. The window size now increases as depicted earlier (the window size starts from $W_1/2$ and increases till w_p, at which time a buffer overflow takes place and $W(t')$ is set to $w_p/2$, and so on) until another random packet loss takes place at a random time instant $S_2 = X_1 + X_2$. Denote the window size at this time (the time of the second loss) by W_2.

In what follows, we call one period from $w_p/2$ till w_p the free-running period or the 'typical' cycle (i.e. free from random loss effects). Note that the second

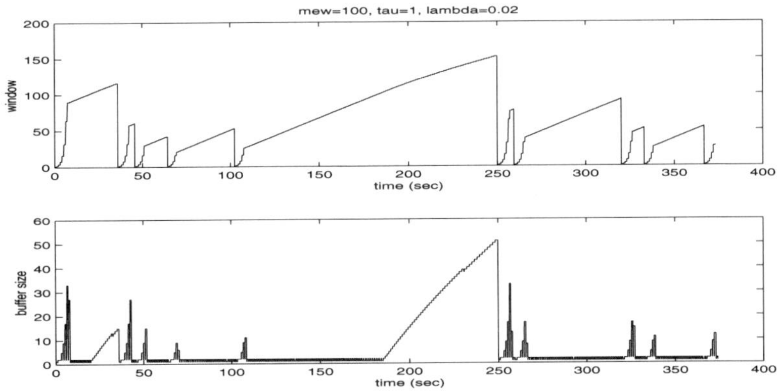

FIGURE 5. Window and buffer size evolution in with random loss of packets for a TCP-T session with $\mu = 100$, $\tau = 1.0$, $\beta = 0.8$ and $\lambda = 0.02$

random loss event can happen before the occurrence of any 'typical' cycles. The window size $W(t')$ is a semi-Markovian stochastic process, because the window size evolution after a random loss (except for its starting value which is half of that just before the random loss) is statistically independent from the window size evolution before the random loss. Further, since $\{X_1, X_2, \dots\}$ are independent and identically distributed (IID), the window sizes $\{W_1, W_2, \dots\}$ (window sizes just before the random loss) form a finite state Markov Chain (i.e. the embedded Markov Chain of the semi-Markov process $W(t')$) [**9, 10**].

3.2. Analysis Assumptions and Notations. With the above model of random loss, two quantities associated with the just defined Markov Chain are of interest
(1) $E[N|W_1 = w_1]$, the expected number of packets successfully transmitted before another random packet loss occurs, given that the most recent random loss took place at w_1; (2) The conditional probability $P[W_2 = w_2|W_1 = w_1]$ (denoted for convenience by P ; the probability that the next *random* loss takes place at $W_2 = w_2$ given that the previous *random* loss took place at $W_1 = w_1$.

Before we attempt to evaluate the above two quantities of interest, we make an approximation for TCP-R, similar to the approximation previously invoked for the channels without random loss. We ignore the first cycle of TCP-R and assume that the TCP session starts with window size $w_p/2$ instead of starting with a window size of 1. This approximation should have a negligible effect on the average throughput, due to two reasons ; (1) A source with an infinite infinite number of packets was assumed; hence the transient behavior (slow start) at the beginning of the connection is expected to be negligible, even for the case of random loss; (2) The duration as well as the number of packets sent during this slow start phase is low (recall that slow start is actually 'fast'). In other words, this transient behavior disappears very fast. We comment on the effect of this approximation in the results section following the analysis.

In the analysis that follows, two ranges of β are considered separately, $\beta < 1$ and $\beta > 1$, and expressions for $E[N|W_1]$ and $P[W_2|W_1]$ are found for each of the two ranges.

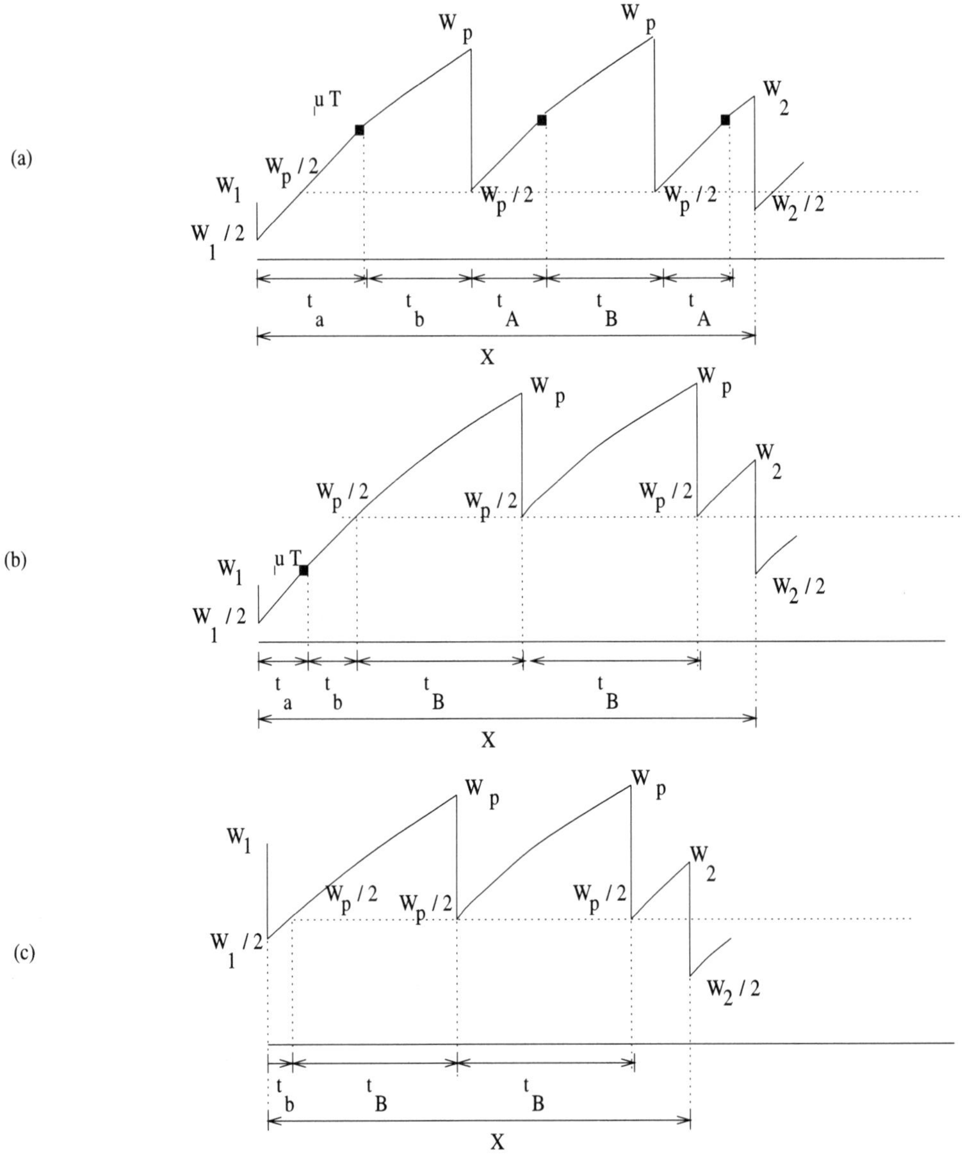

FIGURE 6. A sketch of sample functions of window size in the presence of random loss for TCP-R ; (a) $\beta < 1$ ($\mu T > w_p/2 > W_1/2$), (b) $\beta > 1$ ($w_p/2 > \mu T$) and $W_1/2 < \mu T$ and (c) $\beta > 1$ ($w_p/2 > \mu T$) and $W_1/2 > \mu T$

Define,
N_a is the number of packets sent during the linear congestion avoidance phase in the first TCP cycle just after a random packet loss at a window size $W_1 = w_1$.
N_b is the same as N_a but for the quadratic phase.
N_A is the number of packets sent during the linear congestion avoidance phase of one of the typical cycles.

N_B is the same as N_A but for the quadratic phase.
$N_p = N_A + N_B$ is the number of packets sent in a typical cycle.
The corresponding durations of time where the above number of packets is transmitted (time is counted since the beginning of the phase referenced) are t_a, t_b, t_A, and t_B respectiveley. Define,

$$t_1 = t_a + t_b,$$

$$t_p = t_A + t_B.$$

The values of N and t with capitalized subscripts (A, B) do not depend on W_1 and can be computed using (2.3)-(2.6) as in the case of channels without random packet loss. On the other hand, the values with small letter subscripts (a, b) refer to the cycle of TCP-R just after a random loss at a window size W_1. Those values depend on the value of W_1, and can be computed also by (2.3)-(2.6) by substituting by the appropriate initial and final values of the window size as shown in Figure 6.

For notaional convenience, define,

$$\begin{aligned}
T_a(n) &= T/2\sqrt{w_1^2 + 8(n+1)} - T/2w_1 \\
T_{b_1}(n) &= t_a + (n+1)/\mu \\
T_{b_2}(n) &= (n+1)/\mu \\
T_A(n) &= t_1 + jt_p + T/2\sqrt{w_p^2 + 8(n+1)} - T/2w_p \\
T_{B_1}(n) &= t_1 + t_A + jt_p + (n+1)/\mu \\
T_{B_2}(n) &= t_1 + jt_p + (n+1)/\mu \\
T_a(w_1, w_2) &= T(w_2 - w_1/2) \\
T_{b_1}(w_1, w_2) &= t_a + \frac{w_2^2 - (\mu T)^2}{2\mu} \\
T_{b_2}(w_1, w_2) &= \frac{w_2^2 - (w_1/2)^2}{2\mu} \\
T_A(w_1, w_2) &= t_1 + jt_p + T(w_2 - w_p/2) \\
T_{B_1}(w_1, w_2) &= t_1 + jt_p + t_A + \frac{w_2^2 - (\mu T)^2}{2\mu} \\
T_{B_2}(w_1, w_2) &= t_1 + jt_p + \frac{w_2^2 - (w_p/2)^2}{2\mu}
\end{aligned}$$

where, $0 \leq n < \infty$ is an incremental counter that is initialized at the beginning of each phase (linear or quadratic) and $0 \leq j < \infty$ is an invremental counter that refers to the $j + 1^{st}$ typical cycle after the first atypical cycle following a random packet loss.

The above values represent absolute times of occurances of certain events assuming that a packet loss takes place at a window size $W_1 = w_1$ and no subsequent random packet losses take place. Hence, these times, conditioned on knowing the value of w_1, are deterministic.

For example, $T_A(n)$ represents the time at which, assuming the last random packet loss took place at $W_1 = w_1$, the transmission of packet n will take place in the $(j + 1^{st})$ cycle after the first atypical cycle. $T_A(w_1, w_2)$ represents the time at which the window size will reach w_2 in the $(j + 1^{st})$ cycle after the first atypical cycle.

Finally, the subscript $_1$ for T_b and T_B is used when $\beta < 1$. When $\beta > 1$, both parameters, with $_1$ and $_2$ subscripts will be used, depending on $w_1 < 2\mu T$ or $w_1 > 2\mu T$, as shown in Figure 6.

3.3. $\beta < 1$. $E[N|W_1]$ can be written in terms of the complementary distribution function of N as

$$E[N|W_1 = w_1] = \sum_{n=0}^{\infty} Pr[N > n|W_1 = w_1] \tag{3.1}$$

Using (2.3)-(2.6) and Figure 6, the summation in (3.1) can be written in terms of the p.d.f. of X as a sum of terms of the form $F(a)$ where $F(x)$ is the complementary distribution function of X and $F(a) = Pr[X > a]$.

$$\begin{aligned} E[N|W_1 = w_1] &= \sum_{n=0}^{N_a-1} F[T_a(n)] + \sum_{n=0}^{N_b-1} F[T_{b_1}(n)] \\ &\quad + \sum_{j=0}^{\infty} \sum_{n=0}^{N_A-1} F[T_A(n)] \\ &\quad + \sum_{j=0}^{\infty} \sum_{n=0}^{N_B-1} F[T_{B_1}(n)] \end{aligned} \tag{3.2}$$

In a similar fashion, $P[W_2|W_1]$ can be written in terms of the complementary distribution function $F(x)$. Let $F(a,b) = F(a) - F(b)$. Then by using (2.3)-(2.6) (refer to Figure 6(a) for notations),

$$P = \begin{cases} 0 & 0 < w_2 < w_1/2 \\ F(T_a(w_1, w_2), T_a(w_1, w_2 + 1)) & w_1/2 < w_2 < w_p/2 \\ F(T_a(w_1, w_2), T_a(w_1, w_2 + 1)) & \\ + \sum_{j=0}^{\infty} F(T_A(w_1, w_2), T_A(w_1, w_2 + 1)) & w_p/2 < w_2 < \mu T \\ F(T_{b_1}(w_1, w_2), T_{b_1}(w_1, w_2 + 1)) & \\ + \sum_{j=0}^{\infty} F(T_{B_1}(w_1, w_2), T_{B_1}(w_1, w_2 + 1)) & \mu T < w_2 < w_p \end{cases} \tag{3.3}$$

Some remarks on the derivation above are appropriate. First, notice that when a random loss happens at $W_1 = w_1$, the window size immediately after the loss is set to $w_1/2$; then the TCP algorithm increases the window size, possibly passing through events of buffer overflow (depending on the time to next random loss), until another random loss occurs. This means that the window size in the interval between the two random loss events never takes a value less than $w_1/2$. This explains the value of $P[W_2|W_1]$ being zero for $0 < w_2 < w_1/2$. Second, if the next random loss takes place at $W_2 = w_2$ such that $w_1/2 < w_2 < w_p/2$, then this is equivalent to the event that the next random loss takes place somewhere between the time it reaches w_2 (as described by (2.3)) and the time it should have left w_2 to increase to $w_2 + 1$. This explains the second term. Finally, note that if the next random loss takes place at a w_2 such that $w_p/2 < w_2 < w_p$, it implies that the loss occurs somewhere between the time it reaches w_2 (as described by (2.3) and (2.5)) and the time the size is incremented from w_2 to $w_2 + 1$ in the first cycle just after a random loss at $W_1 = w_1$, or in a subsequent cycle that is $j \geq 0$ typical cycles away from the first cycle after the random loss. This explains the last two terms in the previous equation.

3.4. $\beta > 1$. As for the case of $\beta < 1$, using (2.3)-(2.6) and Figure 6, the summation in (3.1) can be written in terms of the p.d.f. of X. However, we have to differentiate between the case when $W_1/2 < \mu T$ and $W_1/2 > \mu T$. Recall for the non-random-loss case, $w_p/2 > \mu T$, and hence the typical cycles will always consist of a quadratic evolution between $w_p/2$ and w_p. However, for the first cycle after the loss occurring at $W_1 = w_1$, the cycle starts with window size $w_1/2$. If $w_1/2 < \mu T$, the window size will have a linear growth from $w_1/2$ until μT and then a quadratic evolution (unless another random loss takes place) until w_p. On the other hand, if $w_1/2 > \mu T$, the window size will have a quadratic growth from $w_1/2$ without going into a linear phase first. Figure 6(b) and 6(c) show a schematic of a sample function of the two cases.

1. $w_1 < 2\mu T$

$$E[N|W_1 = w_1] = \sum_{n=0}^{N_a-1} F(T_a(n)) + \sum_{n=0}^{N_b-1} F(T_{b_1}(n)) + \sum_{j=0}^{\infty}\sum_{n=0}^{N_p-1} F(T_{B_2}(n)) \tag{3.4}$$

$$P = \begin{cases} 0 & 0 < w_2 < w_1/2 \\ F(T_a(w_1, w_2), T_a(w_1, w_2+1)) & w_1/2 < w_2 < \mu T \\ F(T_{b_1}(w_1, w_2), T_{b_1}(w_1, w_2+1)) & \mu T < w_2 < w_p/2 \\ \sum_{j=0}^{\infty} F(T_{B_2}(w_1, w_2), T_{B_2}(w_1, w_2+1)) & w_p/2 < w_2 < w_p \end{cases} \tag{3.5}$$

2. $w_1 > 2\mu T$

$$E[N|W_1 = w_1] = \sum_{n=0}^{N_b-1} F(T_{b_2}(n)) + \sum_{j=0}^{\infty}\sum_{n=0}^{N_p-1} F(T_{B_2}(n)) \tag{3.6}$$

$$P = \begin{cases} 0 & 0 < w_2 < w_1/2 \\ F(T_{b_2}(w_1, w_2), T_{b_2}(w_1, w_2+1)) & w_1/2 < w_2 < w_p/2 \\ \sum_{j=0}^{\infty} F(T_{B_2}(w_1, w_2), T_{B_2}(w_1, w_2+1)) & w_p/2 < w_2 < w_p \end{cases} \tag{3.7}$$

3.5. Steady State Distribution of the Window Size. The average packet transmission rate R of a TCP session under the current model is given by

$$R = \frac{E[N]}{E[X]} \tag{3.8}$$

where $E[N]$ is the average number of packets successfully sent in an inter-loss duration, and E[X] is the average time between two random losses (which is equal to $1/\lambda$).

$E[N]$ is given by

$$E[N] = \sum_{W=0}^{w_p} E[N|W]\pi(W) \tag{3.9}$$

where π is the steady state distribution of the MC.

It does not appear feasible that a closed-form expression for π given $P[W_2|W_1]$ can be found. Accordingly, MATLAB^{TM} routines for solving the eigenvalue problem were used to compute π for different values of λ, μ, τ and B, assuming an exponential distribution for the inter-loss times in the following section.

4. Analysis and Simulation for Exponential Random Loss

For an exponential inter-loss distribution with average rate λ, $F(a) = e^{-\lambda a}$ and $F(a,b) = e^{-\lambda a} - e^{-\lambda b}$. By substituting in (3.2)-(3.7), we obtain

4.1. $\beta < 1$.

$$\begin{aligned} E[N|W_1 = w_1] &= \sum_{n=0}^{n_a-1} e^{-\lambda(T/2)\sqrt{w_1^2+8(n+1)}-w_1} \\ &\quad + \frac{e^{-\lambda t_1}}{1-e^{-\lambda t_p}} \sum_{n=0}^{n_A-1} e^{-\lambda(T/2)\sqrt{w_p^2+8(n+1)}-w_p} \\ &\quad + e^{-\lambda/\mu}\frac{1-e^{-(\lambda/\mu)n_B}}{1-e^{-(\lambda/\mu)}}\left(e^{-\lambda t_a} + \frac{e^{-\lambda t_1}}{1-e^{-\lambda t_p}}e^{-\lambda t_A}\right) \end{aligned} \tag{4.1}$$

$$P = \begin{cases} 0 & 0 < w_2 < w_1/2 \\ e^{-\lambda T(w_2 - w_1/2)}(1-e^{-\lambda T}) & w_1/2 < w_2 < w_p/2 \\ e^{-\lambda T w_2}(1-e^{-\lambda T})(e^{\lambda T w_1/2} + \frac{e^{-\lambda(t_1 - w_p/2)}}{1-e^{-\lambda t_p}}) & w_p/2 < w_2 < \mu T \\ e^{\lambda\frac{(\mu T)^2}{2\mu}}(e^{-\lambda w_2^2} - e^{-\lambda(w_2+1)^2})(e^{-\lambda t_a} + \frac{e^{-\lambda(t_1+t_A)}}{1-e^{-\lambda t_p}}) & \mu T < w_2 < w_p \end{cases} \tag{4.2}$$

4.2. $\beta > 1$.

1. $w_1 < 2\mu T$

$$\begin{aligned} E[N|W_1 = w_1] &= \sum_{n=0}^{n_a-1} e^{-\lambda(T/2)(\sqrt{w_1^2+8(n+1)}-w_1)} \\ &\quad + \frac{e^{-\lambda(t_a+1/\mu)}}{1-e^{-\lambda/\mu}}\left(1-e^{-(\lambda/\mu)n_b} + e^{-\lambda t_b}\frac{1-e^{-(\lambda/\mu)n_p}}{1-e^{-\lambda t_p}}\right) \end{aligned} \tag{4.3}$$

$$P = \begin{cases} 0 & 0 < w_2 < w_1/2 \\ e^{-\lambda T(w_2 - w_1/2)}(1-e^{-\lambda T}) & w_1/2 < w_2 < \mu T \\ e^{-\lambda(t_a - \frac{(\mu T)^2}{2\mu})}(e^{-\lambda w_2^2} - e^{-\lambda(w_2+1)^2}) & \mu T < w_2 < w_p/2 \\ \frac{e^{-\lambda(t_1 - \frac{(w_p/2)^2}{2\mu})}}{1-e^{-\lambda t_p}}(e^{-\lambda w_2^2} - e^{-\lambda(w_2+1)^2}) & \mu T < w_2 < w_p/2 \end{cases} \tag{4.4}$$

2. $w_1 > 2\mu T$

$$E[N|W_1 = w_1] = \frac{e^{-\lambda/\mu}}{1-e^{-\lambda/\mu}}\left(1-e^{-(\lambda/\mu)n_b)} + e^{-\lambda t_b}\frac{1-e^{(-\lambda/\mu)n_p}}{1-e^{-\lambda t_p}}\right) \tag{4.5}$$

$$P = \begin{cases} 0 & 0 < w_2 < w_1/2 \\ e^{\lambda\frac{(w_1/2)^2}{2\mu}}(e^{-\lambda w_2^2} - e^{-\lambda(w_2+1)^2}) & w_1/2 < w_2 < w_p/2 \\ \frac{e^{-\lambda(t_b - \frac{(w_p/2)^2}{2\mu})}}{1-e^{-\lambda t_p}}(e^{-\lambda w_2^2} - e^{-\lambda(w_2+1)^2}) & w_p/2 < w_2 < w_p \end{cases} \tag{4.6}$$

The average throughput for different values of μ, τ, β and λ are shown in Figures 7 and 8, and discussed in the following section. Note that the above expressions for $E[N|W_1]$ simplify in the extreme cases of $\lambda \to 0$ (non-random loss) and $\lambda \to \infty$ (extremely heavy loss rate). Taking the limit of $\lambda E[N|W_1]$ as $\lambda \to 0$ yields n_p/t_p for $\beta < 1$ (where $n_p = n_A + n_B$) and μ for $\beta > 1$ which agrees with the expressions deduced in [**3**] and summarized earlier in (2.7)-(2.10). Taking the limit of $\lambda E[N|W_1]$ as $\lambda \to \infty$ yields zero as expected.

In the simulations, we considered the same set-up described in the system model in Section 2 *without approximation*. Recall that in the analysis, we approximated the TCP window evolution by neglecting the slow start phase which is expected to contribute to any (small) deviation between the analysis and the simulation results in Figures 7, 8.

5. Results and Concluding Remarks

The main results that can be deduced from the throughput behavior depicted in Figures 7, 8 are:

(1) Consider a link with a given loss rate λ and a bandwidth-delay product $\mu\tau$. The results show that increasing the buffer size (i.e. increasing β) does *not* always increase the throughput. For channels with high loss rate, increasing the buffer size has no positive effect on the throughput; however for channels with low loss rates, increasing the buffer size increases the throughput considerably.

(2) For low loss rates, faster channels (higher μ) have higher throughput. However (contrary to what may be expected) for moderate to high loss rates, slower channels have higher throughput. The explanation for this is simple though perhaps not transparent. Recall that for channels without random loss, the throughput is given by $\frac{n_p/t_p}{\mu}$. For channels with random loss, the throughput is given by $\frac{\lambda E[N]}{\mu}$. The expression in the numerator is the average transmission rate. Now, for the case of no random loss, increasing μ increases n_p significantly and hence the average transmission rate as well as the throughput increase. Similarly, for low random loss rates, increasing μ increases $E[N]$ significantly and hence both average rate and throughput increase. On the other hand, for moderate to high loss rates, increasing μ does not increase the number of packets successfully transmitted proportionately (due to the effect of random loss); hence the average transmission rate increases but the throughput actually decreases.

One practical interpretation of this result for the Internet relates to a user's dial-up modem connection to a server. Purchasing a faster modem would increase the average transmission rate, but may not be economically justifiable in the case of moderate-to-high loss rate channels since the proportion of the used bandwidth (i.e, throughput) for the new faster modem is less than that for the slower (and hence, less expensive) one.

(3) The results from the analysis of the proposed model of random loss matches closely with the simulations. However, the effect of neglecting the slow start phase at the beginning of a TCP-R session results in a deviation between the simulation and analysis results. Note the following : (a) For a given channel (i.e. bandwidth-delay product), the deviation (between the simulation and the analysis results) for low loss rates is small. This is because the slow start phase duration is sufficiently small such that the window size reaches $w_p/2$ in a very short time (compared to the average time to the first random loss) corroborating our approximation. in the

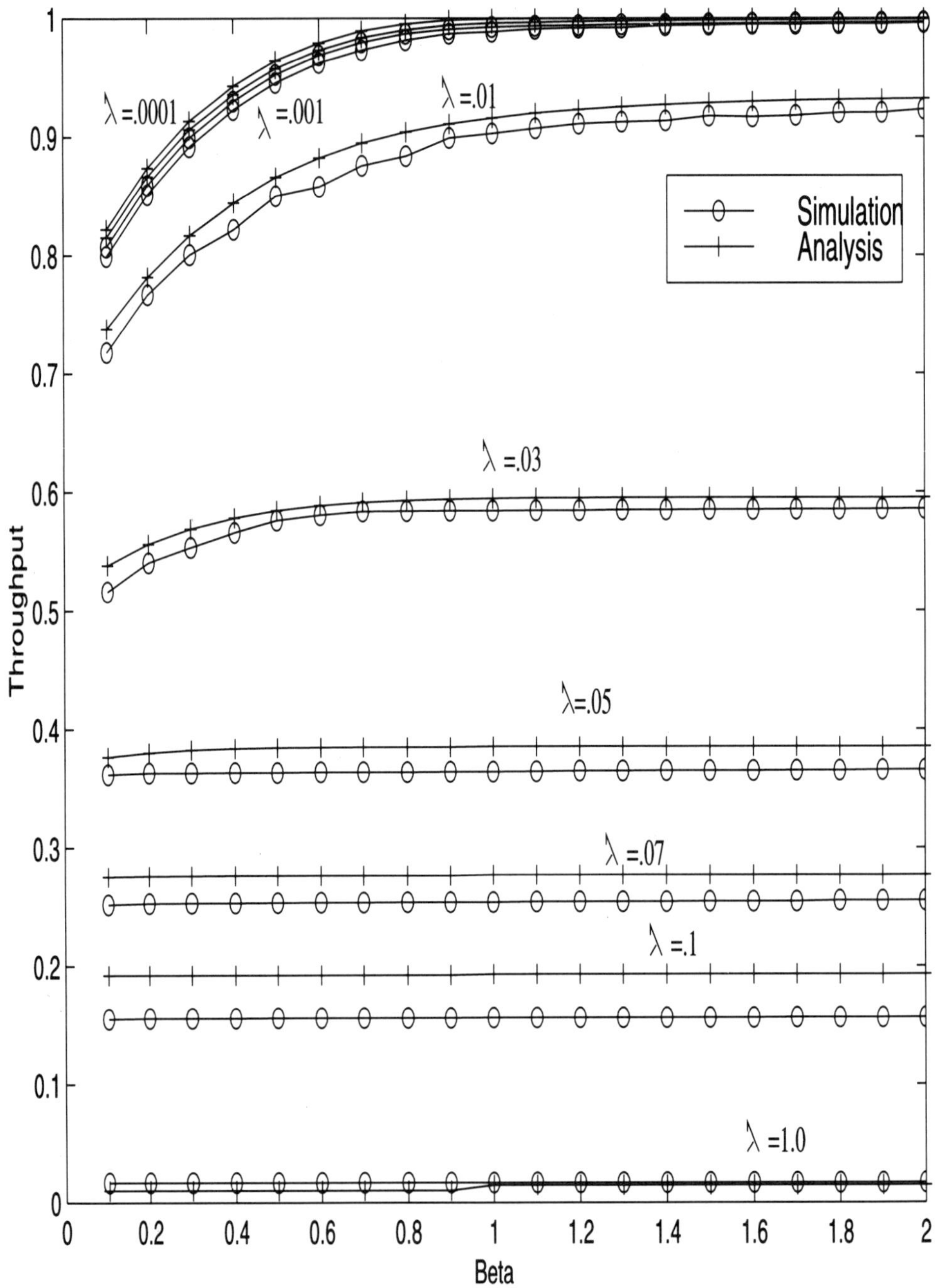

FIGURE 7. $\mu = 100, \tau = 1.0$

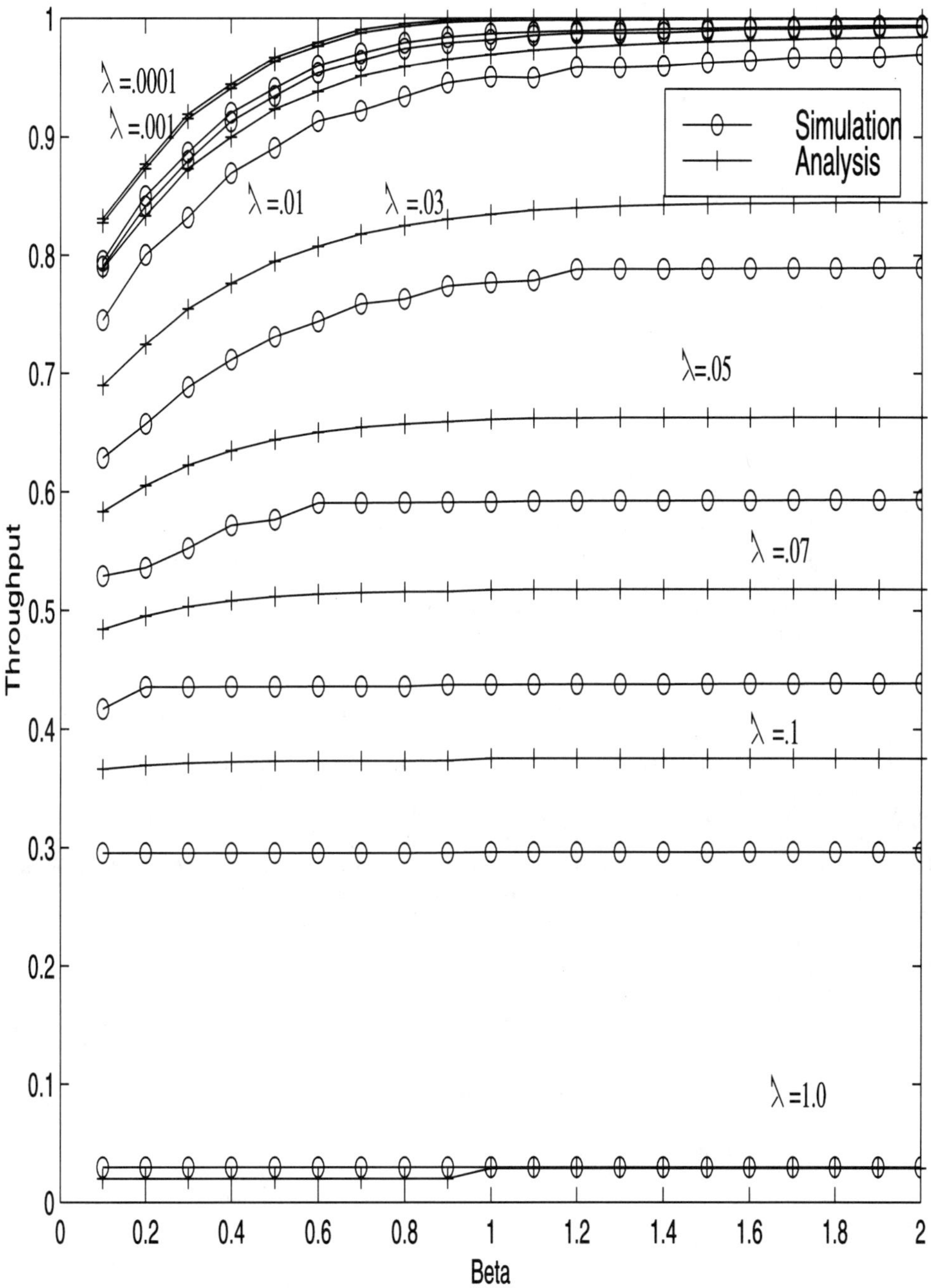

FIGURE 8. $\mu = 50, \tau = 1.0$

analysis. As λ increases, so does the deviation since it becomes increasingly probable that the first random loss takes place early in the slow start phase, thereby precipitating a congestion avoidance phase with an initial window size that is considerably smaller than $\frac{w_p}{2}$ as assumed in the approximation. Consequently, the simulated throughput (on the average) is lower than that predicted by analysis, most noticeably for moderate values of random loss. For heavy loss rates, the deviation decreases again since the approximate window size quickly decreases from its starting value of $w_p/2$ to that (i.e., the true) in the simulations.

(b) The deviation is larger for lower bandwidth-delay product. Recall that the reason for neglecting the slow start phase in the first cycle of TCP-R is that the duration as well as *the number of packets sent* during this phase is low. However, this becomes increasingly inaccurate as w_p decreases. Since $w_p = (\mu\tau + 1)(\beta + 1)$, decreasing $\mu\tau$ (the band-width delay product) decreases w_p and hence increases the deviation for the same value of β.

References

[1] S. Shenker, L. Zhang, and D. D. Clark, "Some observations on the dynamics of a congestion control algorithm," *Computer Communications Review*, pp. 30-39, Oct 1990.

[2] T. J. Ott, J. H. B. Kemperman, and M. Mathis, "The stationary behavior of ideal TCP congestion avoidance," 1996.

[3] T. V. Lakshman, and U. Madhow, "The performance of TCP/IP for networks with high bandwidth-delay products and Random loss," *IEEE/ACM Transactions on Networking*, vol. 5, no. 3, June 1997.

[4] T. V. Lakshman, and U. Madhow, "Window-based error recovery and flow control with a slow acknowledgment channel: a study of TCP/IP performance," *Proceedings of IEEE INFOCOM '97.*

[5] S. Savari, and E. Telatar, "The behavior of certain stochastic processes arising in window protocols," *preprint*, July 1998.

[6] V. Jacobson, "Congestion avoidance and control," *Proceedings of ACM SIGCOMM ' 88*, August 1988.

[7] V. Jacobson, "Modified TCP congestion avoidance algorithm," *Message to end2end-interest mailing list*, April 1990, *URL ftp://ftp.ee.lbl.gov/email/vanj.90apr30.txt.*

[8] J. Bolot, "End-to-end packet delay and loss behavior in the internet," *Proc. ACM SIGCOMM'93.*

[9] S. M. Ross, "Stochastic processes," New York: Wiley, 1983.

[10] R. G. Gallager, "Discrete stochastic processes," Boston : Kluwer, 1996.

[11] S. Floyd, "TCP and successive fast retransmits," February 1995, *URL ftp://ftp.ee.lbl.gov/email/vanj.90apr30.txt.*

Department of Electrical Engineering, University of Washington, Box 352500, Seattle WA 98195

E-mail address: hussein@ee.washington.edu

Department of Electrical Engineering, University of Washington, Box 352500, Seattle WA 98195

E-mail address: azizoglu@ee.washington.edu

Department of Electrical Engineering, University of Washington, Box 352500, Seattle WA 98195

E-mail address: roy@ee.washington.edu

DIMACS Series in Discrete Mathematics
and Theoretical Computer Science
Volume **52**, 2000

Tracking of Multi-level Modulation Formats for DS/CDMA Systems in A Slowly Fading Channel

Ali F. Almutairi, Scott L. Miller, and Haniph A. Latchman

ABSTRACT. This paper studies the behavior of the minimum mean-squared error (**MMSE**) receiver in a frequency-nonselective slow fading channel when multi-level modulation formats, like 16QAM, are used in direct sequence code-division multiple access (**DS-CDMA**) systems. A detailed study of the behavior of the MMSE receiver in a slowly fading channel is shown. A modified MMSE receiver structure is proposed and a study of three channel tracking methods is presented.

1. Introduction

Traditionally, higher level modulation has been used to achieve higher bandwidth efficiency. The price for higher bandwidth efficiency is paid in terms of the required signal to interference plus noise ratio (SINR) to achieve the same bit error rate (BER). Frequently, BPSK is often used as an underlying modulation format with the MMSE receiver. The MMSE receiver, which has been described in detail in [**11**], [**5**], [**6**], is a single user receiver known for its low complexity and ease of implementation. In addition, the MMSE receiver is known to be near-far resistant.

The use of multi-level modulation formats, like 16QAM, lead to some interesting research problems. As with unspread systems, any time a multilevel modulation format is used, it becomes necessary to carefully track the fading. In this work, we propose to track the fading process through the use of periodically spaced pilot symbols and linear prediction. Previous work has shown that decision directed adaptive MMSE receiver have a difficult time operating in fading channels [**2**]. In a fading channel, the received signal strength can dip to very low levels for fairly long periods of time. During that time, decisions made by the MMSE filter are very unreliable. When these unreliable decisions are used to adapt the MMSE filter, the filter can lose track of the desired signal. Pilot symbols are useful in two different ways. First, the MMSE receiver could be updated on pilot symbols only and remove the decision directed mode and hence eliminate the above mentioned problem. Second, pilot symbols can be used to provide channel estimates that can help the receiver to track the channel.

1991 *Mathematics Subject Classification.* Primary 94A13, 94A05; Secondary 94A40.

This paper is organized as follows. In Section 2, An overview of DS-CDMA and the MMSE receiver is given. In Section 3, the system model is described. In Section 4, the behavior of the MMSE receiver in a fading channel is studied. In Section 5, a modified receiver structure is proposed. In addition, three tracking techniques and a comparison simulation study are described. This paper concludes in Section 6 with some observations and a discussion of future work in this area.

2. Overview of DS-CDMA and The MMSE Receiver

Unlike other multiple-access techniques such as frequency division multiple-access (**FDMA**) and time division multiple-access (**TDMA**) where the channel is divided into subchannels and each user is assigned to one of the available sub-channels, CDMA is a digital communication multiple access technique in which the channel is not partitioned in frequency or time but each user is given a distinct spreading sequence to access the channel. In general, in CDMA systems, spreading is accomplished by either direct sequence (DS) or frequency-hopping (FH). In this work, we chose the first method as a mean of spreading . The literature is rich with many outstanding papers about CDMA systems icluding papers such as, ([**9**], [**12**], [**13**],and [**3**]) just to mention a few. In DS-CDMA, data symbols of duration T_S of each user are multiplied by unique narrow chips of duration T_c. The chip rate is N times the symbol rate where N is the spreading gain.

Due to the presence of interference caused by other users, the performance matched-filter type receiver (which is optimum for a single user in additive white Gaussian noise (AWGN) channel) degrades substantially. The performance of the conventional receiver was analyzed in [**10**] and [**14**]. A major defficiency of the conventional receiver is its inability to mitigate what is called the near-far problem. The near-far problem occurs when the received signal of the desired user is overwhelmed by the interfering signals of the other users.

To improve the performance of the CDMA system in the presence of multiple access interference (MAI), and to mitigate the near-far problem, several receivers with different degrees of complexity and performance have been developed. The MMSE receiver is a suboptimum receiver which is known to be near-far resistant. In addition, the MMSE receiver does not require a priori knowledge of certain side information like the code sequence and the carrier frequency of the desired user. This information can be obtained through adequate training if the MMSE is implemented in its adaptive form. practical adaptive algorithms such as the least-mean-square (LMS) and recursive least-square (RLS) can be used to obtain the tap weights of the filter. The performance of the MMSE receiver in an AWGN is presented in [**1**], [**11**], [**5**], [**6**],and [**7**] and in a fading channel in [**2**] and [**8**]. To understand the advantages of the MMSE receiver, we need to describe briefly how it works.

The MMSE receiver is shown in Figure 1. The received signal which consists of the desired user's signal, MAI, and Gaussian noise is fed at the chip rate into the equalizer until the N-tap delay line becomes full. After one symbol time, the equalizer content are correlated with the tap weight vector, **a**, and the result of this correlation is used to make a decision about which symbol was sent. These tap weights are updated every symbol interval to minimize the mean square error between the output of the filter and the desired output. In practice, the filter is trained for a reasonable period of time by a known training sequence to reach a

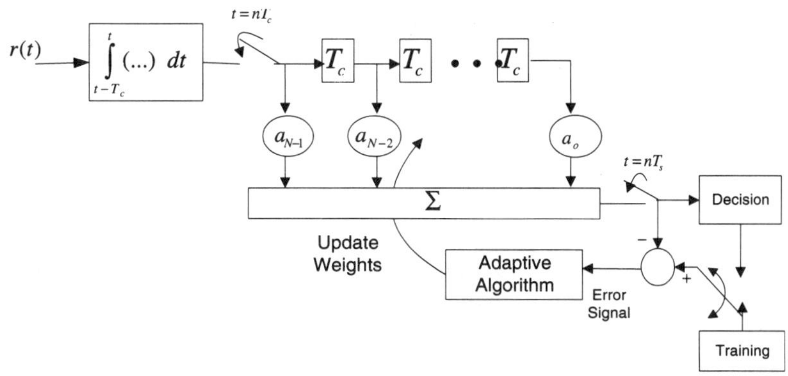

FIGURE 1. The MMSE receiver.

tap weight vector that is close to the optimum weights. After the training period, the receiver switches to decision feedback mode. It has been shown in [**2**] that the decision directed mode proves to be troublesome in a fading channel. In deep fades, with the MMSE structure shown in Figure 1, incorrect decisions are fed back to the receiver and this causes the MMSE receiver to lose track of the desired signal. A modified MMSE receiver structure to overcome this problem was described in [**2**] for a BPSK modulation format.

3. System Model

In this section, a CDMA system model based on the MMSE receiver is described. Each user is assumed to have a unique spreading waveform $c_i(t)$. It is assumed that all users transmit asynchronously over the same channel with carrier frequency w_o. There are K active users in the system. The modulated signal of the jth user can be written as

$$\begin{aligned} s_j(t) &= \mathrm{Re}\{\sqrt{2p_j}d_j(t)c_j(t)e^{jw_ot}\} \\ &= \mathrm{Re}\{g_j(t)e^{jw_ot}\} \end{aligned} \tag{3.1}$$

where $g_j(t)$ is the complex envelope of $s_j(t)$, p_j is the transmitted power, and $d_j(t)$ is a complex baseband signalling format with symbol interval T_s. The waveform $c_j(t)$ is assumed to be in the polar form with chip interval T_c. Therefore, the processing gain N is equal to T_s/T_c. The bandpass received signal at the receiver is given by

$$r(t) = \mathrm{Re}\{\sum_{j=1}^{K} \alpha_j(t)e^{j\theta_j(t)}g_j(t-\tau_j)e^{jw_ot}\} + n(t) \tag{3.2}$$

The variables τ_j, α_j, θ_j are the propagation delay, and the amplitude and phase of the fading process for the jth user. The process $n(t)$ is a real AWGN process with a spectral density of $N_o/2$. The fading amplitude is Rayleigh-distributed while the fading phase is uniformly-distributed. It is assumed, without loss of generality, that user 1 is the desired user and its propagation delay is assumed to be 0. In addition, it is assumed that the fading process of each user varies at a slow rate so that the amplitude and the phase of the fading process can be assumed constant over the duration of a symbol. In addition, the chip matched filter has a scale factor of $\sqrt{2p_1}\,T_c$. Based on these assumptions, the contents of the equalizer are given by

$$\mathbf{r}(m) = d_1(m)\alpha_1(m)e^{j\theta_1(m)}\mathbf{c}_1 + \sum_{j=2}^{K}\sqrt{\frac{p_j}{p_1}}\alpha_j(m)e^{j\theta_j(m)}\Big[d_j(m)\widetilde{\mathbf{f}}_j(l,\delta) + d_j(m-1)\widetilde{\mathbf{g}}_j(l,\delta)\Big] + \mathbf{n}(m) \tag{3.3}$$

Here the vector $\mathbf{n}(m)$ consists of elements which are independent zero-mean complex Gaussian random variables whose real and imaginary parts have variances of $N/(2E_s/N_o)$ where E_s is the average energy per symbol. The delay of the jth user is given by $\tau_j = l_jT_c + \delta_j$ where l_j is an integer and $0 \le \delta_j < T_c$. The vectors $\widetilde{\mathbf{f}}_j$ and $\widetilde{\mathbf{g}}_j$ are defined as follows

$$\widetilde{\mathbf{f}}_j(l,\delta) = \frac{\delta}{T_c}\mathbf{f}_j(N-l-1) + \left(1-\frac{\delta}{T_c}\right)\mathbf{f}_j(N-l)$$

$$\widetilde{\mathbf{g}}_j(l,\delta) = \frac{\delta}{T_c}\mathbf{g}_j(N-l-1) + \left(1-\frac{\delta}{T_c}\right)\mathbf{g}_j(N-l)$$

where

$$\mathbf{f}_j(l) = (\mathbf{c}_j^{(l)} + \widehat{\mathbf{c}}_j^{(l)})/2$$

$$\mathbf{g}_j(l) = (\mathbf{c}_j^{(l)} - \widehat{\mathbf{c}}_j^{(l)})/2$$

$$\mathbf{c}_j^{(l)} = (c_{j,N-l}, c_{j,N-l+1}, ..., c_{j,N-1}, c_{j,0}, c_{j,1}, ..., c_{j,N-l-1})^T$$

$$\widehat{\mathbf{c}}_j^{(l)} = (-c_{j,N-l}, -c_{j,N-l+1}, ..., -c_{j,N-1}, -c_{j,0}, c_{j,1}, ..., c_{j,N-l-1})^T$$

We define the following terms to be used in the next section. The autocorrelation matrix, $\mathbf{R}$, of the equalizer contents is defined as $\mathbf{R} = E\left[\mathbf{r}(m)\mathbf{r}^H(m)\right]$ and the correlation between the desired user response and the received signal is given by $\mathbf{P}_i = E\left[d_i^*(m)\mathbf{r}(m)\right]$. The minimum mean-squared error $J_{\min}$ is given by [**4**]

$$J_{\min} = \sigma^2_{\hat{d}_i} - \mathbf{P}_i^H\mathbf{R}^{-1}\mathbf{P}_i \tag{3.4}$$

where $\sigma^2_{\hat{d}_i}$ is the variance of the estimate of the desired data when the MMSE receiver is operating in decision feedback mode.

4. The MMSE Receiver Behavior in A Fading Channel

In this section, we study the behavior of the MMSE receiver in a fading channel when a multilevel modulation format is used. Since tracking the phase and magnitude of the fading is essential for successful demodulation of a multilevel modulation format, we will study the ability of the present structure of the MMSE receiver to track these fading parameters. In [**2**], the performance of the MMSE receiver in a frequency nonselective fading channel has been evaluated when a BPSK modulation format is used. It has been shown that the MMSE has a difficult time tracking the channel variation due to the fact that during deep fades, unreliable decisions are fed back to the LMS algorithm. This will cause the MMSE receiver to lose lock on the desired signal or it may lock onto another interfering signal.

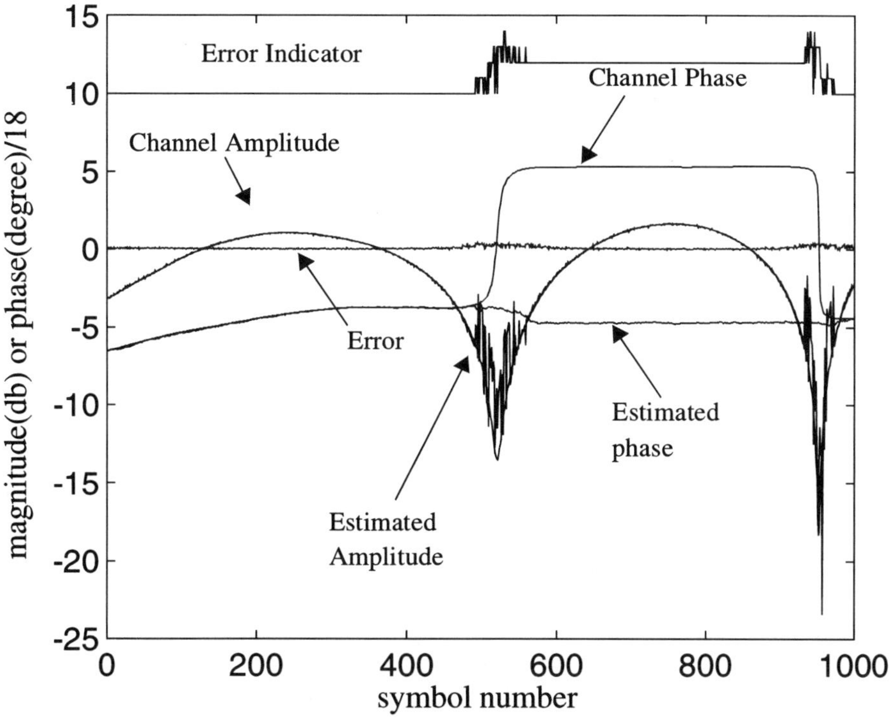

FIGURE 2. The MMSE behavior in a fading channel in decision directed mode.

In this paper, we assume a slow fading environment with a mobile speed of 5 mph, a frequency band of 900 MHz, and a data rate of 9600 bps. This will result in a Doppler frequency of 6.7 Hz. Figure 2 demonstrates the behavior of the present MMSE structure in a slowly varying Rayleigh fading channel for a single user using 16QAM modulation. As expected, the figure shows the inability of the receiver to track the magnitude and phase of the fading process when the desired user goes into deep fades. The phase estimate in Figure 2 represents the MMSE receiver's estimate of the phase based on the receiver coefficients. In a single user case, if the MMSE is doing its job of tracking the channel variation, the phase of the MMSE filter coefficients is equal to the opposite value of the phase of the channel. The amplitude estimate is calculated from the value of the filter output. It is clear from Figure 2 that the MMSE receiver does a good job in tracking the amplitude variation of the fading channel except during the deep fade period. On the other hand, the receiver does a poor job in tracking the phase of the fading process. In fact, the receiver ends up locked 180° out of phase to the desired user after the deep fade period is over. Differential detection may be considered to solve this problem, but differential encoding will not solve a more practical problem, that is when the MMSE receiver locks on to other interfering signals. Figure 3 shows that in training mode, the MMSE receiver always tracks phase and amplitude of the fading channel well. This shows that the decision-directed mode of operation of the MMSE receiver is a disadvantage to its performance in this environment. Therefore, if there is a technique by which we can feed back reliable decisions to

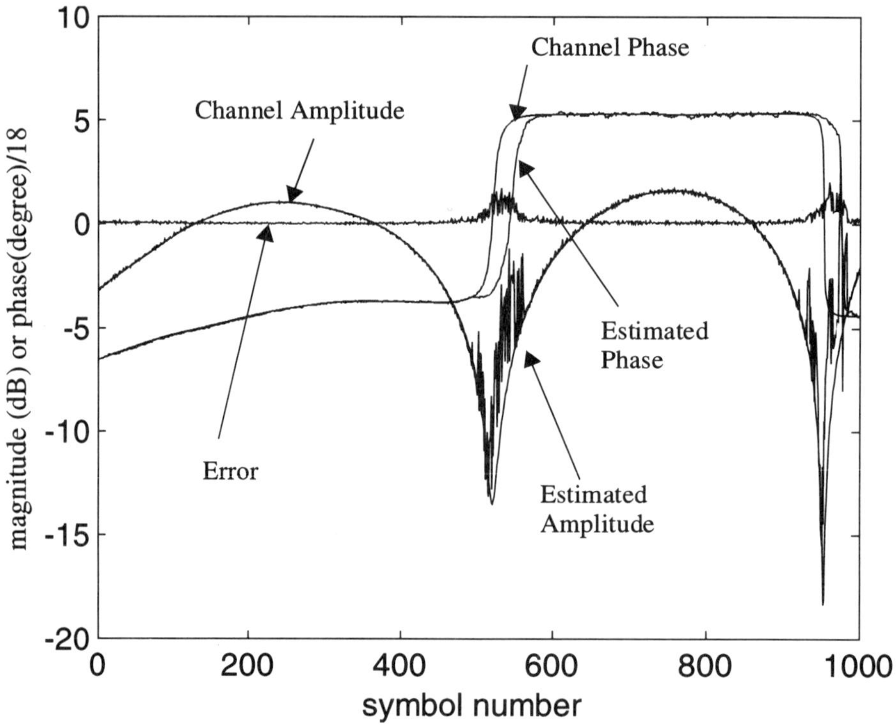

FIGURE 3. The MMSE behavior in a fading channel in a training mode

the adaptive algorithm, the LMS in this case, then the MMSE will perform in an acceptable manner. This is part of the motivation for using periodic pilot symbols to provide a reliable feedback for the LMS and this will be discussed in the next section.

5. Tracking Techniques in A Fading Channel

Having shown the inability of the present MMSE structure to work in the environment described in the previous section, we now consider modification of the MMSE receiver to be capable of demodulating multilevel modulation scheme in a fading environment. In [**2**], a modified MMSE structure for one-dimensional (BPSK) modulation is presented. For the rest of this paper, the same simulation environment is used except that we have 3 users in the system instead of a single user. In this section, we will present a more general modified MMSE structure capable of demodulating a wide range of digital modulation formats. First, since the errors due to the phase of the fading process are dominant, we need to eliminate this phase variation from the input to the adaptive filter. In addition, to eliminate the problem of the MMSE receiver locking to other user's phases, we need to take the real and imaginary part of the input to the adaptive filter. The modified structure is shown in Figure 4. This structure assumes an estimate of the amplitude and phase of the fading process are available at the receiver. It has been shown by [**2**] that phase compensation is an effective method of improving the MMSE receiver performance in a fading channel. In [**2**] a phase estimate is obtained by using a linear

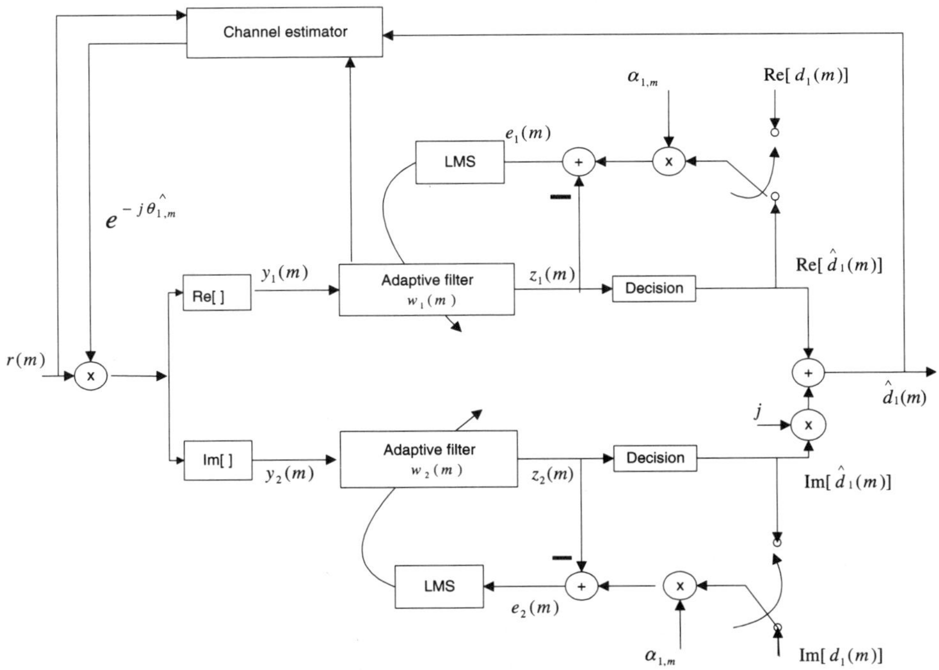

FIGURE 4. The modified MMSE structure.

predictor. In our case, since we are dealing with multilevel modulation, 16QAM, amplitude and phase compensation are needed to improve the performance of the MMSE receiver. We studied the capabilities of three techniques in tracking the fading amplitude and phase. These techniques are based on pilot symbols and/or linear prediction. The first tracking technique uses the decision out of the MMSE to form an estimate of the phase of the desired user's fading parameters using linear prediction. The channel estimation based on this technique is shown in Figure 5. This technique is presented in some detail in [**2**] for a CDMA system with BPSK modulation. It worked fairly well for BPSK modulation but not in the case here where 16QAM modulation is used. Thats motivite the search for a better tracking method. We will now summarize the procedure used to obtain channel estimates using linear prediction.

From Figure 4, the output of the filter output, $z(m)$, when $\mathbf{r}(m)$ is the input, is given by

$$z(m) = d_1(m)\alpha_{1,m}e^{j\theta_{1,m}}\mathbf{a}^T\mathbf{c}_1 + \tilde{\mathbf{n}} \tag{5.1}$$

A noisy estimate of the fading process can be given by

$$\begin{aligned}\beta(m) &= \frac{z(m)}{d_1(m)\mathbf{a}^T\mathbf{c}_1} \\ &\approx \alpha_{1,m}e^{j\theta_{1,m}}\end{aligned} \tag{5.2}$$

In a decision-directed mode, $d_1(m)$ is replaced by $\hat{d_1}(m)$. The linear prediction can be formulated by the following.

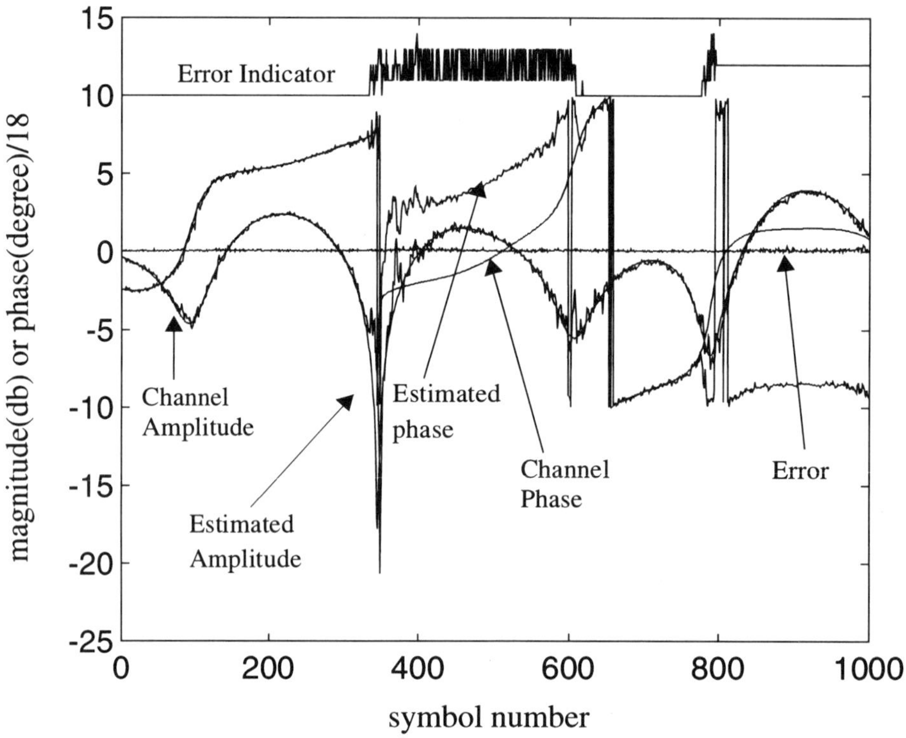

FIGURE 5. Channel tracking using linear prediction.

The Lth order linear prediction of the fading channel is given by

$$\hat{\beta}(m) = \sum_{i=1}^{L} \hat{a}_i \beta(m-i) \tag{5.3}$$

The optimum coefficients of the linear predictor are given by

$$\hat{\mathbf{a}} = \mathbf{C}^{-1}\mathbf{v} \tag{5.4}$$

The expressions for $\mathbf{C}$ and $\mathbf{v}$ for the single user case are given in [**2**] as

$$\mathbf{C} = \mathbf{B} + \left(\frac{E_b}{N_0}\right)^{-1}\mathbf{I} \tag{5.5}$$

where $\mathbf{B}$ is a L x L matrix whose elements are given by

$$[\mathbf{B}]_{i,j} = R_c((i-j)T_s) \tag{5.6}$$

$$[\mathbf{v}]_i = R_c(iT_s) \tag{5.7}$$

and $R_c(\tau)$ is the autocorrelation function of the fading process and is approximated by

$$R_c(\tau) = 1 - (\pi f_D \tau)^2 \tag{5.8}$$

The phase and the amplitude estimates of the fading process out of the linear predictor are then used to remove the phase of the desired user fading from the input of the modified MMSE receiver and to scale the decisions from the modified MMSE

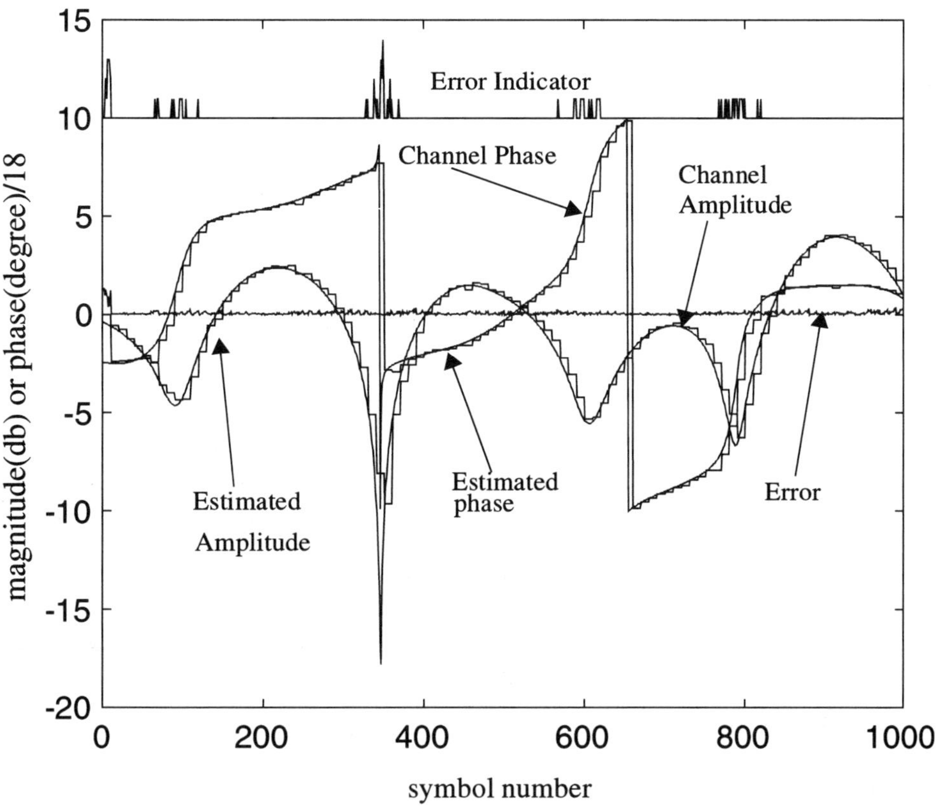

FIGURE 6. Channel tracking using pilot symbols.

receiver, respectively. The use of this technique in [**2**] for a BPSK modulation resulted in efficient tracking of the fading process.

The second tracking technique is based on pilot symbols. The result of tracking the fading channel using this technique is shown in Figure 6. In this technique, pilot symbols, known by the receiver, are sent periodically (every 10th symbol for the case reported in Figure 6). The MMSE receiver uses these pilots to obtain an estimate for the fading process in the same manner as in Eqn. 5.2. The fading parameters obtained by this estimate are used in demodulating the desired user's signal until the next pilot symbol is received and a new estimate is made.

We propose the use of pilot symbols for two reasons. First, pilot symbols can be used to periodically train the MMSE and prevent the MMSE filter from feeding back wrong decisions. The second reason for using pilot symbols is to aid the receiver in estimating the channel fading condition. The fading parameters obtained by this estimate are used in demodulating the desired user's signal until the next pilot symbol is received and a new estimate is made. Obviously, this technique is suitable for fading slowly channel and may not work well for a rapidly fading channel.

The third approach is a combination of the first and second techniques. The tracking of the fading channel using this technique is shown in Figure 7. In this case, channel estimates are made by feeding back a linear prediction of the previous channel estimates, some of which were based on pilot symbols.

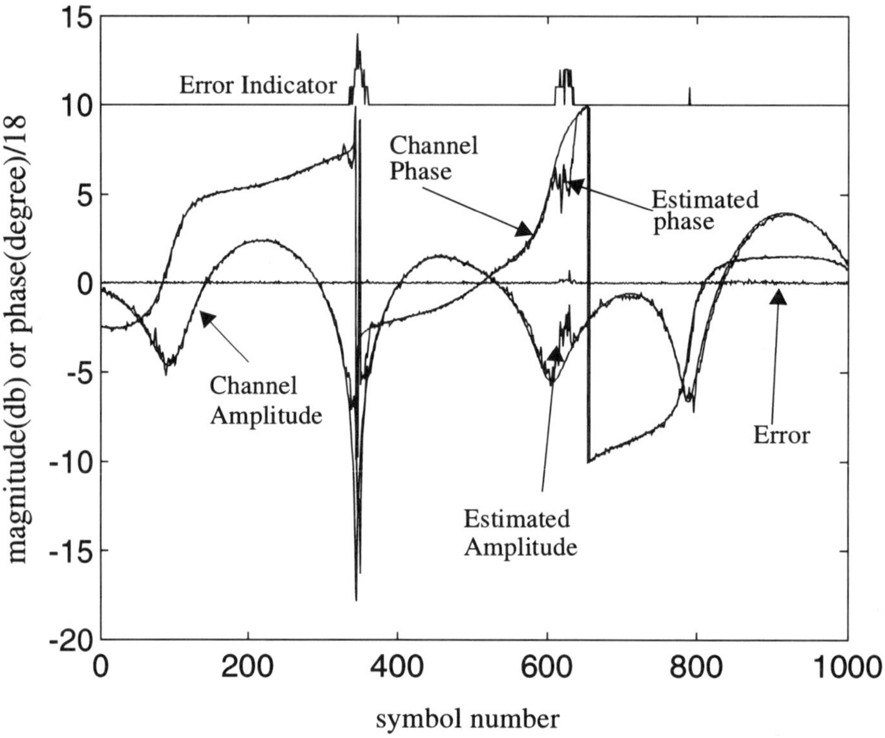

FIGURE 7. Channel tracking using pilot symbols and linear prediction

By comparing Figures 5, 6, and 7, one can conclude that the third technique has better tracking capabilities than those of the other techniques. The good performance of the third technique can be attributed to two reasons. First, the use of pilot symbols provide the MMSE receiver with a reference that helps the receiver not to lose lock on the desired user. Second, using the linear predictor, estimates are made for every received symbol. This gives the linear predictor recent past channel estimates to predict the channel conditions. Figure 6 demonstrates that the MMSE receiver can be updated based on pilot symbols only. This is interesting since the poor performance of the MMSE receiver in a fading channel is often due to the feeding back of unreliable decisions to the adaptive algorithm during deep fades.

6. Conclusion

In this paper, we have investigated the performance of the MMSE receiver in a fading channel when 16QAM modulation is used. We demonstrated the inability of the present MMSE structure to operate in a fading channel for one- and two-dimension modulation formats. We presented a general structure of the MMSE receiver which can operate effectively in a fading channel. This structure is based on estimating the phase and the magnitude of the fading process. Estimates of the fading process are made by using linear prediction and periodic pilot symbols. Besides being used for channel estimates, pilot symbols can be used for periodic reference for the the adaptive algrothm.

This work illustrates the feasibility of employing a multilevel modulation like 16QAM in MMSE receiver based CDMAsystems operating in a fading channel and taking advantage of the higher bandwidth efficiency offered by this type of modulation. Although a multilevel modulation, like 16QAM, will need a higher SINR to operate effectively, this can be achieved by increasing the transmitter power and taking advantage near-far resistance properties of the MMSE receiver.

References

1. Majeed Abdulrahan, Asrar U. H. Sheikh, and David D. Falconer, *Decision feedback equalization for CDMA in indoor wireless communications*, IEEE Journal on Selected Areas in Communications **12** (1994), no. 4, 698–706.
2. Afonso N. Barbosa and Scott L. Miller, *Adaptive detection of DS/CDMA signals in fading channels*, IEEE Transactions on Communications **46** (1998), no. 1, 115–124.
3. C.E. Cook and H.S. Marsh, *An introduction to spread spectrum*, IEEE Communication Magazine (1983), 8–16.
4. Haykin.S, *Adaptive filter theory*, Printice Hall, 1996.
5. Upamanyu Madhow and Michael L. Honig, *MMSE interference suppression for direct-sequence spread-spectrum CDMA*, IEEE Transactions on Communications **42** (1994), no. 12, 3178–3188.
6. Scott L. Miller, *An adaptive direct-sequence code-division multiple-access receiver for multiuser interference rejection*, IEEE Transactions on Communications **43** (1995), no. 2/3/4, 1746–1755.
7. Charles N. Pateros and Gary J. Saulnier, *Interference suppression and multipath mitigation using an adaptive correlator direct sequence spread spectrum receiver*, Proceedings IEEE International Conference on Communications, 1992, pp. 662–666.
8. ———, *Adaptive correlator receiver performance in fading multipath channels*, Proceedings 43rd IEEE Vehicular Technology Conference (Secaucus, NJ), 1993, pp. 746–749.
9. Raymond L. Pickholtz, Donald L. Schilling, and Laurence B. Milstein, *Theory of spread-spectrum communications—a tutorial*, IEEE Transactions on Communications **COM–30** (1982), no. 5, 855–884.
10. Michael B. Pursley, *Performance evaluation for phase-coded spread-spectrum multiple-access communication - Part I: System analysis*, IEEE Transactions on Communications **25** (1977), no. 8, 795–799.
11. Predrag B. Rapajic and Branka S. Vucetic, *Adaptive receiver structures for asynchronous CDMA systems*, IEEE Journal on Selected Areas in Communications **12** (1994), no. 4, 685–697.
12. Robert A. Scholtz, *The spread spectrum concept*, IEEE Transactions on Communications **8** (1977), no. COM–25, 748–755.
13. William F. Utlant, *Principles and possible application to spectrum utilization and allocation*, IEEE Communication Society Magazine (1978), 21–31.
14. K. Yao, *Error probability of asynchronous spread spectrum multiple access communication systems*, IEEE Transactions on Communications **COM-25** (1977), 803–809.

Laboratory for Information Systems and Telecommunications, University of Florida, 458 ENG, Building 33, PO BOX 116130, Gainesville, FL 32611-6130, Phone: (352)392-2584, Fax: (352)392-0044

E-mail address: `almutair@ecel.ufl.edu`

Department of Electrical Engineering, Texas A&M University, College Station, TX 77843-3128

E-mail address: `miller@bashful.tamu.edu`

Laboratory for Information Systems and Telecommunications, University of Florida, 458 ENG, Building 33, PO BOX 116130, Gainesville, FL 32611-613

E-mail address: `latchman@list.ufl.edu`

DIMACS Series in Discrete Mathematics
and Theoretical Computer Science
Volume **52**, 2000

An Adaptive Concurrency Control Protocol for Mobile Transactions

Elisa Bertino, Elena Pagani, and Gian Paolo Rossi

ABSTRACT. Wireless networks are becoming an essential component of any infrastructure for distributed data management. Unlike wired hosts, mobile hosts can be temporarily unreachable as consequence of their movements across different cells, their energy limitation or unavailable wireless channels. Mobility forces mobile hosts to alternate connected and disconnected work. When connected, they perform personal communications and access shared data and services from fixed hosts. When disconnected, they can process locally cached data. Transaction management in such an environment poses several issues and requires extensive modifications to techniques and algorithms developed for wired networks. One relevant issue is to devise caching strategies according to which data can be cached at mobile hosts to enable disconnected work. When working in disconnected mode, a mobile host performs updates on its own cached copy of data. Such updates are then applied to the data stored at the fixed hosts when the mobile host reconnects. We refer to such transactions as mobile transactions. To guarantee database consistency proper concurrency control mechanisms must be adopted to serialize mobile transactions. Two possible approaches are the pessimistic approach, under which only one mobile transaction at the time may have a cached copy of the data, and the optimistic approach, under which all mobile transactions may have a cached copy of the same data. Both approaches have severe shortcomings, that are discussed in the paper, and no one is better than the other under all application patterns. In this paper, we propose an adaptive concurrency control protocol that dynamically modulates the number of data copies that can be concurrently cached by mobile hosts. We evaluate the performance of our protocol through simulation. In particular, the evaluation shows that our protocol helps in minimizing energy expenditure by mobile hosts - an important requirement in mobile computing systems.

1. Introduction

Wireless networks are becoming an important component of a system of interconnected networks. The most rapidly growing segment of personal computers, which includes both portable computers and Personal Digital Assistants (PDAs), can find in the wireless technology the infrastructure to access data and services

1991 *Mathematics Subject Classification.* Primary 68M10, 68M20, 68P15.
Key words and phrases. mobile host networks, concurrency control, replicated databases.

available at the wired part of the Internet. Although such communication infrastructure represents an important opportunity to deploy a variety of new distributed applications (see, for instance, [**20**]), the interconnection between wired and wireless networks requires adaptations at different layers of the communication protocol stack, from the routing protocols up to the application support protocols.

While the development of wireless networks poses several challenging issues to telecommunication research [**16, 5**], several issues also arise with respect to data management research [**14, 6**]. In particular, as one of the main application areas for mobile computing systems refers to ubiquitous information use and modification, proper techniques for transaction management must be devised. Transaction management has already been widely investigated in the area of database systems (both centralized and distributed) [**12, 1**]; problems and their solutions are today well understood. Transaction management in mobile computing systems poses, however, more difficult issues because of the narrow bandwidth of the wireless communication channels, the limited power supply and the changing locations of mobile units. Such characteristics require extensive modifications to existing transaction management techniques. In particular, whereas in database systems over wired networks, the key performance metric is the response time, in wireless networks both response time and energy expenditure of mobile units are important [**10**]. As a consequence, algorithms that were designed to optimize response time only may not be any longer adequate when both response time and energy expenditure must be optimized.

To better focus our discussion, we consider the case of a transactional environment in which mobile hosts (MHs) access databases located on fixed hosts (FHs)[1]. Here, the mobility introduces four main issues: (1) Mobile hosts may need to work in *disconnected* mode, or may happen to be temporarily disconnected. As a consequence, transactions may last longer. (2) MHs are free to move among wireless networks whose technology, traffic conditions and bandwidth (ranging from Kbps to Mbps) can be very different. Thus, the network resource can be sometimes scarce. (3) The battery life-time represents an extremely critical resource for all mobile devices. The communications over the wireless links are the main responsible of the battery consumption and, therefore, the packet exchange between mobile and fixed hosts should be minimized. (4) A very large number of potential users can access the databases from thin, mobile, chip devices. Scalability is one of the primary requirements.

Unfortunately, those four issues entail contrasting requirements with respect to a transaction management system. The need of supporting disconnected work and the characteristics of the communication infrastructure (issues (1) and (2) above) require the use of a caching mechanism according to which data can be stored at MHs, thus enabling MHs autonomy. When working in disconnected mode, a MH performs updates on its own cached copy of data. Such updates are then applied to the data stored at the FHs when the MH reconnects. We refer to such transactions as *mobile transactions*. To guarantee database consistency proper concurrency control mechanisms must be adopted to serialize mobile transactions. Two possible approaches are the pessimistic approach, under which only one mobile transaction at the time may have a cached copy of the data, and the optimistic

[1]Note that other architectures have been devised such as the multi-hop architectures, where all the hosts in the system are mobile. Such architectures are out of the scope of the present paper.

approach, under which all mobile transactions may have a cached copy of the same data.

The *pessimistic* approach, adopted for instance in [**6**], avoids conflicting updates by preventing transactions from executing operations that might *collide*; this is generally achieved through the use of *locks* on data items. Such an approach minimizes the use of critical resources at the price, however, of high latency time, and low throughput. Furthermore, sudden disconnections of MHs can cause *orphan* locks and connected work can be penalized by disconnected work. On the other hand, the *optimistic* approach [**18**] has the opposite behavior by allowing the MHs to concurrently cache data item copies according to their needs. Such an approach increases transaction throughput and reduces latency time. Moreover, when the number of concurrent accesses to the same data items is low, such as in a file system, an optimistic approach fully enables disconnected work. Possible conflicts can be repaired upon reconnection [**23, 15**] through the use of transaction recovery protocols. However, the presence of a large number of MHs (issue (4) above) may drastically increase the number of concurrent transactions and, thus, the frequency of conflicts. When the number of conflicting transaction increases, MHs are repeatedly involved in the recovery of aborted transactions. Transaction recovery can be very expensive in terms of both local processing power and network resources. In this case, disconnected work can be penalized by connected work.

It is thus clear that when a transaction management system must support mobility, by ensuring at the same time scalability and resource usage optimization, with good performance, the adoption of a single concurrency control policy may result in severe shortcomings. Different policies must be adopted that dynamically range between the optimistic and the pessimistic approaches according to the dynamics of the competing transactions, as defined by the access patterns, and of the system resources status.

In this paper, we propose an *Adaptive Concurrency Control Protocol*, henceforth also called *ACP*, which allows the system to modulate the number of copies of data items that are concurrently cached at MHs. By dynamically varying such number, the system is able to modulate the number of concurrent operations on data items and thus to control the number of transaction aborts when critical resources have to be saved.

The paper is organized as follows. In Section 2, we describe the system model and the application environment which are considered throughout the paper. In Section 3, we describe the proposed adaptive protocol. The performance evaluation of the protocol, obtained by means of simulation, are given in Section 4. Finally, Section 5 compares our approach with related work, whereas Section 6 presents some concluding remarks and outlines future research.

2. System model

In our work, we have used a general architecture of a mobile computing system which we show in Figure 1 [**11**]. The architecture includes a number N of mobile hosts (MHs) accessing data items located at fixed hosts (FHs). FHs are interconnected through a wired network in which a subset of the nodes (the *Mobile Support Stations*, MSSs) are equipped with a wireless interface. An MSS supports communications among MHs within a *cell* and among MHs in other cells by relying to the remote MSS. A cell is the area in which the signal generated by the MSS can

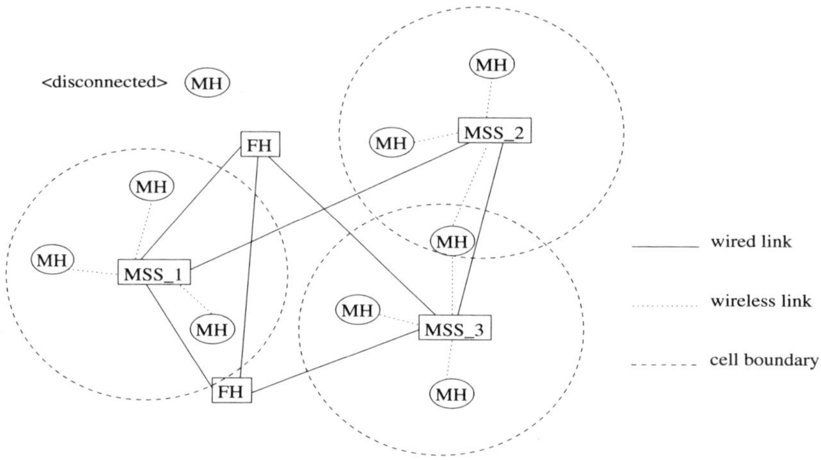

FIGURE 1. A general architecture of a mobile computing system

be received by MHs. Communications within a cell are broadcast and MHs filter the packets according to their destination address. A MH can communicate with another MH in the same cell only by relying to the local MSS.

A MH can *move* across cells, thus changing the *wireless* network topology over time. When the MH_m moves from the cell of MSS_1 to the cell of MSS_2, a *handoff* procedure is activated between MSS_1 and MSS_2 to transfer the state information about MH_m to MSS_2. A *mobility assumption* [**2, 3**] is required to ensure system liveness: MH_m remains in a cell the time needed to complete the handoff procedure and to allow the local MSS to deliver undelivered packets to MH_m; this way, message starvation is avoided.

MHs may also be temporarily *disconnected*. Disconnections occur as the consequence of *planned* events (e.g. to save energy) or of *sudden* events (e.g. moving to uncovered zones, energy exhaustion or crash failure).

We assume that all the MHs have comparable processing and memory resources. The MHs are mainly constrained by the short lifetime of the batteries, that are heavily affected by the communication over the wireless links. To reduce the waste of energy, a MH can enter the *doze mode*: a doze MH has only the network interface active. As soon as a message addressed to the MH is received, the MH is awakened to revert to the normal operation mode. This is an important facility that should be exploited by transaction management protocols designed for a mobile environment.

We consider a database as a collection of *data items*. A *version number* (vn) is associated with each data item. Applications can execute two kinds of operations on the data items: *transactions* and *queries*. Only transactions commit and they can commit if all the accessed data items have the version number of the most recent database copy. We make the assumption that an application performing a query can tolerate stale data; hence, the version numbers of the data items accessed by the query are not checked and the query does not have to commit.

The database, or part of it, can be replicated at the various FHs. We refer to such replication as *primary replication* as opposed to *secondary replication* (or *cached replication*) that refers to cache copies stored at MHs. Primary replicas are kept consistent by using one of the available *multiple copy update* protocols designed for distributed database systems [**9, 21**]. We assume that a MH is able to access the nearest *Replica Manager* (RM) whenever it needs to access a data item. Such an access can be transparently achieved by redirecting the MH request packet from the *home RM*, i.e. the RM where MH firstly logged on, to the current RM. Redirection can be performed by MSSs through tunneling techniques, which are often used to support communications between fixed and mobile stations (e.g. Mobile-IP, [**22**]). If the nearest RM of a MH changes as the consequence of mobility and before the termination of the MH's transactions, RMs communicate to exchange the transaction status of MHs that are moving across cells. This way, MHs are guaranteed that the nearest RMs they access are always aware of their transaction operations. For sake of simplicity, we assume, in the following, that RMs coincide with MSS stations.

To allow disconnected work, MHs *cache* a copy of the selected data items, together with their version numbers. We make the assumption that the cache is large enough to contain the needed data items and that data are cached before disconnection occurs. We use the term *disconnected transaction* to indicate a transaction processed on a disconnected MH.

3. The adaptive concurrency control protocol

In this section, we first outline the mobile transaction management protocol that we have adopted in our work. We then present the adaptive concurrency control mechanism, according to which data items copies are cached at MHs. The protocol is mainly based on a dynamically varying parameter, referred to as *number of releasable copies* (*rc* for short) which determines, for each data item, the number of cached copies of the data item that can be concurrently released to MHs. Such parameter can range between a lower and an upper bound. Heuristics for setting *rc* as well as its lower and upper bounds are also presented in this section. Table 1 lists, for the convenience of the reader, the acronyms and variables more frequently used in this paper. For each acronym and variable, the table reports a brief explanation of its meaning.

3.1. Mobile transaction management protocol. The transaction management protocol that we have used in our work is shown by the diagram in Figure 2. It is similar to the protocol proposed in [**15, 19**].

A MH_a that must process a transaction T_a requests to the nearest RM the data items that T_a requires. The RM replies with the copy of the requested data items according to the adopted concurrency control policy. These data items are cached at MH_a. MH_a may process T_a either while connected or while disconnected. In the former case, either the optimistic or the pessimistic approach can be adopted. If the optimistic approach is chosen, the *write-back* mechanism can be used to inform a MH that its cache is inconsistent[2], whereas if the pessimistic approach is chosen locks have to be acquired on the data items before accessing them. In the latter case (i.e., disconnected execution), MH_a records in a *log file* the updates transaction T_a makes to the cached data items. Upon successful completion, T_a *locally commits* at MH_a. We say that a transaction *locally commits* when it commits on the disconnected MH that processed it. Note that a locally committed transaction may still abort. Upon reconnection of the MH, the log-file is transferred to the

[2]See [**15**] for a discussion about the possible optimization of the write-back policy in the mobile environment.

TABLE 1. Notation and terminology

Acronym	Meaning
ACP	Adaptive Concurrency Control Protocol
MH	mobile host
FH	fixed host
MSS	mobile support station
RM	resource manager
$as(T)$	abort set of transaction T
$vn(X)$	version number of data item X
$rc(X)$	number of releasable cache copies of data item X
$out(X)$	number of currently cached copies of data item X
$cc(X)$	"commit counter" for X
LB	lower bound for rc
UB	upper bound for rc
R	number of RMs in the system
N	number of MHs in the system
n	number of MHs without resource problems
$r = n/N$	ratio of MHs without resource problems
m	constant $\simeq N$

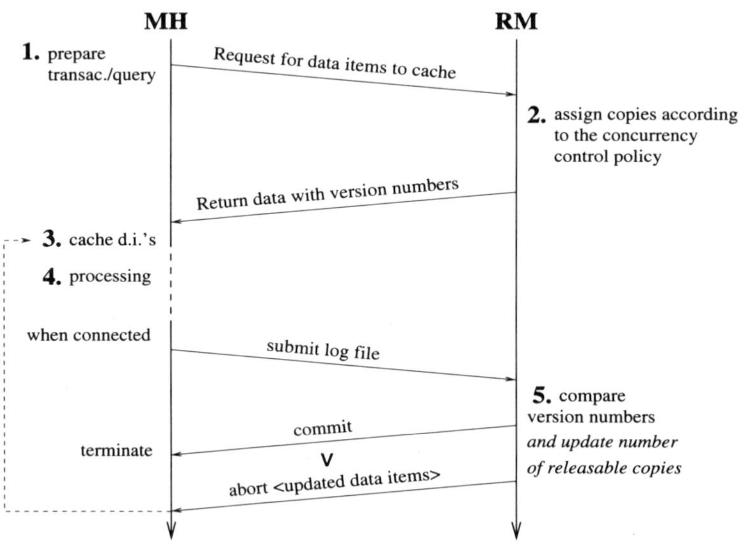

FIGURE 2. Mobile transaction management protocol

nearest RM, which forwards it to the other RMs having copies of the modified data items. All RMs check for conflicts with other possible concurrent transactions by comparing the version numbers of the data items. If no conflict occurs, all RMs redo the logged operations and the disconnected transaction *globally commits*. When a disconnected transaction globally commits, the version numbers of all the written data items are incremented. The durability of locally committed transactions is not guaranteed until they globally commit. All RMs have a consistent view of the data items and of their vn's. A RM notifies a MH whether its transaction committed or aborted. In the case of abort, a MH decides the recovery actions to undertake according to specific application requirements. To enable possible automatic re-execution of the transaction, the abort notification packet transports the copy of the current version of the transaction's data items [**19**]. The abort of a disconnected transaction T may cause the *cascading abort* of all the other disconnected transactions that locally committed after T on the same MH (see paragraph 3.2.1).

In some situations, an RM may unilateraly decide to abort a disconnected transaction. Such an abort, that we refer to as *forced abort*, is necessary in our approach to handle disconnected transactions that after a reasonable amount of time do not "connect" again. The amount of time should be large enough to let the majority of transactions complete and force the abort of a transaction only under exceptional circumstances. We refer to such amount of time as *execution timeout*. We elaborate more on its use in subsection 3.2.

Figure 3 shows the state diagram for transactions in our transaction management protocol. In the graphical representation, the state "forced to abort" denotes a forced abort of the transaction by its nearest RM because of the execution timeout has expired. In the graphical representation, an arc connecting the dashed rectangle to the "forced to abort" state denotes that a transition can occur between any state inside the rectangle and the "forced to abort" state. A RM may indeed force the abort of a transaction running at a disconnected MH, independently from the transaction status, of which the RM is unaware. The MH will be notified about the transaction abort only upon reconnection. It is important to notice that a forced abort has the same effect as a global abort with respect to data held by the aborted transaction and to transaction recovery. However, in our adaptive protocol, as we will show in the following subsection, we need to distinguish between global aborts due to concurrency control conflicts and aborts due to execution timeout expiration.[3]

Moreover, in the description of our protocol, we will use the notion of *transaction abort set*. Such a set contains the identifiers of data items for which a given transaction experienced conflicts, due to concurrent accesses, which resulted in the abort of the transaction. Let T be a transaction: its abort set is denoted as $as(T)$. For example, consider the disconnected transaction performed by MH_i in Figure 6, let us indicate it as T_i: T_i globally aborts because of conflict on the data item X; hence, $as(T_i) = \{X\}$.

3.2. Overview of the ACP protocol. The ACP protocol, used at step 2 of the transaction management protocol (Figure 2), is based on maintaining for each data item X, a parameter, denoted as $rc(X)$, indicating the number of *releasable copies* of X. Releasable copies refer to copies of X that can be cached at different MHs.

The problem of adapting to changing conditions in a mobile transactional environment can be therefore stated as the problem of dynamically modifying rc values for the various data items in the database according to the changing conditions of the environment. To this purpose, two aspects must be considered. First, the amount of the releasable copies rc of a data item should be determined according to the current access pattern. If the observed rate of conflicting transactions per data

[3]For simplicity of presentation, we assume that global aborts are only due to concurrency control conflicts. Of course, transactions may globally abort for other reasons. Our assumption can, however, be easily removed.

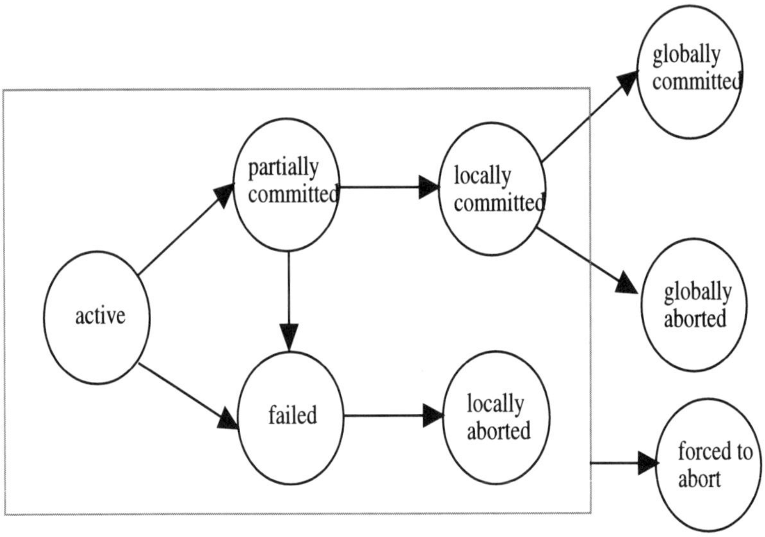

FIGURE 3. Transaction state diagram

item is low, then RMs should quickly shift towards an optimistic policy by incrementing rc. When the conflict rate grows, RMs should reduce rc accordingly, thus behaving more pessimistically. Second, for a given access pattern, the current value of rc should reflect the system status (in terms of on-board and network resources). If resources are lacking, it is necessary to control rc accordingly. The design of the adaptive mechanism must combine the above requirements into a single policy that controls the number of copies that can be concurrently cached at MHs. We elaborate on these issues in the following subsections.

To support an adaptive behavior between a pessimistic policy and an optimistic policy, the rc parameter for a data item X is associated with an interval (LB, UB) denoting the range of values that rc can assume. In particular, LB is the lower bound corresponding to a pessimistic policy, and UB is the upper bound corresponding to an optimistic policy. (Note that different data items can have different rc). By dynamically regulating rc according to the current rate of abort notifications per data item, we can address the first of the above requirements. LB is the second dynamically changing parameter of the ACP protocol. If the reported status of the resources is not critical, LB can be kept high to emphasize the optimistic nature of the policy. As resources conditions get critical, LB is reduced, although, in our simulation experiments, we never set LB to 1. By properly modulating LB, we are able to adapt the concurrency control policies to varying conditions in the resources and thus to address the second of the above requirements.

In the remainder of this section, we describe in some detail the mechanisms adopted to determine the values of the variables rc and LB.

3.2.1. *The adaptive mechanisms of the ACP protocol.* To support the ACP protocol, each RM maintains a table with an entry for each data item, local at the RM. An entry of such a table at RM_i contains for a data item X, located at RM_i, the following information:

- the version number of X, denoted as $vn(X)$;

- the current number of copies of X, denoted as $rc_i(X)$, that can be concurrently released by RM_i;
- the number of currently cached copies of X, denoted as $out_i(X)$, released by RM_i;
- the commit counter for X, denoted as $cc_i(X)$. $cc_i(X)$ is incremented by RM_i whenever a transaction T, which accesses X and is performed by the MH using RM_i, globally commits. In the event of global abort of T due to a conflict on X, $cc_i(X)$ is set to 0.

Note that $vn(X)$ is a global variable which is consistently maintained among all RMs which store a copy of X. rc, out and cc are local to each RM and are initialized respectively to LB, 0 and 0. The total amount of copies that the system can release at the time t for data item X is obtained by adding the $rc(X)$ values of each RM at time t. Similarily, we can obtain the total amount of outstanding copies of X at time t. The value for $rc(X)$ depends on the current history of successful global commits and varies according to the following heuristics, which is implemented by each RM.

Heuristics for rc. Let X be a data item, $rc_j(X)$ and $out_j(X)$ be, respectively, the number of releasable copies and currently cached copies of X at RM_j. Then, $rc_j(X)$ is incremented by one when the number of consecutive globally committed transactions accessing X is equal to the maximum number of copies of X that can be concurrently cached, whereas when a single transaction T_j accessing X globally aborts because of a conflict on X, $rc_j(X)$ is set to LB. The way we dynamically adapt rc does not result in any additional cascading aborts. It only affects new transactions, whose execution is delayed, whereas ongoing transactions normally complete. If T_j globally aborts and $X \notin as(T_j)$, then $rc_j(X)$ is left unchanged. If, by contrast, we were to update $cc_j(X)$ and $rc_j(X)$, as if T_j had committed, there would have been the risk that too many copies of X would become releasable when the locality in accessing the data items changes, thus increasing the probability of aborts.

In Figure 4 we outline the algorithm executed by RM_j on receiving the request to globally commit T_j. By adopting this heuristics, the values of rc cannot grow too fast and the concurrency degree on a given data item also depends on the variable LB, thus allowing for the status of the system resources. Figure 5(a) shows the rc profile for given data item X according to the transaction commit history.

Each RM storing a copy of X, can locally maintain $rc(X)$ by exploiting its global knowledge about the results of the global commit procedure.

As long as $out_i(X) < rc_i(X)$, RM_i can satisfy requests for copies of X to cache. Since some transactions may last indefinitely and never release their copies of data items, the number of available copies may happen to remain indefinitely lower than rc. To cope with this event, the RM assigns an execution timeout to each transaction and forces the transaction to abort in the event of expiration of such timeout. Each MH is made aware of execution timeout for all its local transactions. This is important to allow the MH to locally discontinue a transaction after the expiration of the timeout. Because of the stringent requirements concerning energy expenditure it is important that MHs avoid useless computations by discontinuing transactions that eventually will be aborted by the RM upon reconnection. A proper dynamic tuning of the execution timeout is therefore relevant. In the current version, we assign the same execution timeout to all transactions.

ALGORITHM 3.1. **Transaction_Commit_Request**

INPUT: A request for globally committing a transaction T_j
OUTPUT: YES, if T_j globally commits
NO, otherwise

METHOD:

1. execute global commit protocol
2. **if** T_j globally commits **then**
    reply $\leftarrow$ YES;
    $\forall$ X accessed by T_j:
        $out_j(X) \leftarrow out_j(X) - 1$;
        $cc_j(X) \leftarrow cc_j(X) + 1$;
        if $cc_j(X) = rc_j(X)$ **then**
            $rc_j(X) \leftarrow rc_j(X) + 1$;
            $cc_j(X) \leftarrow 0$;
   otherwise /* abort */ :
    reply $\leftarrow$ NO;
    $\forall$ X accessed by T_j:
        if $X \in as(T_j)$ **then**
            $rc_j(X) \leftarrow LB$;
            $cc_j(X) \leftarrow 0$;
        else do nothing;
3. **return** reply

FIGURE 4. An algorithm for processing global commit requests

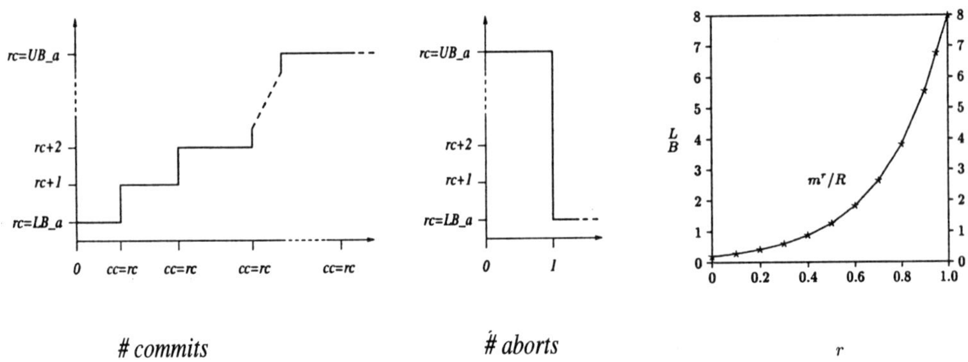

(a) (b)

FIGURE 5. (a) Profile of rc vs. the result of global commit. (b) Value of LB with $m = N = 40$ vs. different values of r.

The following example illustrates the approach.

EXAMPLE 3.2. Figure 6 (a) shows RM_a at a given time t. RM_a released two out of three copies of data items X and Y. When the disconnected transaction T_i of MH_i tries to globally commit, it aborts because it reports an old vn for X. An updated copy of X is sent to MH_i together with the abort message. On the contrary, the transaction performed by MH_k globally commits. In Figure 6 (b), we show how the system changes after the events described in Figure 6 (a). RM_a sets $cc_a(X)$ to 0 and decreases $rc_a(X)$ to the current value of the lower bound (in this

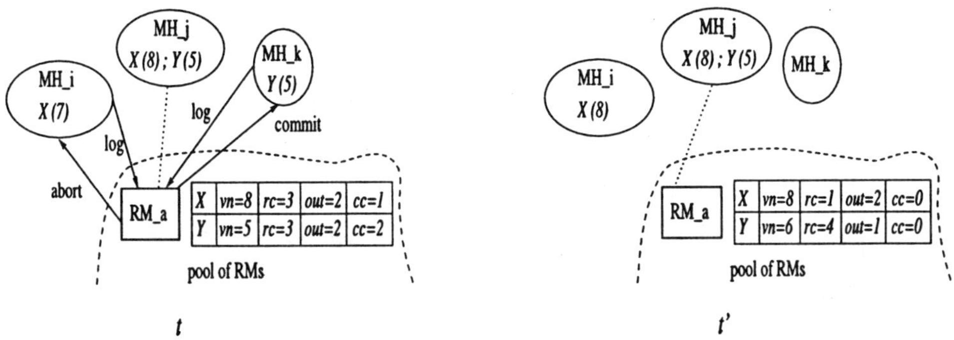

FIGURE 6. Data structures maintained by RMs: (a) at time t; (b) situation at time $t+1$

ALGORITHM 3.3. **Data_Copy_Request**

INPUT: A request for a cache copy of data item X

OUTPUT: YES, if the requested copy can be granted
NO, otherwise

METHOD:

1. **if** $out_j(X) < rc_j(X)$, **then**:
   reply $\leftarrow$ YES;
   $out_j(X) \leftarrow out_j(X) + 1$;

   otherwise:
   reply $\leftarrow$ NO;
   add request to the hold-back queue

2. **return** reply

FIGURE 7. An algorithm for processing data item requests

example, $LB = 1$). The version number of the data item Y is incremented, as well as $cc_a(Y)$. Since $cc_a(Y)$ becomes equal to $rc_a(Y)$, $rc_a(Y)$ can be incremented and $cc_a(Y)$ is reset to 0. Should a second transaction T_i' be running on MH$_i$ and should T_i' depend on T_i, then T_i' is forced to abort after the abort of T_i.

If a copy cannot be granted, the transaction request is queued into an *hold-back queue*. The MH locally running such transaction can enter the doze mode until some copy is released[4]. While waiting, a MH can move across cells. The implemented *handoff* procedure contains the exchange of any queued request being generated by the moving MH. Note that possible deadlock situations that can arise can be solved by the RMs according to usual distributed deadlock algorithms [**17**].

Figure 7 outlines an algorithm, executed by RMs, for handling data copy requests from mobile transactions.

3.2.2. *Dimensioning UB and LB.* If conflicting transactions are infrequent, disconnected transactions can globally commit and the values of rc can grow accordingly. In this event, to enable the system to behave optimistically, rc should

[4]Unless, of course, the MH has other work to do.

be able to grow until the highest concurrency degree is reached. In the current implementation of ACP, UB_i for the RM_i is set to the number of MHs for which RM_i is the nearest RM (in our case, the number of MHs in the cell of RM_i; membership information are available at each MSS).

The value of LB determines the lower bound to which rc is reset initially or whenever an abort occurs. It determines the nature of the system, optimistic or more controlled, and should be dynamically modified according to:

1. the reported status of the critical resources, such as the wireless bandwidth and battery charge;
2. the application (and the user) behavior in case of abort. In fact, if the system behaves optimistically, MHs have a relevant role in the transaction recovery, while if it behaves pessimistically, the control is left to RMs and MHs can be awaken up when the transaction globally committed. Users or application processes could be wishing to indicate their preferences.

We assume that each MH periodically sends control information about points 1 and 2 to its current RM. For sake of simplicity, this ACP implementation uses a single bit, the *status-bit*, which is piggybacked on the *Request* packet to report whether or not the MH has resource problems. RMs use *status-bits* to compute LB, whose value must be equal for all the RMs, that is, all RMs use the same LB, to ensure that a MH in a critical area has equal opportunities of a MH plenty of resources. If, by contrast, RMs in a critical area were to use a LB value lower than the one used by the other RMs, their MHs would have less probability of obtaining the required data items and would starve. To this purpose, RMs periodically exchange the received *status-bits* and compute the ratio $r = n/N$, where n is the number of MHs that *do not* report resource problems and N is the total amount of MHs in the system. Let R be the number of RMs in the system. We compute the lower bound $\boxed{LB \text{ as } m^r/R}$, where m can be set close to N. We have chosen this function for computing LB because it shows a sort of 'slow start'. It forces LB to grow slowly when most MHs have problems, while it leaves LB to grow quickly otherwise, as shown in Figure 5(b), where the values of LB are reported for $m = N = 40$, $R = 5$ and for different values of r.

The combined use of the simple heuristics to compute rc and LB satisfies the adaptation requirements of mobile transactions. In fact, if the system suffers from lack of resources and the rate of aborts is low, high performance can be achieved by letting rc to grow. However, as soon as the rate of abort increases, rc remains close to LB, thus enabling resources to be saved. On the contrary, if resources are available, LB is high and the optimistic nature of the system is emphasized.

4. Performance evaluation

4.1. Simulation environment. We measured the performance of the described protocol by means of simulation techniques; the adopted simulation package reproduces in details a single-hop network architecture up to the transport layer.

The simulated network consists of a number G of MHs and of 5 MSSs; we have evaluated the performance for different values of G. We assume that the wired network topology is completely connected. Initially the MHs are uniformly distributed in the five cells. They randomly move at a predefined speed, so that each MH enters a new cell about every second. We consider a wired network bandwidth of 10 Mbps and a wireless network bandwidth of 2 Mbps. These parameters describe a wired LAN with wireless LANs attached to the MSSs. At the transport layer we adopted the reliable multicast protocol described in [**2, 3**]. It guarantees that packets are delivered to MHs in spite of their current locality.

Each MH has a backlog of transactions and queries; a new operation on the database is generated as soon as the previous one has terminated. All the generated operations are processed in disconnected mode; we assume that during a disconnection only one transaction can be processed. After having received all the requested data items a MH disconnects. The duration of a disconnection is modeled with a random variable having a negative exponential distribution. At the end of the disconnection, the MH reconnects to a randomly chosen MSS. By varying the amount of transactions that access the same set of data items we can obtain different access patterns. For instance, a condition in which transactions have a high level of competition over a set of data items can be modeled by forcing 80% of the transactions to work on 20% of the data items. We assume that the database contains 100 data items, and that each query or transaction accesses at most 10 of them. According to the selected access pattern, 70% of the data items being accessed by a transaction are read, while 30% are written. The simulation tick is 1 msec. We assume that the packet size is 3 Kbytes; hence, it takes about 1.5 msec. to traverse a wireless link and about 0.3 msec. to traverse a wired link. The total simulation time was 50000 ticks, that corresponds to 50 sec. of simulated time.

4.2. Simulation results. Simulations have been utilized to obtain preliminary indications about the capability of the ACP to identify the trade-off between the need of saving system resources and performance. To this purpose, performance indexes have been observed under different access patterns and system conditions. We have measured:

1. the throughput, that is, the number of transactions processed by the system in the time unit;
2. the latency time, that is, the elapsed time between the start of a disconnected transaction and its global commit;
3. the number of aborts for transaction, which is expected to grow with the increase in the concurrency degree;
4. the resource utilization. To this purpose, we measured the number of exchanged messages over the wireless links and the time a MH spends (i) to process a transaction, (ii) to communicate with its RM, (iii) to wait for RM replies. The number of messages exchanged between MH and RM depends on the protocol and on the amount of aborts. If this number is low, it is possible to save the channel bandwidth and the energy. Furthermore, if the protocol behaves pessimistically, MHs wait a longer time for the RM replies, thus allowing MHs to enter a *doze* mode.

Figure 8 shows the behaviors of ACP in a general condition. In fact, we assume that only 30% of the MHs report bandwidth or battery problems and, for this reason, LB can be set and maintained quite high. However, by assuming that 80% of the transactions randomly access 20% of the data items, the high number of aborts can become the cause of resource problems. It is interesting to analyze how ACP behaves under these conditions by dynamically adapting rc and LB variables. Figure 8 shows the throughput (a), the number of aborts (b) and the

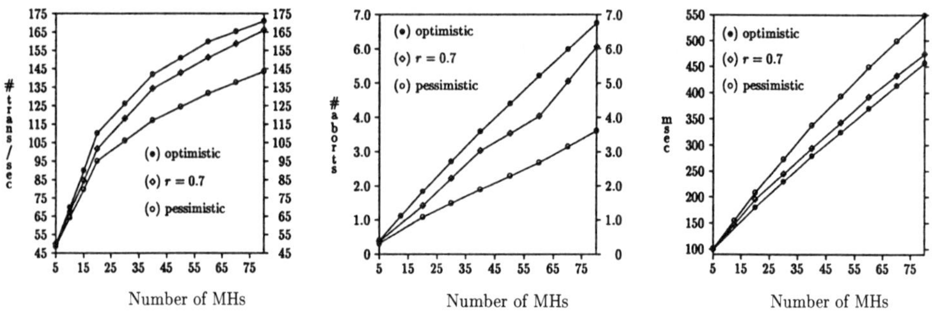

(*a*) Throughput (*b*) Number of aborts (*c*) Latency time

FIGURE 8. Performance results of ACP with 80% of the transactions accessing 20% of the data items.

latency time (c) of ACP compared with the optimistic and pessimistic[5] approaches. In Figure 9, the composition of the MH time utilization is reported. Figure 8 (a) shows that, under the considered conditions, ACP behaves closely to the optimistic approach. This behavior, which is confirmed by the results in Figure 8(b) and 8(c), depends on the values of LB which are maintained fairly high. It is clear that, by adopting different heuristics for LB, it is possible to obtain a protocol behavior which is more aggressive or more conservative according to the need of privileging performance or resource saving. The results in Figure 8 also indicate the good scaling capabilities of ACP. However, the reported results highlight a tendency of the protocol to approximate the behaviour of the optimistic approach with the increase in the number of MHs in the system. This event could not be acceptable in large systems where LB would grow too much, thus generating an increase of aborts and the consequent reduction of the benefits in terms of resource saving (see, for instance, in Figure 9 the computing-time bar for 80 MHs[6]). This tendency of LB to grow can be compensated, for instance, by acting on the variable m, which is used to determine LB.

Figure 9 shows that ACP succeeds in achieving energy saving with respect to the optimistic approach and, at the same time, it allows a higher number of concurrent transactions than the pessimistic approach. When the number of abort increases, under the optimistic approach, MHs have to reload the updated items and re-execute their transactions. By contrast, under ACP, MHs wait to obtain the required data items. Hence, the use of ACP lets MHs spend less time in computing and exchanging messages.

The throughput is influenced by the different settings of LB. To analyze these effects, we have considered two different system conditions (Figure 10). In the first,

[5]We have actually simulated a nearly pessimistic approach in which each RM grants up to 1 copy per data item, so that at most 5 copies are concurrently cached in the considered system (against 1 copy of a pure pessimistic approach). This is the consequence of the fact that we decide not to include, in the current version of our algorithm, any agreement protocol amongst RMs to decide which one of them can release the copy.

[6]These measurements are also affected by the discrete growing of LB, whose values are approximated to the lower or the higher integer digit.

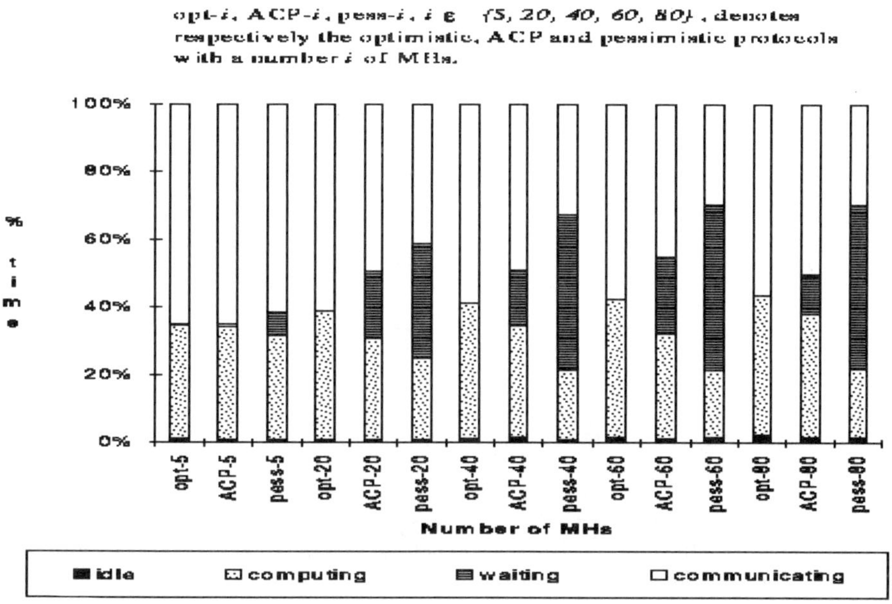

FIGURE 9. MH time utilization.

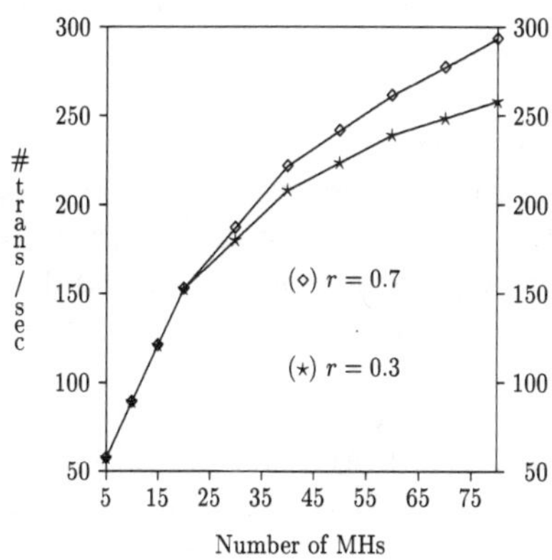

FIGURE 10. Throughput for different values of LB.

70% of the MHs are supposed to report resource problems; in the second, only 30% of MHs has resource problems. 60% of the transactions access 40% of data items; this setting has been chosen to reduce the interference of aborts in the experiments. Figure 10 shows that the throughput is nearly the same for systems of few MHs because, under these conditions, LB assumes nearly the same values. With the growing of the group membership, LB assumes different values and the throughput adapts to the different system conditions. This indicates that, by modulating LB, it is possible to properly determine the optimistic or pessimistic behaviors of ACP.

Unlike the parameter LB, rc has been introduced to dynamically adapt the number of concurrent copies of a data item according to changing access patterns.

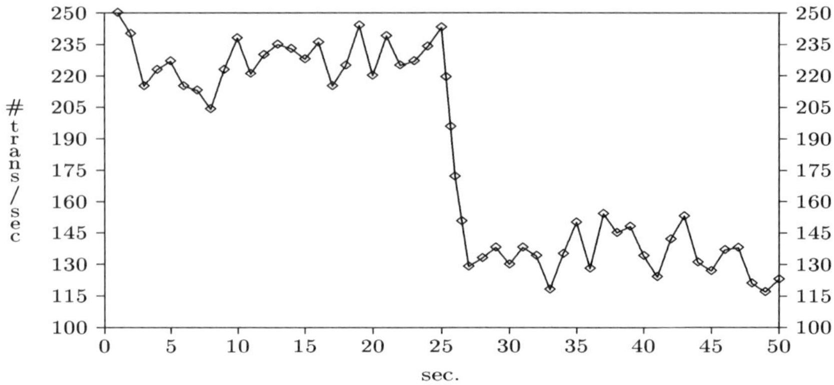

FIGURE 11. Throughput over time for changing access pattern.

To analyze this capability, in Figure 11, we report the throughput over the simulation time for a system of 40 MHs and LB set according to 30% of reported resource problems. The simulations ran a sequence of two different access patterns: in the first 25 seconds of simulated time, 60% of the transactions randomly access 40% of data items; in the second 25 seconds of simulated time, the locality of the accesses changes and 80% of the transactions access 20% of data items. Figure 11 clearly indicates how the throughput fluctuates according to the dynamics of rc and how quickly ACP adapts rc to changing access patterns. ACP requires less than 2.5 seconds to adapt to the new locality.

The combined analysis of the results reported in Figure 10 and 11 is quite encouraging and, although different heuristics can be envisaged, indicates that the use of both the parameters rc and LB is instrumental in making the ACP protocol dynamically adapting to changing access patterns and resource situations. Note that the use of only one of those parameters is not sufficient for such purpose. In fact, from the results in Figure 8 and 10, it is clear how the throughput can be affected by LB to reflect the status of the resources. rc alone would not work properly. Finally, we observe that ACP can approximate the performance of an optimistic approach, but it never behaves as a pure optimistic approach. This behaviour is due to a specific design choice; in fact, measurements indicate that an excessive amount of aborts can be the source of a bad resource utilization even when no apparent resource problem is reported.

5. Related work

Several areas are related with our work. First of all, concurrency control algorithms have been widely investigated in the area of both centralized and distributed databases. Despite, however, the large number of approaches proposed, most actual systems use locking-based mechanisms. In general, such mechanisms are not adaptive, even though some systems provide lock escalation and de-escalation techniques to dynamically convert fine-grained locks into coarse-grained locks and viceversa [**12**]. However, the concurrency control policy is always pessimistic and based on locking data before using them. A notable exception is represented by the GemStone object-oriented database system [**8**] providing two different concurrency control policies; a pessimistic one, based on locking, and an optimistic one, under which data are cached at the client and when the client tries to commit a validation test is performed. However, the choice between those two policies is left to the applications. The system makes no automatic choice neither provides any adaptive behavior, as in our approach.

Concurrency control algorithms devised specifically for mobile computing systems have also been developed. In particular, the approach proposed in [**15, 19**] to support disconnected operations in the framework of the Coda file system uses an optimistic approach, as such it can be supported as a special case of our approach. Similarly, the approach proposed in [**6**] uses a pessimistic approach and thus can be supported as a special case of our approach. A relevant shortcoming of such an approach is that it does not tolerate sudden disconnections. Both approaches differ from ours in that they only support a single policy (either optimistic or pessimistic) and do not provide any adaptive behavior.

Another relevant body of research deals with dynamic data replication approaches in distributed database systems [**24, 4**]. Those approaches develop algorithms that are able to dynamically change the data replication according to the read-write patterns of data access operations, so to optimize performance. Even though such approaches may appear relevant to mobile computing systems, they are not applicable to the problem of deciding the optimal of copies that can be cached at MHs. Those approaches assume that an host can get a missing data item from another host, when such item is not available locally. Cost functions are then formulated accordingly; in particular, the fact that an host has a certain data item may impact the access costs by other hosts. In a mobile system, the fact that a MH has a copy of a data item does not impact the access costs for the other MHs, since these MHs have anyhow to require the data item from the MSS. Moreover, those approaches seek to minimize the communication costs only, as all algorithms designed for distributed data management over fixed networks. Further, even though those approaches dynamically replicate data, they still use a single concurrency control protocol, usually based on the 2-phase locking approach or quorum approaches. As such, they do not support adaptive concurrency control protocols. It is, however, important to notice that such adaptive data replication protocols could be used to control primary data replication, that is, data replication at RMs. However, such replication is orthogonal to cache replication for MHs.

The problem of copy replication on MHs has been briefly addressed in [**13**], as a special case of a more general data replication algorithm. The approach proposed is based on an optimistic policy. Thus the same data can be freely replicated at the various MHs. Whenever a MH performs a local modification, it notifies the RM, which then is in charge of invalidating the copies that all other MHs may have. Such an approach has the serious shortcoming that in the presence of high conflict rates, the various MHs will have to frequently re-load data, which involves high energy expenditure as well high network usage. In this respect, our approach is more effective because it reduces the number of MHs that can get a data item if there is a high probability of conflict over such data item.

Finally, a recent approach has been proposed supporting adaptive data broadcast over mobile computing systems [**10**]. The goal of this approach is to develop algorithms able to dynamically determine which information have to be broadcast and how often according the request patterns of mobile users. Even though the goal of this work is very different from ours, still it is important to point out that in this approach, as in ours, both response time and energy expenditure are a main concern.

6. Concluding remarks

In the transactional systems of the near future, users will use mobile hosts to access data items located at both fixed and mobile stations. Mobile hosts alternate connected and disconnected work and are free to arbitrarily move across disomogeneous wireless cells. Mobility is changing the focus of transaction management protocols from the need to optimize response time only to a new one where response time and energy/bandwidth expenditure must be both optimized and where disconnected work is performed on cached copies of the data items.

This paper considers the problem of transaction management in such an environment and proposes an adaptive concurrency control protocol that dynamically modulates the number of data copies that can be concurrently cached by mobile hosts. The introduction of adaptive mechanisms is required to address the novel problem issued by mobility: to achieve optimal performance and, at the same time, to save the resources of the network and the mobile hosts. As the initial simulation runs indicate, this is a promising approach which, however, is open to further analysis in several directions such as the definition of new heuristics to dimension rc and LB, and the way to adaptively set the execution timeout.

However, when the amount of MHs scales from hundreds to thousands, the 'system-controlled' adaptation of the concurrency degree is likely to become not sufficient to solve the problem alone. Under these conditions, a set of 'application-' and 'system-controlled' mechanisms should be placed side by side to consider, for instance, application specific wishes, MH location awareness or pre-fetching techniques of data items. When the information provided by the applications can be supported by the proper system parameters, new more efficient approaches to manage mobile transactions could be devised.

References

[1] Recovery in Database Management Systems, V. Kumar, and M. Hsu eds., Prentice-Hall, Upper Saddle River (New Jersey), 1998.

[2] Acharya A., Badrinath B. R., *"A Framework for Delivering Multicast Messages in Networks with Mobile Hosts"*, ACM/Baltzer Journal of Wireless Networks, Special Issue on *Routing in Mobile Communication Networks, Vol. 1* (1996), 199–219.

[3] ———, *"Delivering Multicast Messages in Networks with Mobile Hosts"*, Proc. 13th Intl. Conf. on Distributed Computing Systems (May 1993), 292.

[4] Awerbuch B., Bartal Y., Fiat A., *"Optimally-competitive Distributed File Allocation"*, Proceedings of the 25th Annual ACM STOC (Victoria (B.C., Canada), 1993).

[5] Badrinath B. R., Acharya A., Imielinski T., *"Impact of Mobility on Distributed Computations"*, ACM Operating Systems Review, Vol. 27, No. 2 (Apr. 1993), 15–20.

[6] Barbará D., Garcia-Molina H., *"Replicated Data Management in Mobile Environments: Anything New Under the Sun ?"*, Proc. IFIP Conf. on Applications in Parallel and Distributed Computing (Caracas, Venezuela, Apr. 1994).

[7] Bertino E., Pagani E., Rossi G. P., *"Fault-Tolerance and Recovery in Mobile Computing Systems"*, **"Recovery in Database Management Systems", ch.24, Eds. V. Kumar and M. Hsu** (Prentice-Hall, 1998), 661–696.

[8] Breitl R. et al., *"The GemStone Data Management System"*, In *Object-Oriented Concepts, Databases, and Applications*, W.Kim, and F.Lochovsky eds., Addison-Wesley, Reading (Massachusetts), 283–308.

[9] Ceri S., Pelagatti G., Distributed Databases: Principles and Systems, McGraw-Hill, 1984.

[10] Datta A., VanderMeer D., Celik A., Kumar V., *"Broadcast Protocols to Support Efficient Retrieval from Databases by Mobile Users"*, ACM Transactions on Database Systems (1998).

[11] Dunham M.H., Helal A., *"Mobile Computing and Databases: Anything New?"*, ACM SIGMOD Record, Vol. 24 (Dec. 1995), 5–9.

[12] Gray J., Reuter A., Transaction Processing: Concepts and Techniques, Morgan-Kaufmann Publishers, San Mateo (California), 1993.

[13] Huang Y., Wolfson O., *"Object Allocation in Distributed Databases and Mobile Computers"*, Proc. 10th Intl. Conf. on Data Engineering (Houston, TX, Feb. 1994), 20–29.

[14] Imielinski T., Badrinath B. R., *"Mobile Wireless Computing: Challenges in Data Management"*, Communications of the ACM, Vol. 37, No. 10 (1994), 18.

[15] Kistler J. J., Satyanarayanan M., *"Disconnected Operation in the Coda File System"*, ACM Trans. on Computer Systems, Vol. 10, No. 1 (Feb. 1992), 3–25.

[16] Kleinrock L., *"Nomadic Computing - An Opportunity"*, ACM Computing Communication Review, Vol. 25, No. 1 (Jan. 1995), 36–40.

[17] Knapp E., *"Deadlock Detection in Distributed Databases"*, ACM Computing Surveys, Vol. 19, No. 4 (Dec. 1987), 303–328.

[18] Kung H. T., Robinson J. T., *"On Optimistic Methods for Concurrency Control"*, ACM Trans. on Database Systems, Vol. 6, No. 2 (Jun. 1981), 213–226.

[19] Lu Q., Satyanarayanan M., *"Isolation-Only Transactions for Mobile Computing"*, ACM Operating Systems Review, Vol. 28, No. 2 (Apr. 1994), 81–87.

[20] Nixon P., Cahill V., eds., *"Technologies for a Distributed Society"*, IEEE Internet Computing, Vol. 2, No. 1 (1998).

[21] Ozsu T., Valduriez P., Principles of Distributed Database Systems, Prentice-Hall, 1991.

[22] Perkins C., *"Mobile Network Computing through Mobile-IP"*, IEEE Internet Computing, Vol. 2, No. 1 (1998).

[23] Pitoura E., Bhargava B., *"Maintaining Consistency of Data in Mobile Distributed Environments"*, Proc. 15th Intl. Conf. on Distributed Computing Systems (Vancouver, British Columbia, Canada, Jun. 1995), 404–413.

[24] Wolfson O., S. Jajodia, *"An Adaptive Data Replication Algorithm"*, ACM Transactions on Database Systems, Vol. 22, No. 2 (Jun. 1997), 255–314.

Computer Science Dept., Università degli Studi di Milano, via Comelico, 39. I-20135 Milano, Italy

E-mail address: `bertino@dsi.unimi.it`

Computer Science Dept., Università degli Studi di Milano, via Comelico, 39. I-20135 Milano, Italy

E-mail address: `pagae@dsi.unimi.it`

Computer Science Dept., Università degli Studi di Milano, via Comelico, 39. I-20135 Milano, Italy

E-mail address: `rossi@dsi.unimi.it`

DIMACS Series in Discrete Mathematics
and Theoretical Computer Science
Volume **52**, 2000

Supporting Adaptive-QOS over Multiple Time Scales in Wireless Networks

Javier Gomez and Andrew T. Campbell

ABSTRACT. For wireless channels, interference mitigation techniques are typically applied at the packet transmission level. In this paper, we present an adaptive-QOS framework that also responds to impairments over multiple time scales that are present at the flow/session level. Our framework is based on three different mechanisms that operate over distinct adaptation time scales. At the packet transmission time scale, a channel predictor determines whether to transmit a packet or not depending on the state of the wireless channel. At the packet scheduling time scale, a compensator credits and compensates flows that experience bad link quality. Over even longer time scales an adaptator regulates flows taking into account the ability of wireless applications to adapt to changes in available bandwidth and channel conditions. In this paper, we argue that to effectively support QOS across wireless links there needs to be interworking or integration between the predictor, compensator and adaptator. We achieve this by deploying an arbitrator that coordinates the operation of each mechanism in response to environmental factors, scheduling state and flow semantics.

1. Introduction

There has been considerable discussion in the mobile networking research community about the most suitable service model for the delivery of mobile multimedia services over wireless networks. One school of thought believes that the radio can be engineered to provide wireline type hard-QOS assurances, (e.g., guaranteed delay or constant rate services). Another school argues that the wireless link cannot be viewed in this manner because of the inherent time-varying environmental factors found in radio communications (e.g., fading). In this case, wireless services lend themselves to more adaptive-QOS approaches [**Kat94**] or better than best-effort service paradigms [**NJZ97**].

We take our lead from the adaptive camp and propose a packet-based adaptive-QOS framework for application and channel dependent quality of service control. Our approach incorporates adaptation techniques for packet scheduling and application-level rate control taking into account wireless channel conditions and the ability of application level flows/sessions to adapt to these conditions over multiple time scales. In this paper, we argue that an adaptive-QOS paradigm is suitable for the delivery of voice, video and data to mobile devices.

The most prominent characteristics associated with wireless networks is the extraordinary premium placed on bandwidth and power efficiency as well as the use of unrelialable transmission links. Existing protocols for wireline networks are limited in their ability to deal with these issues; they are generally designed to provide specific services with little ability to adapt to highly time-varying conditions associate with wireless networks. What is required is an appropriate set of adaptive protocols that pass state information across layers in an effort to cope with this variability.

In this paper, we introduce an adaptive-QOS model that is founded on the notion of exchanging state information between mechanisms capable of responding to time-varying wireless characteristics. These mechanisms operate over three distinct time scales and include a *predictor*, *compensator* and *adaptator*. An *arbitrator* monitors the state of each component coordinating their operation in an integrated manner. Channel prediction allows the arbitrator to defer transmission to mobile devices experiencing fading conditions. Channel prediction, however, does not compensate mobile devices that have previously experienced 'outages' due to poor channel conditions. To overcome this problem, an arbitrator interworks with a compensator (based on channel state dependent packet scheduling [**BBKT97**]) to deliver enhanced throughput to mobile devices. The compensator attempts to resolve unfairness experienced by different spatially distributed receivers and operates on the packet scheduling time scale. When persistent fading conditions exceeds the operational range of the compensator, the arbitrator activates an adaptator module. The adaptator is designed to operate over longer time scales and takes into account application profiles (e.g., packet priorities within a flow/session) in the case of severe channel conditions or variations in available bandwidth. Ideally an adaptive-QOS model should be used in conjunction with adaptive modulation/coding techniques and other interference mitigation techniques (e.g., smart antennas, multiuser detection, power control) in order to achieve optimum performance and a high degree of adaptive-QOS integration.

In this paper, we present analytical analysis of the predictor, compensator and adaptator modules operating over IP networks supporting IEEE 802.11 last hop wireless LANs. The paper is organized as follows. In Section 2, we present an overview of the adaptive-QOS model. In Section 3, we describe our channel predictor followed by a description of a compensator scheme in Section 4. In Section 5, we discuss an adaptator mechanism that supports application-specific adaptation. The adaptive-QOS model has been implemented using existing wireless LAN technology (e.g., IEEE 802.11) and the ns simulator [**Ngu98**].

2. Adaptive-QOS Model

Network dynamics in wireless networks are the result of several different systems interactions operating over multiple time scales. These time scales range from received signal strength variations operating in the order of nanoseconds, to deep fade situations or bandwidth variations occurring anywhere between hundreds of milliseconds to minutes. It is well known that several mechanisms such as modulation, forward error correction, automatic repeat request, interleaving, etc., are useful in dealing with fast radio channel impairments at the packet transmission level. It is unclear, however, which measures are the most appropriate when channel

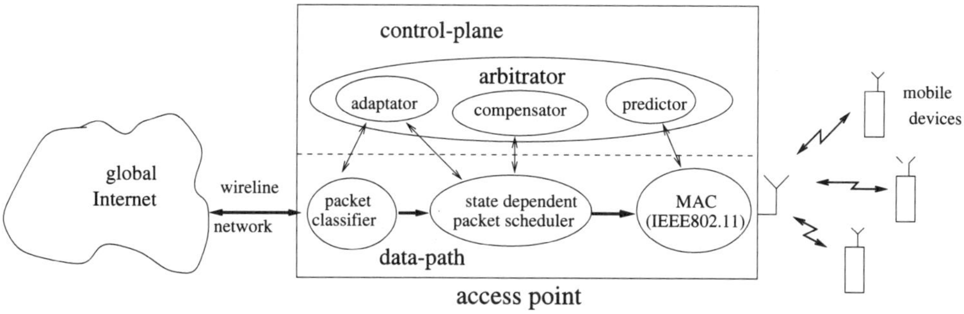

FIGURE 1. The Controlled-QOS Model

impairments become severe and go far beyond the operational range of these mechanisms. The adaptive-QOS model attempts to take this time-varying behavior into account by operating over three distinct time scales to respond to changing network conditions. The main controller of our QOS-adaptive model is an arbitrator present at each wireless access point.

Figure 1 illustrates our adaptive-QOS model. The model comprises a data path, which includes a packet classifier, state dependent packet scheduler and MAC access. In addition to the data path a control plane supports a number of QOS mechanisms that support the data path; these include:

- *an arbitrator*, which coordinates the predictor, compensator and adaptator. Before a packet can be transmitted, the arbitrator requests the predictor to test the state of the wireless link. Depending on the state of the channel, the arbitrator will either agree to transmit the packet or hold it in a buffer and trigger the compensator to 'credit' the flow/session. When a flow's buffer is about to overflow, the arbitrator will set the right filter in the adaptator to drop low priority packets;
- *a predictor*, which probes the wireless channel between the access point and mobile devices to determine the current state of a wireless channel before a packet can be transmitted by the scheduler. The probing mechanism is based on the IEEE 802.11 request-to-send (RTS) and clear-to-send (CTS) packet pair but it could be implemented with any other arbitrary packet pair exchange. If an RTS-CTS probe detects the channel is in a 'bad' state, then the packet remains queued in the scheduler for later transmission and the flow-state is 'credited'. If the channel is detected to be in a 'good' state the packet is transmitted [**FSS98**];
- *a compensator*, which is operational at the packet scheduling time scale. Channel prediction allows the arbitrator to defer scheduled transmission to a receiver in a bad channel state until the fading period is over; thus it can proceed with the transmission of packets to other receivers that are in a good channel state. Channel prediction does not, however, provide mechanisms to compensate mobile devices that deferred transmission in the past. The compensator is able to 'credit' mobile devices experiencing fast and slow fading channel conditions and 'compensate' the same flows when the link becomes good. At the same time the compensator keeps packet delay variation bounded and attempts to achieve fairness among all active flows. Our compensator is built around a deficit round robin (DRR) scheduler introduced in [**GCM98**]; and

• *an adaptator*, which comprises two components: (i) a buffer controller which operates at a slower time scale than prediction and compensation; and (ii) a regulator, which performs end-to-end rate control [**Sch92**] over longer time scales. Both components are based on the insight that adaptation is application-specific. The buffer controller is suited to drop semantically less important packets while responding to changes in the available bandwidth either due to persistent channel conditions or new flows being established by mobile devices. The adaptator sets appropriate dropping marks in the buffer based on the different priorities within each flow and the long term average measurements of the channel. While the buffer controller tries to maintain 'good' quality over short intervals (e.g., when the buffer is about to overflow due to a deep fade), the regulator performs longer-term adaptation that reacts to long-term observed conditions that are experienced in the network.

In our framework, applications specify their flows as having a minimum bandwidth requirement and a number of enhancement layers. The base layers are treated at a higher priority than the enhancement layers by the buffer controller. Both priority and delay information can be carried in each packet using in-band fields such as the differentiated services codepoint (DSCP [**KNB97**]) or lightweight signaling techniques (INSIGNIA [**LC98**]).

While the adaptive-QOS model has been designed to operate over a variety of radios our implementation is focused on the IEEE 802.11 standard [**WMSW97**], [**P8097**] that operates between 1-20 Mbps. The IEEE 802.11 standard operates in two modes: (i) Distributed Coordination function (DCF) where mobile to mobile communication is established using collision sense multiple access with collision avoidance (CSMA/CA) with or without the RTS-CTS option; and (ii) Point Coordination mode (PCF) where an access point provides a centralized controller for contention free communications. IEEE 802.11 is optimized to support best-effort IP delivery using DCF and real time flows using PCF. To support a channel predictor capability based on the RTS-CTS probe we have modified the network simulator (NS-2) IEEE 802.11 code suite [**Ngu98**] to support this new feature in PCF mode. The access point operates as central scheduler for both up/down link communications in this case.

3. Predictor

Channel prediction allows a transmitter to probe the state of the wireless channel before transmitting a packet. If the predictor detects that the channel is in a 'bad' state, the packet remains queued in the scheduler for later transmission and the flow-state is 'credited'. If the channel is detected to be in a 'good' state then the packet is transmitted [**FSS98**]. Previous work on channel prediction either assumes that the state of the channel or the duration of bad link periods are known in advance [**FSS98**], [**ESZ98**], [**SLS97**]. In practice, however, the state of wireless links cannot be entirely predicted. Our main motivation in this section is to use an analytical framework to investigate bounds and utility of the approach.

3.1. Operation. In what follows, we discuss our approach to channel prediction. To estimate the channel state, we have implemented a simple hand-shake protocol based on the well known RTS-CTS probing mechanism. RTS-CTS as a channel predictor was proposed in [**FSS98**], however, no analytical or simulation results concerning the performance of such an approach have been discussed in the literature. Our channel predictor operates as follows. Before the start of packet transmission to a mobile device a short probing RTS packet is sent to the designated receiver. The mobile device responds by sending the CTS packet as an acknowledgment to the RTS. If the CTS packet is received intact the channel state is assumed to be good. If, on the other hand, the CTS does not arrive after a given timeout then the channel state is considered to be in a bad state. The assumption is that the RTS or CTS could have been corrupted, lost or incorrectly received because of degrading channel conditions manifest as increased bit errors and loss of signal at the receiver.

In IEEE 802.11, RTS-CTS is used in the DCF mode to compensate for the hidden terminal problem, which can lead to a very high numbers of collision in the channel for heavy traffic loads. However, even if RTS-CTS fails because of channel errors, the transmitting mobile device will always assume the problem was caused by hidden terminals and will back-off before trying again. During the PCF operation, the access point is able to acquire the channel before any of the mobile devices in its coverage area. Therefore, there is no need to use RTS-CTS to prevent collisions. Any packet received in error in the PCF mode is unambiguously the result of channel conditions. The predictor we have implemented works in PCF mode to verify the state of the channel. In IEEE 802.11/PCF mode the access point always initiate transmission for both downlink (transmitting the packet) or uplink (polling a mobile) communications. Therefore, RTS-CTS can be used in both downlink/uplink transmissions. As a means to differentiate between up/down link operations we use RTS-CTS for the downlink and request to receive (RTR) and clear to receive (CTR) for the uplink.

3.2. Analysis. A two state Markov model can be used to model the good and bad states of a wireless channel [**ZR96**]. Transmission of packets during good state periods assures error free delivery. On the other hand, during a bad period the packet will be received in error. This assumption simplifies the analysis and is realistic for IEEE 802.11 where no Forward Error Correction (FEC) is used [**P8097**]. The transition between states occur at discrete time instances according to the transition rates. Rather than using a single set of transition rates for a particular channel model, we analyzed the performance of the channel predictor for a wide range of rates.

Table 1 shows all the possible outcomes of RTS, CTS, DATA and ACK events for one transmission. Note that uplink analysis is similar using the RTR-CTR pair. Any packet transmitted can be received error-free (0) or in error (1). If both RTS and CTS packets are received correctly, the state of the channel is predicted as error-free, otherwise the channel is predicted in error. Depending on the reception of the DATA and the ACK packets the transmission is evaluated in the same way as the predictor.

Let $1/\lambda$ and $1/\gamma$ be the average time the channel is in good and bad states, respectively. The transition matrix of the markov model [**ZR96**] is as follows:

$$P = \begin{pmatrix} P(0|0) & P(1|0) \\ P(0|1) & P(1|1) \end{pmatrix} = \begin{pmatrix} 1-\lambda & \lambda \\ \gamma & 1-\gamma \end{pmatrix} \tag{3.1}$$

With the steady state probability of the channel being in bad/good state given by:

$$\pi_1 = \lambda/(\lambda+\gamma); \qquad \pi_0 = 1 - \pi_1 \tag{3.2}$$

RTS	0	0	1	0	0	0	0	1	1
CTS	0	1	*	0	0	1	1	*	*
prediction	0	1	1	0	0	1	1	1	1
DATA	0	0	0	1	0	1	0	1	0
ACK	0	0	0	*	1	*	1	*	1
transmission	0	0	0	1	1	1	1	1	1

TABLE 1. Legend: 0 : error-free, 1 : error, * : timeout

The probability that the channel prediction is accurate (P_C) is equal to the probability that RTS, CTS, DATA and ACK packets are received error-free ($P(pre = 0, tra = 0)$) plus the probability that predictor (RTS, CTS) and transmission (DATA, ACK) are received in error ($P(pre = 1, tra = 1)$), see table 1; then:

$$P_C = P(pre = 0, tra = 0) + P(pre = 1, tra = 1) \tag{3.3}$$

If the channel is currently in one of the two states, with κ being the transition rate to the other state, the probability that the channel will remain in that state for x more seconds is equal to $e^{-\kappa x}$. Now let rts, cts $data$ and ack be the size in bytes of RTS, CTS, DATA and ACK packets, respectively. Before the transmission of CTS, DATA and ACK packets in 802.11 the transmitter should wait for a short inter frame space (SIFS) [**P8097**]. If the speed in bytes/sec of the wireless local area network (WLAN) is C then $P_{(pre=0,tra=0)} = P(tra = 0|pre = 0)P(pre = 0)$, where P(pre=0) can be approximated by $\pi_0 e^{-(\frac{rts+cts}{C}+SIFS)\lambda}$, therefore:

$$P_{(pre=0,tra=0)} \approx \pi_0 e^{-(\frac{rts+cts+data+ack}{C}+3SIFS)\lambda} \tag{3.4}$$

This represents the probability that the channel is in a good state at the beginning of RTS transmission and remains in a good state for a period longer than the reception of the corresponding ACK. In this equation we neglected the case in which the channel changes from good to bad and from bad to good state during a SIFS interval. In the same way a good approximation for $P_{(pre=1,tra=1)}$ under realistic conditions (e.g., where good channel periods are much longer compared to bad channel periods), is:

$$P_{(pre=1,tra=1)} \approx \pi_1 e^{-(\frac{rts+cts}{C}+SIFS)\gamma} + \pi_0(1 - e^{-(\frac{rts+cts}{C}+SIFS)\lambda}) \tag{3.5}$$

This equation is the sum of two components, the first component represents the probability that the channel is in a bad state at the beginning of the RTS transmission and remains in a bad state for a period at least longer than the beginning of the DATA packet transmission. The second term represents the probability when the channel is in a good state at the beginning of RTS transmission but changes to a bad state before the beginning of the DATA packet transmission.

The RTS-CTS probe introduces a small overhead in the protocol in PCF mode. For mobile devices experiencing continuous fading, the predictor will provide enhanced throughput. In contrast, mobile devices experiencing a consistently good link will receive little benefit from the use of the prediction probe; the downside being the penalty of sending the probe for each packet transmission. Based on channel prediction the packet scheduler operates under the assumption that the predicted channel state is accurate.

Because channel prediction can avoid unwarranted multiple retransmissions to receivers in a bad channel state, throughput is greatly enhanced. Channel prediction, however, does not provide any compensation mechanism for receivers that have deferred transmission in the past due to a bad channel state [**BBKT97**]. Although receivers in a good channel state can benefit from the deferred transmission of receivers in a bad channel state, they are not typically re-compensated after the state of the channel of the deferred receiver becomes good.

4. Compensator

To overcome the potential unfairness of mobiles devices experiencing different channel conditions, our compensator uses a modified version of deficit round robin (DRR) [**SV95**] to 'credit' and 'compensate' flows. Transmission of data packets in DRR is controlled by the use of quantum (Q) and deficit counters (DC) [**SV95**]. The quantum accounts for the number of allocated bytes to each flow for transmission during each round, whereas the deficit counter keeps track of the transmission-credit history for each flow. A "round" is defined as the process of visiting each queue in the scheduler once. At the beginning of each round, a quantum is added to the deficit counter for each flow. The scheduler visits each flow comparing the size of the deficit counter with the size of the packet at the head of the queue. As long as the packet size is smaller than the deficit counter, a packet will be transmitted and the deficit counter reduced by the packet size. When the packet size is bigger than the deficit counter, the scheduler will maintain the deficit value in a flow-state table for the next round and move to serve the next flow in a round robin order. As long as the quantum size is larger than the maximum packet size the system is work-conserving [**SV95**].

An equal allocation of the link is achieved when the quantum size for all flows is the same. Making the quantum size for some flows different leads to weighted round robin (WRR), which allows a proportional share of the link according to the weights given to each flow [**SV95**]. For example, if three flows have a similar quantum (e.g., equal to 100), they all will receive 1/3 of link bandwidth. If $Q_1 = Q_2 = 100$ and $Q_3 = 200$, the share of the link would be $\frac{1}{4}$ $\frac{1}{4}$ and $\frac{1}{2}$, respectively.

4.1. Operations. We modify the weighted round robin algorithm to achieve fairness in the presence of location dependent fading conditions by introducing a compensation counter (CC) that is maintained for each receiver. For each round, αCC_i extra bytes (if the compensation counter for $flow_i$ is positive) are allocated, where α is a value between 0 and 1. Each time αCC bytes are used to compensate the flow, its compensation counter is decreased by the same amount. It should be noted that if a compensation counter for a receiver is positive then the session will get αCC more bytes for transmission in one round than other sessions with nonpositive compensation counters. This compensate receivers sessions which have been deferred in previous rounds. To this end, even if the channel has estimated a bad state (and hence data packets are not transmitted) the deficit counter for the receiver is decreased by the quantum size. In return for the decrease, the compensation counter of the session is adjusted by a quantum size increase of the same amount[1]. It is important to clarify that the compensation process realizes two goals:

[1]The actual compensation may vary between 0 and the quantum size according to the observed load of the system as discussed later.

- it determines how many bytes to credit a flow after the channel predictor diagnoses a bad channel; and
- it determines which portion of the credit is used for compensation of a flow in each round.

Considering the former goal, it is intuitive to credit by Q bytes every time transmission is deferred. When the system is heavily loaded this is a good solution as we discussed below. However, when the system is lightly loaded the rate at which the round robin scheduler serves a flow is faster than the worst case (e.g., under full load). Crediting by Q bytes at this rate will over-credit flows leading to unfairness for newly arriving flows over the long term. Therefore we credit flows according to the load of the system providing less credit in light loaded systems and a full quantum size credit for heavily loaded systems. If n flows are registered with the central scheduler (each $flow_i$ with a weight Q_i), the load of the system is defined as the ratio of the sum of Q_i for active flows [2] (Q_i^{Act}) and the total capacity of the system in each round (denoted hereafter as G). The definition of G can be considered arbitrary but has to be consistent. For example if G is set to 1000 and a particular flow gets a 15 percent share of the link, the quantum size for that flow should be set to 150. Let $CC_i(k)$ be the compensation counter of flow i in round k then, if $flow_i$ deferred transmission in round k, the compensation counter in round $k+1$ will be:

(4.1)

$$CC_i(k+1) = \begin{cases} CC_i(k) + \Big((\sum_{j=1}^{n} Q_j^{Act} - Q_i)/(G - Q_i)\Big)Q_i & if \quad G > Q_i \\ \\ CC_i(k) & if \quad G = Q_i \end{cases}$$

Only when $G = \sum_{i=1}^{n} Q_i^{Act}$, is the system operating at full load and the compensation given to $flow_i$ is equal to Q_i. When $\sum_{i=1}^{n} Q_i^{Act} = Q_i$, only $flow_i$ is active with a compensation of zero.

Now we analyze the second goal discussed above; that is, how many bytes of the credit should be used for compensation in one round. It is desirable to compensate a flow that is behind schedule as soon as is possible. This means adding CC bytes to DC in one round no matter what the size of CC is. The problem with this approach is that the latency for flows is likely to be sensitive to the amount of compensation that is given to a particular flow in each round specially during loaded periods. In order to bound the latency it is necessary to bound the maximum compensation that a flow acquires in a single round. Similar to [**ESZ98**], we bound the maximum amount of bytes $flow_i$ transmit in one round to a constant parameter DC_i^{max} even under loaded conditions.

Let $\sum_{i=1}^{n} Q_i^+$ be the sum of quantums for flows having positive compensation counters (e.g., $Q_i^+ = 0$ if $CC_i = 0$ and $Q_i^+ = Q_i$ if $CC_i > 0$) then the number of bytes available for compensation to $flow_i$ in one round is given by:

$$(4.2) \qquad \alpha CC_i = min\Big[max\Big[(\frac{Q_i^+}{\sum_{j=1}^{n} Q_j^+})(G - \sum_{j=1}^{n} Q_j^{Act}),\ DC_i^{max} - Q_i\Big],\ CC_i\Big]$$

The first term inside the brackets in Equation 4.2 accounts for compensation in the case where unused bandwidth is available. This can be obtained by computing

[2] We consider an 'active' flow to be one that has at least one packet in the scheduler's queue.

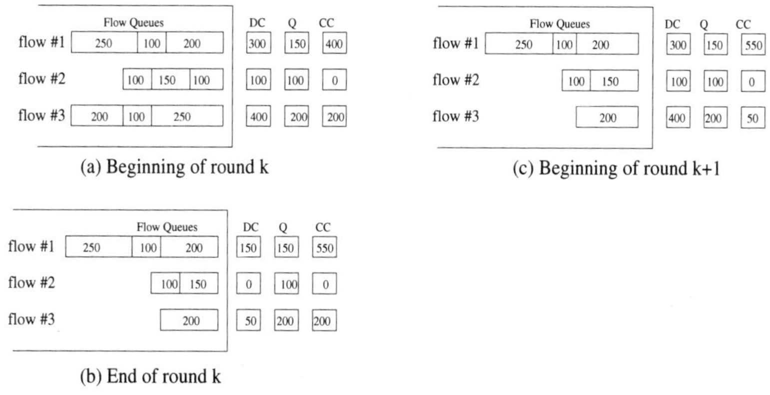

FIGURE 2. Compensator Operation

the available bandwidth and the portion of that bandwidth that corresponds to each flow with a positive CC. The second term $(DC_i^{max} - Q_i)$ accounts for the minimum compensation given to a flow in each round in the case the system is heavily loaded and there is no unused bandwidth available. In both cases the amount of compensation given to $flow_i$ is bounded by CC_i.

An illustration of the scheduler state and the operation of the compensator is shown in Figure 2. A snapshot of the scheduler at the beginning of a round (after the quantum and compensation bytes have been added) is illustrated in part 2(a). Three flows associated with three different mobile devices are active and the sum of the allocated rates is equal to the system capacity (i.e., the system is fully loaded). In this example $DC_i^{max} = 2 * Q_i$ for each $flow_i$. Figure 2 (b) illustrates the state of the scheduler at the end of the round. The following events take place during the round:

• the channel prediction for flow #1 detects a bad channel state and the scheduler defers the transmission of the packet, updates the compensation counter by the quantum size and reduces the deficit counter by the same amount;

• the channel prediction for flow #2 indicates a good channel state and the scheduler transmits the packet reducing the deficit counter by the packet size (normal weighted round robin operation); and

• the channel prediction for flow #3 indicates a good channel state, allowing two packets to be transmitted and the deficit counter to be decreased by the packet size. Figure 2 (c) illustrates the state of the scheduler at the beginning of next round, when Q_i plus αCC_i bytes (if the compensation counter is positive) are added to the deficit counter for each flow i. Note that only a portion of the compensation counter for $flow_1$ and $flow_3$ is added to their respective deficit counters so $DC_i \leq DC_i^{max}$.

4.2. Fairness. Now we discuss the fairness properties of our compensator mechanism under full load conditions. Using the same nomenclature defined in **[SV95]**, let $DC_i(k)$ and $CC_i(k)$ be the value of the deficit counter and compensation counter, respectively, for $flow_i$ at the end of round k. Let $bytes_i(k)$ be the bytes sent by $flow_i$ in round k, and let $sent_i(k)$ be the sum of the bytes sent by $flow_i$ in rounds 1 through k (where $sent_{i,K} = \sum_{k=1}^{K} bytes_i(k)$). Based on the protocol description in Section 4.1 it follows:

$$bytes_i(k) + DC_i(k) + CC_i(k) = Q_i + DC_i(k-1) + CC_i(k-1) \tag{4.3}$$

The fairness property of DRR is analyzed in [**SV95**]. In order to prove fairness for our compensator we must consider the scenario when the mobile device first defers transmission due to a bad channel state prediction in some rounds and then when the channel is predicted to be in a good state and compensation is provided to the mobile device. Assume that $DC_{i,0} = CC_{i,0} = 0$ are the initial conditions for round 1 letting the predictor diagnose a bad channel state for the next N rounds. The fact that the mobile device defers transmission for rounds $k = 1$ through $k = N$ implies that $sent_{i,N} = 0$. It immediately follows that:

$$DC_i(k+1) = DC_i(k) = DC_i^{max} - Q_i \;\; ; \;\; 1 \le k \le N \tag{4.4}$$

Using this result in Equation 4.3 we get:

$$CC_i(k+1) = Q_i + CC_i(k) \;\; ; \;\; 1 \le k \le N \tag{4.5}$$

Now let us assume that the predictor diagnoses good channel state for rounds $k = N+1$. In this case the mobile device transmits packets and will be compensated for the previous rounds that it deferred transmission. Then the amount of bytes of compensation given in one round is:

$$CC_i(k-1) - CC_i(k) = (DC_i^{max} - Q_i - DC_i(k-1)); \;\; k > N \tag{4.6}$$

Since $DC_i(k-1) = DC_i^{max} - bytes_i^{pred}(k-1)$, Equation 4.6 can also be written as:

$$CC_i(k-1) - CC_i(k) = bytes(pred)_i(k-1) - Q_i \;\; ; \;\; k > N \tag{4.7}$$

Where $bytes^{pred}$ indicates the dependency of the number of bytes transmitted successfully based on the accuracy of the prediction (e.g., $bytes^{pred} = 0$ if the packet was corrupted by channel errors not detected by the predictor). Since conditions at round N are: $CC_i(N) = NQ$ and $DC_i(N) = DC_i^{max} - Q_i$, we get:

$$CC_i(k) = kDC_i^{max} - bytes_i^{pred}(k) - ... - bytes_i^{pred}(N) \;\; ; \;\; k > N \tag{4.8}$$

or:

$$CC_i(k) = kDC_i^{max} - sent_i^{pred}(k) \;\; ; \;\; k > N \tag{4.9}$$

Clearly, the compensation of $flow_i$ will occur as long as $CC_i(k)$ remains positive and will stop when equal to zero. The ideal bytes allocated to flow i in WRR after k rounds under normal conditions (persistent good channel conditions) is $sent_i(k) = kQ_i$ [**SV95**]. Subtracting this from Equation 4.9, it follows that as soon as $CC_i(k)$ equals zero, the flow reaches its ideal bandwidth allocation (e.g., the flow has been fully compensated).

The mobile device can only transmit data after round N when the channel is predicted to be in a good channel state. Since $bytes_i^{pred}(k)$ is always smaller than DC_i^{max}, then as long as $DC_i^{max} > Q_i$ the flow will reach its bandwidth allocation. The main implication of this analysis is that even if the mobile device experiences a deep fade, fairness can be reached as long as the channel recovers in the future. Fairness in practical situations, however, does not hold when channel prediction fails and the packet is transmitted and corrupted by channel errors which cannot be anticipated by the predictor. In this case, the accuracy of the predictor plays a critical role in the operation of adaptive-QOS wireless systems.

The choice of DC^{max} is a design parameter. Choosing a small DC^{max} will reduce the latency bound but increase flow compensation time. On the other hand, choosing a large DC^{max} increases the latency bound during periods of heavy load but decreases the compensation time. Since only a fraction of CC is used for compensation, CC can become large without affecting the latency bound of flows in the system. Because of this we do not limit the maximum size of the compensation counter.

4.3. Delay Analysis. The latency bound provided by normal WRR is given by $\sum_{i=1}^{n} Q_i/C$ [**SV95**], where C represents the transmission speed when there are n flows in the scheduler[3]. An interpretation of this equation is that a small packet arriving at the head of the queue can be delayed by the quantum size of the other flows in the scheduler. In our case, the quantum size could be bigger than the default size (Q) when compensation bytes are added; therefore the latency bound becomes:

$$LatencyBound = \frac{\sum_{i=1}^{n} DC_i^{max}}{C} \tag{4.10}$$

This equation does not consider the delay associated with the RTS-CTS packet exchange. This delay is in the order 3 msecs for 802.11 operating at 2 MBPS [**P8097**]. For small packets this delay can generate a large overhead. The value of DC^{max}, which is also translated to how fast flows recover their share of the link has a direct impact on the latency bound at which a flow can probe the state of the channel. It is important to note that this latency bound does not represent the worst case packet delay but the worst case channel prediction delay. Because it is out of the scheduler's control how long the channel is in a bad state, the best the scheduler can do is to bound the time between channel predictions for each flow.

Ideally the system should attempt to probe the channel as soon as is possible if the channel is in a bad state. Experimental results show [**BBKT97**], however, that fading periods are usually correlated. Therefore, waiting for some time before testing the channel again may be intuitive. On the other hand, waiting too long to test the channel may lead to poor performance. This is because the scheduler can miss periods in which the channel is in a good state and packets could have been transmitted. The fact that WRR visits flows at discrete times (once every round) matches with this 'intuitive' probing timing of the predictor. Determining the optimal interval and time for probing the channel during a fade is still an open research issue which depends on how well the duration of bad periods can be accurately estimated.

The fairness properties of the compensator assumes that buffer space is infinite and packets can remain in the buffer indefinitely. Buffer space is a finite resource, however. If bad channel periods persist and build up the queues, then arriving packets may find the buffer full and be dropped or the application regulated. This observation calls for additional adaptation mechanisms capable of responding to these conditions over longer time scales.

---

[3]This equation is valid only when the quantum size is greater than the maximum packet length, which is a necessary condition in DRR to make the system work-conserving. Otherwise Q_i should be replaced by the maximum packet size.

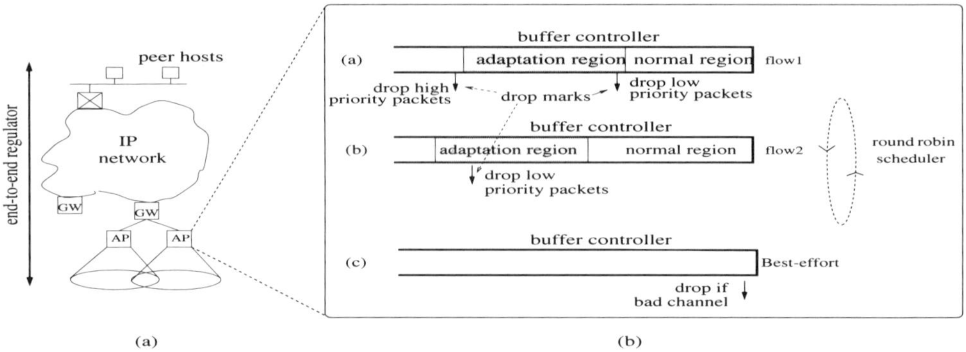

FIGURE 3. Adaptation Mechanism.

5. Adaptator

The final component of our adaptive-QOS model exploits the ability of applications to adapt to channel dependent conditions or variations in available bandwidth over longer time scales. For example, audio and video flows may require discrete or smooth adaptation while some real-time data services may be greedy and capable of responding to any available bandwidth [**BCL98**]. Some applications may be able to tolerate fast time-scale adaptation while others, conversely, may require slow adaptation to available bandwidth conditions rather than instantly reacting to any availability. In either case the wireless access point can respond to these conditions by dropping low priority packets and by controlling the rate of flows over longer time scales.

In what follows, we discuss how QOS information such as delay, priority and multi-resolution semantics can be used to enhance the quality of service delivered to mobile devices. For example, layered audio/video applications can be transmitted using different layers of resolution (e.g., MPEG-2, in response to network conditions [**ACKL98**]). Typically, multi-resolution applications transmit a base layer plus a number of enhancement layers. These applications are capable of gracefully utilizing enhancements layers as bandwidth become available as channel conditions improve.

5.1. Adaptator Operation. While the goal of the compensator is to maintain stability of supporting adaptive real-time flows (e.g., minimum bandwidth assurances), fast time-scale dynamics are also resident. Such dynamics, which translated to application level QOS, can lead to poor performance for continuous media applications. For example, a video sequence in which the received quality is switching between high and low quality because of bandwidth variations due to new sessions or changing link conditions is undesirable for some applications. Subjective tests suggested that many users are susceptible to such changes and a stable even lower quality is sometimes preferred. The observation that adaptation is application-specific motivates the notion of adaptation in wireless network. The adaptator includes two components that support the notion of adaptive wireless services; these are:

- a buffer controller, which operates over the wireless hop; and
- a regulator, which operates on an end-to-end basis.

A typical real time application will use a regulator to adapt the rate of a source to the average bandwidth observed by the network and use it to support the basic stream quality (e.g., base layer). Enhanced quality streams (e.g., enhancement layers) can be transmitted within the available rate seen from the network, or can be transmitted above that rate with the risk that those packets may be dropped before reaching the destination. Here we assume that if congestion or bad channel conditions occur, then enhancement layers should be dropped first. In our adaptive-QOS framework a buffer controller supports this type of operation by partitioning the mobile device's buffer allocation into two regions using dropping marks as illustrated in Figure 3:

•*a normal region*. During normal operations the buffer occupancy is likely to be small reflecting the fact that the channel is in a good state and no burst of data occur; and

•*an adaptation region*. When severe channel degradation occurs or bursty data arrives, the buffer occupancy can reach high levels where packet dropping will likely occur. When this situation occurs, the compensator notifies the adaptator which set the proper filter in the packet classifier to drop low priority packets.

In our adaptator we assume that another protocol running on an end-to end basis regulates the rate of flows over a longer time scale according to the measured network performance (e.g., throughput, packet losses/delay). Using end-to-end regulation in this manner limits the likelihood of persistent high occupancy queues due to congestion.

Figure 3 illustrates a per-mobile buffering scenario at a wireless access point. In this example, two real-time flows are supported by per-mobile queues and buffering with all best effort flows being aggregated into a single queue. Figure 3 (a) illustrates the case when a flow consist of three different priorities; this may for example be associated with a video flow with a base layer and two enhancement layers. In Figure 3 (b), a flow with only two priorities is active; this may correspond to an audio flow with normal and enhanced qualities. Finally in Figure 3 (c), a single buffer is used to aggregate best effort traffic for all mobile devices within a wireless cell. The aggregation of several flows into a single queue leads to the head-of-line problem [**BBKT97**]. In order to avoid this, the adaptator drops the packet at the head of the buffer if the predictor diagnoses a bad channel state.

5.2. Setting the dropping marks. The optimal position of the drop marks illustrated in Figure 3 depends on the average queue size. Without channel prediction, the average queue size depends on several factors that relate to the mismatch between traffic load, link capacity and traffic burstiness. Assuming a regulator operates on an end-to-end basis, the source can match the available rate at the bottleneck node in the network; therefore, small queue sizes should be anticipated (mostly related to jitter in the network). When the predictor is operational, the length of the queue will increase as the length of fade periods increase. If the typical queue size is small then the drop mark should be correspondingly large. This allows the wireless link to operate at a relatively high throughput without having to drop packets. When the average queue size is large, then the drop marks must be correspondingly small. This allows the arbitrator to drop low priority packets earlier, which saves buffer space for high priority packets in case of severe network conditions. Currently, we are working on dynamic techniques for setting up and maintaining these marks.

6. Conclusion

In this paper we have discussed three adaptation components of an adaptive-QOS framework for wireless networks; that is, the predictor, compensator and adaptator mechanisms. We have argued that a systems approach needs to be taken to support the delivery of adaptive services over time-varying wireless networks where multiple time scales come into play. We believe that the predictor, compensator and adaptator mechanisms should work in unison to deliver adaptive services and not in isolation. We argue for a level of integration and interworking managed by an arbitrator that is operational in the access point. Currently we are working on an in depth evaluation of our approach.

References

[ACKL98] O. Angin, A. T. Campbell, M. E. Kounavis, and R. F.-L. Liao. The Mobiware Toolkit: Programmable Support to Adaptive Mobile Networks. *IEEE Personal Communications Magazine*, August 1998.

[BBKT97] P. Bhagwat, P. Bhattacharya, A. Krishna, and S. Tripathi. Enhancing Throughput over Wireless LANs using Channel State Dependent Packet Scheduling. *In Proceedings of the IEEE INFOCOM*, Kobe, April 1997.

[BCL98] G. Bianchi, A. Campbell, and R. Liao. On Utility-Fair Adaptive Services in Wireless Networks. *In Proceedings of International Conference on Quality of Service, IWQoS*, Napa, California 1998.

[ESZ98] T. S. Eugene, I. Stoica, and H. Zhang. Packet Fair Queueing Algorithms for Wireless Networks with Location-Dependent Errors. *In Proceedings of the IEEE INFOCOM*, San Francisco, March 1998.

[FSS98] C. Fragouli, V. Sivaraman, and M. Srivastava. Controlled Multimedia Wireless Link Sharing via Enhanced Class-Based Queuing with Channel-State-Dependent Packet Scheduling. *In Proceedings of the IEEE INFOCOM*, San Francisco, California, March 1998.

[GCM98] J. Gomez, A. T. Campbell, and H. Morikawa. A Systems Approach to Prediction, Compensation and Adaptation in Wireless Networks. *ACM First International Workshop on Wireless Mobile Multimedia (WOWMOM '98)*, Dallas, October 1998.

[Kat94] R. H. Katz. Adaptation and Mobility in Wireless Information Systems. *IEEE Personal Communications Magazine, Vol. 1, No.1, First Quarter*, 1994.

[KNB97] F. Baker K. Nichols, S. Blake and D. Black. Definition of the Differentiaded Services Field (DS Field in the ipv4 and ipv6 headers). *RFC 2474*, 1997.

[LC98] S. Lee and A. Campbell. INSIGNIA: In-Band Signaling Support for QOS in Mobile Ad Hoc Networks. *In Proceedings of Mobile Multimedia Communications, MoMuC*, Belin, October 1998.

[Ngu98] G. Nguyen. Wireless Features in ns Simulator. *http://www.cs.berkeley.edu/ gnguyen/ns/*, 1998.

[NJZ97] K. Nichols, V Jacobson, and L. Zhang. A Two-bit Differentiated Services Architecture for the Internet. *Internet Draft, draft-nichols-diff-svc-arch-00.txt, work in progress*, 1997.

[P8097] P802.11. IEEE Standard For Wireless LAN Medium Access Control (MAC) and (PHY) Specifications, 802.11. *IEEE standard*, November 1997.

[Sch92] H. Schulzrinne. RTP: The real-time Transport Protocol. *MCNC 2nd Packet Video Workshop, vol. 2*, North Carolina, December 1992.

[SLS97] V. Bharghavan S. Lu and R. Srikant. Fair Scheduling in Wireless Packet Networks. *In Proceedings of ACM SIGCOMM*, San Jose, 1997.

[SV95] M. Shreedhar and G. Varghese. Efficient Fair Queueing using Deficit Round Robin. *In Proceedings of ACM SIGCOMM*, Berkeley, California, 1995.

[WMSW97] J. Weinmiller, A. Festag M. Schlager, and A. Wolisz. Performance Study of Access Control in Wireless LANs- IEEE802.11 DFWMAC and ETSI RES 10 HIPERLAN. *Mobile Networks and Applications journal, MONET, Volume 2*, 1997.

[ZR96] M. Zorzi and R. Rao. Error Control Strategies for the Wireless Channel. *IEEE International Conference on Universal Personal Communications, ICUPC*, Cambridge, Massachusetts, 1996.

DEPARTMENT OF ELECTRICAL ENGINEERING AND CENTER FOR TELECOMMUNICATIONS RESEARCH, COLUMBIA UNIVERSITY, NEW YORK, NY 10027

E-mail address: javierg,campbell@comet.columbia.edu

DIMACS Series in Discrete Mathematics
and Theoretical Computer Science
Volume **52**, 2000

Using Self-Stabilization to Design Adaptive Multicast Protocols for Mobile Ad Hoc Networks

Sandeep K. S. Gupta and Pradip K. Srimani

ABSTRACT. Mobile ad hoc networks present unique kinds of challenges in designing and implementing communication protocol in the sense that the underlying topology of the network varies with time due to mobility of the nodes. When the nodes in the system move, existing links often no longer exist and/or new links are created. Any communication protocol designed for such multi-hop ad hoc networks must be able to tolerate such change in network topology that can occur anywhere in the network. Another characteristics of such networks is that bandwidth is a comparatively precious resource and needs to be conserved. Our purpose in the present paper is to explore the possibilities of applying the relatively new paradigm of self-stabilization in distributed fault tolerant algorithm design to tackle the problem of topology change in mobile networks. Self-stabilizing distributed algorithms converge to a global legitimate state in presence of any number of intermittent faults and still use only local knowledge for actions at each node. Our objective is to view the topology change as a change in the node adjacency information at one or more nodes and utilize the tools of self-stabilization to converge to a stable global state in the new network graph. We illustrate the concept by designing a new efficient distributed algorithm for multi-cast in a mobile network that can accommodate any change in the network topology due to node mobility.

1. Introduction

Mobile ad hoc networks are being increasingly used for military operations, law enforcement, rescue missions, virtual class rooms, and local area networks. A mobile multi-hop network consists of n identical mobile hosts (nodes) with unique ids $1, \dots, n$. These mobile hosts communicate among each other via a packet radio network. When a node transmits (broadcasts) a message, the nodes in the *coverage area* of the sender can simultaneously receive the message. A node i is called a *neighbor* of node j in the network if node j is in the *coverage area* of node i. This relationship is time varying since the nodes can and do move. At any given time a node i can correctly receive a message from one of its neighbors, say j, iff j is the only neighbor of i transmitting at that time. A multiple access protocol is used by nodes to get contention-free access to the wireless channel. Since, a limited communication bandwidth is shared between several neighboring nodes the available communication bandwidth between any two neighboring nodes depends upon number of nodes in the vicinity and their communication activity. Further, since nodes are battery-powered they can go to sleep in order to conserve battery power. Hence, in mobile ad hoc network i) old links fail and new links are formed

2000 *Mathematics Subject Classification.* Primary 68W15, 65C05, 68M12, 68M15.

over time due to node movement ii) nodes can be unreachable since they may go to sleep iii) the bandwidth available on a (logical) link between two neighboring nodes varies with time, and iv) the topology may get disconnected.

Multi-cast is an important group communication primitive; it involves sending a message to an arbitrary subset of nodes in the network. It is useful in a large number of applications in a distributed system, where a set of nodes need be informed of a specific event. Multi-cast protocols for distributed systems with a static network have been widely studied. Multi-cast protocols are evaluated in terms of bandwidth consumed in the entire process. A multi-cast message can be send to each member of the multi-cast group separately; this wastes bandwidth since the message may be repeatedly sent over the same communication link. Conceptually, this wastage can be avoided by constructing a minimal cost tree spanning only the members of the multi-cast group in the entire network graph and then forwarding the message from the root of the tree (source of the message) along the tree edges. Unfortunately, the problem of computing the minimal cost multi-cast tree spanning an arbitrary number of nodes (members of the multi-cast group) in a given weighted graph is NP-complete [**S**]; hence, people have concentrated on designing heuristic algorithms for determining near optimal multi-cast trees for given multi-cast groups and analyzing their performance [**D**]. All these studies are made under the assumption that the underlying network topology does not change.

Recently, several protocols for delivering multi-cast messages in mobile networks have been developed [**Z, G**]. These works are also based on the assumption that the node mobility does not alter the network topology, i.e., mobile hosts can directly communicate to the fixed mobile switching stations, which are connected by fixed wired network. Multi-hop radio or the so called ad-hoc [**I, O, W**] networks present another dimension of complexity in the sense that node movements do change the network topology and any protocol for multi-cast or broadcast must provide mechanism to ensure delivery of messages to the recipients even when the topology changes during the process. The constraint of bandwidth conservation becomes even more stringent especially when multiple multi-cast groups are present in the system. Several routing mechanisms for unicast messages in ad-hoc networks have been developed [**R, V**]. Chlamtac and Kutten [**I**] describe a tree based broadcasting in multi-hop radio networks. A general framework for developing reliable group communication primitives is given in [**Q**]. Chiang and Gerla have presented a "best effort" multi-cast algorithm in [**H**]. This algorithm is based on underlying multi-cluster multi-hop packet network [**O**] and uses core based trees for multi-cast framework [**E**]. This algorithm does not guarantee reliability in the sense that messages may be lost if the topology changes during multi-casting of a message. This method is suitable when the rate of topology change is low. Pagani and Rossi have presented a reliable broadcast algorithm in [**W**] which is also based on underlying multi-cluster multi-hop packet network [**O, L**].

Our purpose in the present paper is to propose a new distributed algorithm to maintain the multi-cast tree in a given radio network for a given multi-cast group; our algorithm is fault tolerant (reliable) in the sense that the algorithm automatically can detect occasional link failures and/or new link creations in the network (due to mobility of the hosts) and can readjust the multi-cast tree. Our approach is to use the paradigm of *self-stabilization* for distributed fault tolerance to design the fault tolerant multi-cast tree computation algorithm and thereby to investigate the appropriateness of this paradigm (and thus to use the various algorithms therein) to solve the problems associated with mobile networks (dynamic topology or graph).

2. Mobility Tolerance and Self-Stabilization

2.1. Self-stabilizing Distributed Systems. *Self-stabilization* is an alternate way of looking at system fault tolerance, especially it provides a "built-in-safeguard" against "transient failures" that might corrupt the data in a distributed system. The objective of self-stabilization is (as opposed to mask faults; the common approach to design the fault tolerant systems is to mask the effects of the fault) to recover from failure in a reasonable time and without intervention by any external agency. The concept of self-stabilization was first introduced in [**J**] and the possibility of using this concept for designing fault tolerant algorithms was first explored in [**T**]. One of the goals of a distributed system is that the system should function correctly in spite of intermittent faults. In other words, the global state of the system should ideally remain in a legitimate state. Often, due to node failures or other perturbations, the global state of a distributed system is in some illegitimate state, and is desirable that it reaches a legitimate state without the interference of an external agency. Systems that reach a legitimate state starting from any illegitimate state in a finite number of steps are called self-stabilizing systems [**J, K**]. Recently there has been a spurt of research in designing self-stabilizing distributed graph algorithms for many applications [**N, F, M, A**]; a good survey of self-stabilizing algorithms can be found in [**X**].

2.2. Mobility Tolerant Multi-cast Tree. In our context of computing a multi-cast tree in a network with dynamic topology, each node maintains an adjacency set (set of the nodes adjacent to it). In absence of any mobility, these adjacency sets are constant. However, when any node moves a sufficient distance so that the topology of the network changes, the adjacency set of at least one node will change. *Note that it is neither possible nor necessary that each node of the system will be aware of this change.* In this paper, we assume that the graph is edge-weighted, i.e., each edge is assigned a nonzero positive weight; in a mobile network when a new link comes up, we assume that the relevent nodes can determine the edge weight (depending on available bandwidth, .. etc.). It is possible that many edges in the system graph have the same edge weight or even a single node may have multiple incident edges of same weight. In our algorithms, there are occasions when a node has to choose among a set of edges the one with minimal weight; in case of a tie, we assume that the algorithm uses the end-node identifiers of the edges lexicographically to resolve the tie. Thus, the choice of a minimal edge from among a set of edges is always unique since each node has an unique identifier [**U**]. The proposed algorithm constructs the multi-cast tree in three logical stages. First, it computes the minimal spanning tree (MST) of the new topology; second, it establishes the unique path from each node to the root node (the source node for the multi-cast message) by establishing a predecessor pointer for each node in the tree; third, it determines for each node if the node needs to further propagate the multi-cast message down the tree based on the multi-cast group membership of its successors in the tree. The following features of the algorithm are to be noted: (1) the multi-cast tree maintained by the algorithm is not the minimal multi-cast tree; it simply prunes the MST of the topology to include the necessary nodes to propagate the message all group members; (2) whenever there is a topology change due to mobility of the nodes, the algorithm is automatically triggered (without the need of intervention by any external agency) to construct the new minimal spanning tree of the graph and the subsequent pruning operations. It takes finite amount of time for the system to stabilize to the new tree, nodes may further move during that time, but the mobility rates are low enough to allow the system to stabilize before the next move; (3) the entire process is distributed in nature; no central coordination is necessary.

3. Assumptions for the System Model

We make the following assumptions about the system.

- A data link layer protocol at each node i maintains the identities of its neighbors in some list $neighbors(i)$. This data link protocol also resolves any contention for the shared medium by supporting logical links between neighbors and ensures that a message sent over a correct (or functioning) logical link is correctly received by the node at the other end of that link. The logical links between two neighboring nodes are assumed to be bounded and FIFO. The link layer protocol informs the upper layer of any creation/deletion of logical links using the *neighbor discovery protocol* described below.
- Each node periodically (at intervals of t_b) broadcasts a *beacon* message. This forms the basis of *neighbor discovery protocol.* When a node i receives the beacon signal from a node j which is not in its neighbors list $neighbors(i)$, it adds j to its neighbors list (data structure $neighbors_i$ at node i), thus establishing link (i, j). For each link (i, j), node i maintains a timer t_{ij} for each of its neighbors j. If node i does not receive a beacon signal from neighbor j in time t_b, it assumes that link (i, j) is no longer available and removes j from its neighbor set. Upon receiving a beacon signal from neighbor j, node i resets its appropriate timer.
- When a node j sends a beacon message to any of its neighbors, say node i, it includes some additional information in the message that are used by node i to compute the cost of the link (i, j) as well as other information regarding the state of the node j, as used in the algorithm.
- The links between two adjacent nodes are always bidirectional.
- The topology of the ad-hoc network is modeled by a (undirected) graph $G = (V, E)$, where V is the set of nodes and E is the set of links between neighboring nodes[1]. Since the nodes are mobile, the network topology changes with time. We assume that no node leaves the system and no new node joins the system; we also assume that transient link failures are handled by the link layer protocol by using timeouts, retransmissions, and per hop acknowledgments. Thus, the network graph has always the same node set but different edge sets. Further, we assume that the network topology remains connected. This assumptions holds in mobile ad hoc networks in which the movement of nodes is co-cordinated to ensure that the topology does not get disconnected.
- There is an underlying unicast routing protocol to send unicast messages between two arbitrary nodes in the network.

[1]Note that all links are bidirectional.

4. Fault Tolerant Multi-cast Tree Protocol

4.1. The General Approach. Our protocol for fault tolerant maintenance of the multi-cast tree for a given source node (we call it root node r) and its multi-cast group consists of 3 logical steps: (1) construction of the minimal spanning tree (MST) of the mobile network graph in presence of topology change due to node mobility; (2) once the MST stabilizes for the new topology, rooting the MST with respect to the given source node of the multi-cast message (establishing unique parent pointer for each node in the MST); (3) pruning from the MST the nodes that are not needed to send the message to the multi-cast group members.

(1) MST of the Network in Presence of Topology Change

Our purpose is to design a *distributed* algorithm to construct the MST of a given weighted graph; the requirement is that when the topology changes, the algorithm will be automatically kicked in and the system will recompute the new MST of the changed topology. As stated earlier, whenever there is a change in topology of the network, at least node will be aware of that fact in the sense that the set of neighbor nodes for that node will change. We use similar concepts as developed in the self-stabilizing MST construction in a symmetric graph [**C**]. The minimal spanning tree (MST) of the graph is defined to be a spanning tree of the graph such that the sum of the weights of the edges in the tree is less than or equal to that for all possible spanning trees of the graph.

REMARK 4.1. If the weights $\{w_{ij}\}$ of a graph are unique (distinct), the graph has an unique MST [**P**].

To design a mobility tolerant algorithm for the MST of a graph, we introduce a new characterization of any path in a given graph.

DEFINITION 4.2. α-cost of any path from node i to j is defined to be the maximum of the weights of the edges belonging to the path. $Mcost_{ij}$ is defined as the minimum cost among the α-cost of all possible paths between the nodes i and j.

REMARK 4.3. We call the path, along which $Mcost_{ij}$ is defined, to be the minimum-α path between nodes i and j; this should not be confused with the traditional shortest path between nodes i and j. The shortest path is defined to be the path of minimum length where the length of a path is the sum of the weights of the edges on the path. Most significant difference between the two metrics, α-cost and length, of a path, assuming nonzero positive edge weights, is that when a path is augmented by an additional edge, length must increase while α-cost may remain constant.

REMARK 4.4. Consider a graph G with unique edge weights. An edge (i, j) is in the unique MST if and only if $Mcost_{ij} = w_{ij}$ [**U**]. For details of the proof, see [**U**].

We use Remark 4.1 and Remark 4.4 to develop our algorithm for MST construction. First, we can safely assume the edge weights to be unique; this is no restriction since if not, we can easily add lexicographic information to make them unique [**U**]. Second, if a distributed algorithm can compute the α_{ij} values for all nodes, we can add an additional data structure $Tree\_Neighbors_i$ at each node i that keeps track of the MST edges incident on node i, i.e., $Tree\_Neighbors_i = \{k|\ the\ edge\ e_{ik} \in MST\}$. Computing α_{ij} values for all nodes is not similar to the all pairs shortest path problem since the metric α-cost does not have the desirable properties of the metric length (see Remark 4.3); in order for the algorithm to stabilize, we need to have additional data structures $Mcost_i$ at each node to keep track of the minimal α-cost paths from node i to every other node in the graph. The details of the pseudo-code of the algorithm and the different events are described in the next section.

(2) Rooting the MST at the Multi-cast Source r

Once the MST has stabilized, we need to root the MST at the multi-cast source node r. When the MST has stabilized, each node i knows its neighbors in the MST (the data structure $Tree\_Neighbors_i$ keeps track of that information). In order to root the MST at r, we use two variables $Level_i$ and $Parent_i$ at each node i. The node r knows it is the source node and hence it unilaterally sets its $Level_r$ at 0 and makes its $Parent_r$ Null to indicate that it is the root node. Every other node i looks at all of its neighbors in the MST and sets its $Level_i$ value at one plus the minimum of the level values of its tree neighbors and accordingly sets its parent pointer $Parent_i$.

(3) Pruning the MST to Build the Multi-cast Tree

The multi-cast source node r needs to send the message to the members of the arbitrary multi-cast group. Each node in the network knows whether it is a member of the multi-cast group (IS_Member_i is true). In the rooted MST, each node i can determine if it is a leaf node in the MST (size of the $Tree\_Neighbors_i$ is 1) and if node i does not belong to the multi-cast group, it sets its $Flag_i$ variable to 0; otherwise 1. Any other node i (i is not a leaf node in the MST) will look at all its successors in the MST and will set its $Flag_i$ to 1 iff at least one of its successors either has a Flag of 1 or is a member of the multi-cast group. When this process stabilizes, each node i, when it receives the multi-cast message from its parent in the tree, knows that it needs to forward the message to its successors if $Flag_i$ is 1. Note that the nodes with $Flag_i$ value 1 constitute the internal nodes of multi-cast tree (although not all the nodes in the multi-cast tree are necessarily members of the multi-cast group).

4.2. Complete Pseudo-Code. Each node i executes in an event driven mode – it executes the procedure shown in Figure 1. There are 2 different types of messages in the system as described in subsequent subsections; the actions initiated by the nodes in response to different events are also described as the set of events.

4.2.1. *Data Structures.* We assume there are n nodes (mobile hosts) in the system; the nodes are numbered 1 through n. There is one designated node r which is to be the root of the multi-cast tree (source of the multi-cast message). M is a large positive constant greater than the maximum of the edge weights in the graph. Each node i in the system maintains the following local data structures:

- $\boldsymbol{neighbors_i}$: A set of node identifiers denoting the immediate neighbors of node i.
- $\boldsymbol{Is\_member_i}$: A Boolean variable; the value Is_member_i is `true` if node i is a member of the given multi-cast group otherwise it is `false`.
- $\boldsymbol{W_i[\,]}$: An array of positive real numbers. The value $W_i[j]$ is the edge weight of the link (i, j) where $j \in neighbors_i$.
- $\boldsymbol{Mcost_i[1:n]}$: An array of positive real numbers; $Mcost_i[j]$ is the current estimate of the minimum α-cost path from node i to j. Initially, $Mcost_i[i] = 0$ and $Mcost_i[j] = M$ for all $j \neq i$.

- $\boldsymbol{mcost_{ij}[1:n]}$ $(\boldsymbol{j \in neighbors_i})$: A set of n-tuples; for each $j \in neighbors_i$, $mcost_{ij}$ is a copy of the array $Mcost_j[1:n]$ as stored at node i.
- $\boldsymbol{Hops_i[1:n]}$: An array of positive integers; $Hops_i[j]$ is the number of hops in the minimum α-cost path between node i and j as currently known to node i. $Hops_i[j] = n$ for any j indicates that node i currently does not know of any path between i and j. Initially, $Hops_i[i] = 0$ and $Hops_i[j] = n$ for all $j \neq i$.
- $\boldsymbol{hops_{ij}[1:n]}$ $(\boldsymbol{j \in neighbors_i})$: set of n-tuples; for each $j \in neighbors_i$, $hops_{ij}$ is the local copy of $Hops_j[1:n]$ as stored at node i.
- $\boldsymbol{Tree\_Neighbors_i}$: A set of node identifiers denoting the neighbors of node i are also its neighbors in the MST.
- $\boldsymbol{Parent_i}$: A node identifier; $Parent_i = k$ if node k is the parent of node i in the multi-cast tree with node r as the root node, otherwise $Parent_i = \Lambda$. Initially, $Parent_i = \Lambda$.
- $\boldsymbol{Children_i}$: A set of node ids; $Children_i$ is the set of node i in the multi-cast tree.
- $\boldsymbol{Level_i}$: A positive integer that denotes the level of node i in the multi-cast tree rooted at node r (the core node of the multi-cast tree).
- $\boldsymbol{level_i[1..j]}$: An array to keep a local copy at node i the values of $Level_j$ for each node $j \in neighbors_i$.
- $\boldsymbol{Flag_i}$: A boolean flag which is true when the node i needs to forward a multi-cast message down to its children in the multi-cast tree.
- $\boldsymbol{flag_i[1..j]}$: An array to keep a local copy at node i the values of $Flag_j \vee IS\_Member_j$ for each node $j \in neighbors_i$.
- $\boldsymbol{MH\_Set_i}$: A set of triplets of the form (k, m, h). An element (k, m, h), where $m = Mcost_i[k]$ and $h = Hops_i[k]$ is included in set MH_Set_i if either $Mcost_i[k]$ or $Hops_i[k]$ changed since node i broadcasted the last BEACON message. In case node i has a new neighbor since its last BEACON message, it includes $(k, Mcost_i[k], Hops_i[k])$, $1 \leq k \leq n$, in the MH_Set_i. Node i includes its MH_Set_i in its BEACON messages. After sending out a BEACON message it sets the $MH\_Set_i = \emptyset$.

4.2.2. *Message Types.*

- **BEACON** $(i, MH\_SET, LEVEL, FLAG)$ — is the beacon message from node i; MH_SET is a set of triplets (k, m, h) where $1 \leq k \leq n$, $m = Mcost_i[k]$ and $h = Hops_i[k]$; $LEVEL = Level_i$; $FLAG = Flag_i \vee Is\_Member_i$.
- **MULTICAST** (i, m) — indicates that the sender i wants to propagate the message m down the multi-cast tree.

4.2.3. *System Primitives.* The following primitives are used in the pseudo code given in the next subsection. They are used to simplify the presentation of the algorithm and their functions are described below.

- **set_timer** $(timer, t)$: sets the timer $timer$ for time t. If the timer expires without being reset, the event "*timer* expired" is generated.
- **compute_link_weight** (i, j): computes the weight of link (i, j) based on factors such as the contention a node is experience in communicating with it neighbors, signal-to-noise ratio etc.
- **Add** $(mh\_set, (k, m, h))$: adds the triplet (k, m, h) to the set mh_set. Add() ensures that (k, m, h) is the only triplet in the set with the first component

as k. Hence, before adding (k, m, h) to mh_set, Add() deletes any triplet with the first component as k from mh_set.

4.2.4. *Events.* **Node i wants to Multi-cast a Message**: When a node wants to multi-cast message m it sends a MULTICAST(i, m) to the core node r of the multi-cast group using the underlying unicast routing protocol.

Node i receives MULTICAST message: When a node receives a MULTICAST (j, m) messages it broadcasts it to nodes in $Children_i$ if the value of $Flag_i$ is true.

Beacon timer expired: When the beacon timer expires node i broadcasts BEACON$(i, MH_Set_i, Level_i, Flag_i)$ to its neighbors and sets $MH\_Set_i = \emptyset$.

Node i receives a BEACON message: When a node receives a beacon message BEACON$(j, MH_SET, LEVEL, FLAG)$ from another node j it invokes procedure process_beacon_msgs() which is shown in Fig. 2. If this is a message from a node whose id is not currently in $neighbors_i$ then this indicated that a new node has moved to the vicinity of node i. The procedure *Link_Established* is invoked to update the $neighbors_i$ set. Node i then sets the expiration timer for link (i, j) and computes the weight $W_i[j]$ of link (i, j). Then it updates $mcost_{ij}[k]$ and $hops_{ij}[k]$ using (k, m, h) in the MH_SET and recomputes $Mcost_i[k]$ and $Hops_i[k]$ using the procedure compute_mhcost().

After processing all the triplets in the MH_SET, node i determines its new level in the multi-cast tree rooted at node r and its parent in this rooted tree. Node i computes its level to be one plus the minimum level of all its tree neighbors. Further, the parent node of i is the node which has the minimum level among its tree neighbors. Once node i has determined its parent node $Parent_i$ in the rooted multi-cast tree, it determines its children set $Children_i$. Lastly it computes the new value of $Flag_i$ by OR-ing the $flag_{ij}$ values of all its neighbors.

New link to node j established: When node i comes to know the existence of a new link to node j, it invokes the procedure Link_Established given in Fig. 4. This procedure adds id of node j to $Neighbors_i$ set and adds triplets $(k, Mcost_i[k], Hops_i[k])$ for all $1 \leq k \leq n$ so that node j can learn about the $Mcost_i[]$ and $Hops_i[]$ from the node i's next BEACON message.

Failure of link to node j: In case the timer for link (i, j) expires, node i assumes that link to node j has been lost. In this event node i invokes procedure Link_Failed(). This procedure removes the id of node j from $Negihbor_i$, $Tree\_Neighbors_i$, and $Childern_i$ sets. Further, if node j was its parent node in the multi-cast tree, it sets $Parent_i = \Lambda$. Node i recomputes $Mcost_i[k]$ for each node k in the system and determines its current neighbors in the MST, $Tree\_Neighbors_i$.

5. An Example

We illustrate the working of the algorithm for an example network shown in Figure 6(a). The multi-cast group consists of nodes 0, 1, and 6 and the multi-cast source node is the node 1.

Initially, Level ($\mathcal{L}$) = N (=7), Parent ($\mathcal{P}$) = Λ, Flag ($\mathcal{F}$) = 0 for all nodes. The following table shows the initial status of each node ($MAX = M$, $N = 7$):

Id	Mcost							Hops							TNbrs	$\mathcal{L}$	$\mathcal{P}$	$\mathcal{F}$
	0	1	2	3	4	5	6	0	1	2	3	4	5	6				
0	0	M	M	M	M	M	M	0	N	N	N	N	N	N	{}	7	Λ	0
1	M	0	M	M	M	M	M	N	0	N	N	N	N	N	{}	7	Λ	0
2	M	M	0	M	M	M	M	N	N	0	N	N	N	N	{}	7	Λ	0
3	M	M	M	0	M	M	M	N	N	N	0	N	N	N	{}	7	Λ	0
4	M	M	M	M	0	M	M	N	N	N	N	0	N	N	{}	7	Λ	0
5	M	M	M	M	M	0	M	N	N	N	N	N	0	N	{}	7	Λ	0
6	M	M	M	M	M	M	0	N	N	N	N	N	N	0	{}	7	Λ	0

In the following we show the status of each node after every round. A round is defined as a period of time in which each node in the system receives beacon messages from all its neighbors. Since the state after a round is dependent on the order in which beacon messages are processed by nodes, for the sake of illustration, we have assumed that a node receives beacon message from all its neighbors before it generates its next beacon message. Further, a node generates only one beacon message per round. Tables below give the state of node in a round to be its state just before it generates its beacon message in that round.

For this example four rounds are required for computing the multi-cast tree. Figure 6(b)-(d) illustrate the state of node 0 after rounds 1-4, respectively. Figure 6(e) shows the final multi-cast tree. The state of each node after each round is shown in the following tables.

After round 1:

Id	Mcost							Hops							TNbrs	$\mathcal{L}$	$\mathcal{P}$	$\mathcal{F}$
	0	1	2	3	4	5	6	0	1	2	3	4	5	6				
0	0	3	M	9	1	4	M	0	1	N	1	1	1	N	{1 4 3 5 }	7	Λ	1
1	3	0	10	9	2	4	M	1	0	1	2	1	2	N	{0 2 4 }	0	Λ	1
2	10	10	0	5	10	10	M	2	1	0	1	2	3	N	{1 3 }	1	1	0
3	9	9	5	0	9	9	8	1	2	1	0	2	2	1	{0 2 6 }	2	2	1
4	1	2	10	9	0	4	M	1	1	2	2	0	2	N	{0 1 }	1	1	1
5	4	4	10	9	4	0	7	1	2	3	2	2	0	1	{0 6 }	7	Λ	1
6	7	7	8	8	7	7	0	2	3	2	1	3	1	0	{3 5 }	3	3	1

After round 2:

Id	Mcost							Hops							TNbrs	$\mathcal{L}$	$\mathcal{P}$	$\mathcal{F}$
	0	1	2	3	4	5	6	0	1	2	3	4	5	6				
0	0	2	9	9	1	4	7	0	2	2	1	1	1	2	{4 3 5 }	2	4	1
1	2	0	9	9	2	4	7	2	0	3	2	1	2	3	{4 }	0	Λ	1
2	9	9	0	5	9	9	8	2	3	0	1	3	3	2	{3 }	3	3	0
3	8	8	5	0	8	8	8	3	4	1	0	4	2	1	{2 6 }	4	2	1
4	1	2	9	9	0	4	7	1	1	3	2	0	2	2	{0 1 }	1	1	1
5	4	4	8	8	4	0	7	1	3	3	2	2	0	1	{0 6 }	3	0	1
6	7	7	8	8	7	7	0	2	4	2	1	3	1	0	{3 5 }	4	5	1

After round 3:

Id	Mcost							Hops							TNbrs	$\mathcal{L}$	$\mathcal{P}$	$\mathcal{F}$
	0	1	2	3	4	5	6	0	1	2	3	4	5	6				
0	0	2	8	8	1	4	7	0	2	4	3	1	1	2	{4 5 }	2	4	1
1	2	0	8	8	2	4	7	2	0	5	4	1	2	3	{4 }	0	Λ	1
2	8	8	0	5	8	8	8	4	5	0	1	5	3	2	{3 }	5	3	0
3	8	8	5	0	8	8	8	3	5	1	0	4	2	1	{2 6 }	5	6	0
4	1	2	8	8	0	4	7	1	1	4	3	0	2	2	{0 1 }	1	1	1
5	4	4	8	8	4	0	7	1	3	3	2	2	0	1	{0 6 }	3	0	1
6	7	7	8	8	7	7	0	2	4	2	1	3	1	0	{3 5 }	4	5	0

After round 4:

	Mcost							Hops										
Id	0	1	2	3	4	5	6	0	1	2	3	4	5	6	TNbrs	$\mathcal{L}$	$\mathcal{P}$	$\mathcal{F}$
0	0	2	8	8	1	4	7	0	2	4	3	1	1	2	{4 5 }	2	4	1
1	2	0	8	8	2	4	7	2	0	5	4	1	2	3	{4 }	0	Λ	1
2	8	8	0	5	8	8	8	4	6	0	1	5	3	2	{3 }	6	3	0
3	8	8	5	0	8	8	8	3	5	1	0	4	2	1	{2 6 }	5	6	0
4	1	2	8	8	0	4	7	1	1	4	3	0	2	2	{0 1 }	1	1	1
5	4	4	8	8	4	0	7	1	3	3	2	2	0	1	{0 6 }	3	0	1
6	7	7	8	8	7	7	0	2	4	2	1	3	1	0	{3 5 }	4	5	0

6. Correctness of the Algorithm

The correctness of the proposed algorithm follows from the correctness of the underlying three processes: stabilization of the MST reconstruction process when there is any change in the topology (due to failure of an existing link or creation of a new link caused by mobility of the nodes: (2) once the MST is stabilized, convergence or stabilization of the process of rooting the MST at the multi-cast source node; (3) stabilization of the process of properly setting of the $Flag_i$ variable at each node i to indicate if the multi-cast message needs to be further propagated down to the successors of node i.

Stabilization of the MST construction process directly follows from the correctness proof of the algorithm described in [**C**]. It is to be noted that the data structures at each node as well as the actions taken by each node in [**C**] and in the current algorithm are identical; the only exception is that the algorithm in [**C**] operates in a shared memory paradigm while the proposed algorithm works in a distributed message passing paradigm in the sense that each node first collects the information about the local states of its neighbors (by making local copies at the node) and then takes necessary action to change its own local state. We should also note that it has been shown in [**C**] that the algorithm converges to the new MST after any transient failure (change of topology in our case here) in finite amount of time; no analysis was presented in [**C**] regarding time needed to stabilize after each change. In the present context, we claim that the system will converge to a new MST after each topology change in finite time; but, we also need to evaluate the average time needed for the system to do that in order the algorithm to be useful in real life application. We present some preliminary experimental results (via simulation) to show that the algorithm does indeed converges very fast, taking almost linear (in number of nodes in the system) time to converge.

The correctness of the process of rooting the MST at the multi-cast source node (once the MST is stabilized) directly follows from the correctness of the BFS spanning tree algorithm of [**Y**]. Note that the multi-cast source node r knows that it is the source node and it unilaterally sets its predecessor to be Null and its $Level_r$ to be zero. Every other node checks the *Level* variables of all of its neighbors in the MST and sets its own *Level* variable to one plus the minimum of its neighbors as well as sets its predecessor to its neighbor (in the MST) with the minimum *Level* value. The rest of the argument for convergence is identical to those in [**Y**].

The correctness of the process of setting of the $Flag_i$ variable at each node i in the MST directly follows from the correctness of the self-stabilizing algorithm of [**B**]. Note that once the MST is stabilized and the MST has been rooted at the source node r (each node knows its unique predecessor in the tree), each node can determine if it is a leaf node in the tree. As soon as a node determines that it is a leaf node, it can unilaterally decide its $Flag_i$ variable based on the fact if the node i is a member of the multi-cast group or not. Every other node looks at the $Flag$ variables of its immediate successors in the tree and sets its own $Flag$ variable if at least one of its successors has its $Flag$ True. Thus, the process is essentially a bottom up one anchored at the leaf nodes of the tree; the rest of the argument for convergence is identical to those in [**B**].

7. Preliminary Experimental Results

The proposed algorithm was simulated to study its behavior with respect to how quickly the protocol stabilizes. In this section, we assume that all the processors are involved in the multi-cast which represents the worst case for our algorithm. The simulation is performed in *rounds*. A round is defined as a period of time in which each node in the system receives beacon messages from all its neighbors. The simulation continues to the next round if there is at least one node that has changed state in the current round. Hence, nodes are synchronized at the round boundaries. Since mobility causes some links to fail and establishes other new links, we studied the behavior of the self-stabilizing Multi-cast tree maintenance algorithm with respect to link failures and additions in the network.

The input to the simulation is an initial edge-weighted topology graph. The MST algorithm is run on this graph to get the initial MST. The links which are in the MST are marked as *tree* links and the remaining links are marked as *non-tree* links. Following determination of the initial multi-cast tree, the simulation successively deletes a link from the graph, computes the new multi-cast tree, then adds back the link, and recomputes the multi-cast tree. The simulator records the number of rounds it took for the algorithm to recompute the MST after a link was deleted, called its *d_rounds*, and the number of rounds it took for the algorithm to recompute the MST after the link was reintroduced in the network, called its *a_rounds*. At the end of the simulation, the following statistics are reported over all the link failures and establishments:

- Maximum number of rounds over all the link deletions and additions: $Max_r = max_{e \in G}(max(d\_rounds(e), a\_rounds(e)))$.
- Average number of rounds over all the link deletion and additions: $Avg_r = (\sum_{e \in G}((d\_rounds(e) + a\_rounds(e))/2))/|E|$, where E is the edge set of G.

Additionally, the simulator reports similarly defined statistics for:

- tree link deletions ($Max_{dr}^{(t)}$ and $Avg_{dr}^{(t)}$)
- tree link additions ($Max_{ar}^{(t)}$ and $Avg_{ar}^{(t)}$)
- non-tree link deletions ($Max_{dr}^{(nt)}$ and $Avg_{dr}^{(nt)}$), and
- non-tree link additions ($Max_{ar}^{(nt)}$ and $Avg_{ar}^{(nt)}$)

Table 1 shows the statistics obtained for the algorithm on the graph shown in Figure 7.

The results shown in the table are representative of simulation results obtained for other graphs. The following can be observed from the simulation results we have obtained:

- Recovery from failures of non-tree links and addition of new links requires few rounds.

TABLE 1. Simulation results for the example graph.

$Max_r = 54$	$Avg_r = 19.99$
$Max_{dr}^{(t)} = 54$	$Avg_{dr}^{(t)} = 47.60$
$Max_{ar}^{(t)} = 12$	$Avg_{ar}^{(t)} = 5.87$
$Max_{dr}^{(nt)} = 6$	$Avg_{dr}^{(nt)} = 2.57$
$Max_{ar}^{(nt)} = 5$	$Avg_{ar}^{(nt)} = 2.22$

- Recovery from failures of tree links requires rounds proportional to the size of the network.

8. Conclusion

Our objective in the present paper has been to demonstrate the applicability of the tools of self-stabilization (a relatively new paradigm to design fault tolerant distributed algorithms) to solve the problems associated with dynamic topology change in a mobile multi-hop radio network due to mobility of the host nodes. We have developed a new distributed algorithm for multi-casting a message form a given source node in a mobile network; our algorithms maintains the multi-cast tree in the presence of dynamic topology change of the network. The unique features of the algorithms are: (1) it is distributed in nature; no central coordination is necessary to recompute the underlying minimal spanning tree (MST) of the network; (2) the network nodes take local actions based on local knowledge (the states of its immediate neighbors only); (3) in presence of a topology change, the algorithm is automatically triggered, i.e., at least one node is always aware of the topology change (due to the change of its node adjacency information). We have designed the multi-cast tree based on the MST of the network graph by pruning the nodes properly using the multi-cast group membership. On one hand, we do not get the minimal multi-cast tree for a given multi-cast group; then again computing the minimal multi-cast spanning tree is computationally intractable [**D, S**]. On the positive side, although the algorithm has been developed in the paper based on one multi-cast group, it can be extended in a straightforward way to handle multiple multi-cast groups and each multi-cast group will generate different multi-cast trees, but each of them will use the same underlying minimal spanning tree (thus the reconfiguration due to topology change would be minimal). Our very preliminary experimental results show that the time to converge to a stable state in presence of a topology change is relatively small at least for the examples we have seen; it'd be interesting to conduct elaborate experiments to come to a better understanding of the cost of the algorithm in terms of time to converge as well as to explore to arrive at a theoretical upper and/or lower bound. It'd also be interesting to explore applicability of self-stabilization to solve other problems related to mobile networks.

References

[A] A. Arora, S. Dolev, and M. Gouda. Maintaining digital clocks in step. *Parallel Processing Letters*, 1(1):11–18, 1992.

[B] G. Antonoiu and P. K. Srimani. A self-stabilizing leader election algorithm for tree graphs. *Journal of Parallel and Distributed Computing*, 34:227–232, 1996.

[C] G. Antonoiu and P. K. Srimani. A self-stabilizing distributed algorithm for minimal spanning tree problem in a symmetric graph. *Computers Mathematics and Applications*, 35(10):15–23, October 1998.

[D] N. E. Belkeir and M. Ahamad. Low cost algorithms for message delivery in dynamic multicast groups. In *Proceedings of Int'l Conference on Distributed Computing Systems*, pages 110-117, 1989.

[E] T. Ballardie, P. Francis, and J. Crowcroft. Core based trees (CBT): an architecture for scalable inter domain multicast routing. In *SIGCOMM*, pages 85–95, ACM, 1993.

[F] G. M. Brown, M. G. Gouda, and C. L. Wu. Token systems that self-stabilize. *IEEE Trans. Comput.*, 38(6):845–852, June 1989.

[G] K. Chao and K. P. Birman. A group communication approach for mobile communication. In *Proc. Mobile Computing Workshop*, Santa Cruz, CA, December 1994.

[H] C. C. Chiang and M. Gerla. Routing and multicast in multihop, mobile wireless networks. In *Proceedings of ICUPC '97*, October 1997.

[I] I. Chlamtac and S. Kutten. Tree based broadcasting in multihop radio networks. *IEEE Transactions on Computers*, C-36(10), October 1987.

[J] E. W. Dijkstra. Self-stabilizing systems in spite of distributed control. *Communications of the ACM*, 17(11):643–644, November 1974.

[K] E. W. Dijkstra. A belated proof of self-stabilization. *Distributed Computing*, 1(1):5–6, 1986.

[L] A. Ephremides, J. E. Wieselthier, and J. D. Baker. A design concept for reliable mobile radio networks with frequency hopping signaling. *Proceedings of IEEE*, 1(1):56–73, 1987.

[M] M. Flatebo and A. K. Datta. Two-State self-stabilizing algorithms. In *Proceedings of the IPPS-92*, California, June 1992.

[N] M. Gouda and T. Herman. Stabilizing unison. *Inf. Processing Letters*, 35(4):171–175, 1990.

[O] M. Gerla and J. T. C. Tsai. Multicluster, mobile, multimedia radio network. *ACM Journal on Wireless Networks*, 1(3):255–265, 1995.

[P] E. Horowitz and S. Sahni. *Fundamentals of Computer Algorithms.* Computer Science Press, 1984.

[Q] V. Hadzilacos and S. Toueg. Fault tolerant broadcasts and related problems. In S. Mullender, editor, *Distributed Systems.* Addison-Wesley, New York, 1993.

[R] D. B. Johnson. Routing in adhoc networks of mobile hosts. In *Proceedings of Workshop on Mobile Computing and Applications*, 1994.

[S] R. M. Karp. Reducibility among combinatorial problems. In *Complexity of Computer Computations.* Plenum Press, New York, 1972.

[T] L. Lamport. Solved problems, unsolved problems, and non-problems in concurrency. In *Proceedings of the 3rd Annual ACM Symposium on Principles of Distributed Computing*, pages 1–11, 1984.

[U] B. M. Maggs and S. A. Plotkin. Minimum-cost spanning tree as a path finding problem. *Information Processing Letters*, 26:291–293, January 1988.

[V] C. Perkins and P. Bhagwat. Routing over multi-hop wireless network of mobile computers. In T. Imielinski and H. F. Korth, editors, *Mobile Computing*, pages 182–205. Kluwer Academic Publisher, 1996.

[W] E. Pagani and G. P. Rossi. Reliable broadcast in mobile multihop packet networks. In *Proceedings of MOBICOM 97*, pages 34–42, Budapest, Hungary, 1997.

[X] M. Schneider. Self-stabilization. *ACM Computing Surveys*, 25(1):45–67, March 1993.

[Y] S. Sur and P. K. Srimani. A self-stabilizing algorithm for coloring bipartite graphs. *Information Sciences*, 69:219–227, 1992. Information Sciences.

[Z] A. Acharya and B. R. Badrinath. A framework for delivering multicast messages in networks with mobile hosts. *ACM/Baltzer Journal of Mobile Networks and Applications*, 1:199–219, 1996.

Department of Computer Science, Colorado State University, Ft. Collins, CO 80523

E-mail address: `gupta@cs.colostate.edu`

Department of Computer Science, Colorado State University, Ft. Collins, CO 80523

E-mail address: `srimani@cs.colostate.edu`

```
main()
begin
   case (event) do
      beacon-timer-expired:
         broadcast BEACON(i, MH_Set_i, Level_i, Flag_i ∨ Is_member_i)
         to nodes ∈ neighbors_i
         MH_Set_i = ∅;
         break;
      beacon-msg-received: /* received BEACON(j, MH_SET, LEVEL, FLAG) */
         process_beacon_msg(j, MH_SET, LEVEL, FLAG);
         break;
      multicast-msg-received: /* received MUTLICAST (j, m) */
         if(Flag_i > 0) then
            send MULTICAST (i, m) to Children_i;
         if(Is_Member_i) then
            Deliver the message m to the application layer;
      timer_ij expired:
         Link_Failed(j);
end
```

FIGURE 1. Main Procedure Executed by Node i.

```
Procedure process_beacon_msgs(j, MH_SET, LEVEL, FLAG)
begin
   if(j ≠ neighbors_i)
      Link_Established(j);
   set_timer(timer_ij, time() + t_b);
   W_i[j] = compute_link_weight(i, j);
   /* Update mcost_ij and hops_ij arrays */
   foreach (k, m, h) ∈ MH_SET
      mcost_ij[k] = m;
      hops_ij[k] = h;
      compute_mcost(k);
   end foreach
   /* Compute the tree neighbors */
   Tree_Neighbors_i = ∅;
   foreach(k ∈ neighbors_i) do
      if(W_i[k] = Mcost_i[k])
      then /* (i, k) is a MST link */
         Tree_Neighbors_i = Tree_Neighbors_i ∪ {k};
   end foreach
   level_i[j] = LEVEL;
   if(i = r) /* if node i is the root node of the multicast group */
   then
      Level_i = 0;
      Parent_i = NULL; /* root node has no Parent node*/
   else
      Level_i = n;
      foreach (k ∈ Tree_Neighbors_i)
         if(Level_i > (level_ik + 1))
            Level_i = level_ik + 1;
            Parent_i = k;
      end foreach
   Children_i = Tree_Neighbors_i − Parent_i;
   flag_i[j] = FLAG;
   Flag_i = ⋁_{k∈Children_i} flag_i[k];
end
```

FIGURE 2. Node i receives a BEACON() message from node j.

```
Procedure compute_mcost(k)
begin
    Mcost_i[k] = MAX;
    Hops_i[k] = n;
    foreach((ℓ ∈ neighbors_i) ∧ (hops_iℓ[k] < n))
        cost = max(W_i[ℓ], mcost_iℓ[k])
        if((cost < Mcost_i[k])
        then
            Mcost_i[k] = cost;
            Hops_i[k] = hops_il[k] + 1;
        else
            hops = hops_iℓ[k] + 1;
            if((cost = Mcost_i[k]) ∧ (hops < Hops_i[k]))
                Hops_i[k] = hops;
    end foreach
    if((k ∈ neighbors_i) ∧ (Mcost_i[k] = W_i[k]))
    then
        Tree_Neighbors_i = Tree_Neighbors_i ∪ {k};
    Add(MH_Set_i, (k, Mcost_i[k], Hops_i[k]));
end
```

FIGURE 3. Procedure compute_mcost.

```
Procedure Link_Extablished(j)
begin
    neighbors_i = neighbors_i ∪ {j};
    for k = 1 to n
        Add(MH_Set_i, (k, Mcost_i[k], Hops_i[k]));
    end for
end
```

FIGURE 4. New link to node j established.

```
Procedure Link_Failed(j)
begin
   neighbors_i = neighbors_i − {j};
   Tree_Neighbors_i = Tree_Neighbors_i − {j};
   Children_i = Children_i − {j};
   if (Parent_i = j) then
      Parent_i = NULL;
   for k = 1 to n
      compute_mcost(k);
   /* Compute the tree neighbors */
   Tree_neighbors_i = ∅;
   foreach(k ∈ neighbors_i) do
      if(W_i[k] = Mcost_i[k])
      then /* (i,k) is a MST link */
         Tree_Neighbors_i = Tree_Neighbors_i ∪ {k};
   end foreach
end
```

FIGURE 5. Failure of node i's link to node j.

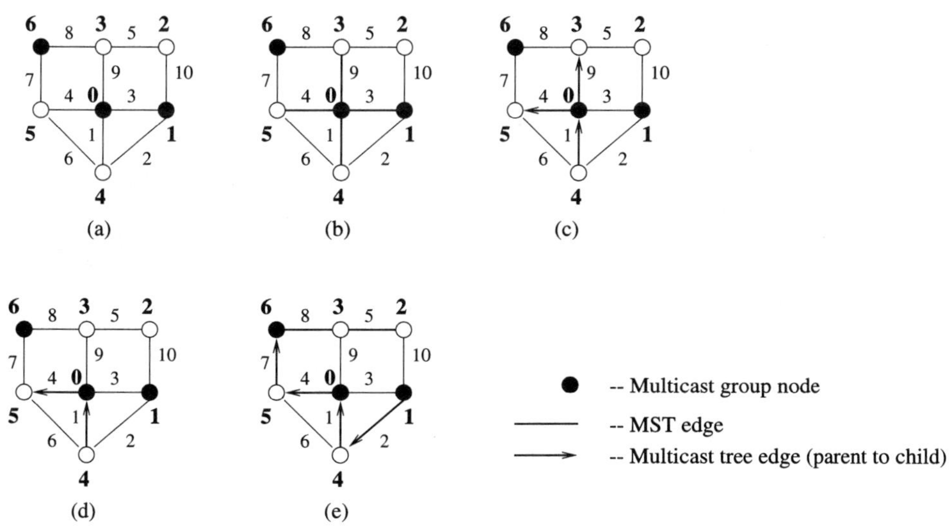

FIGURE 6. Example graph to illustrate the working of the algorithm: a) initial graph; b) node 0's status after round 1; c) node 0's status after round 2; node 0's status after round 3; d) node 0's status after round 4; and e) Multi-cast tree computed by algorithm.

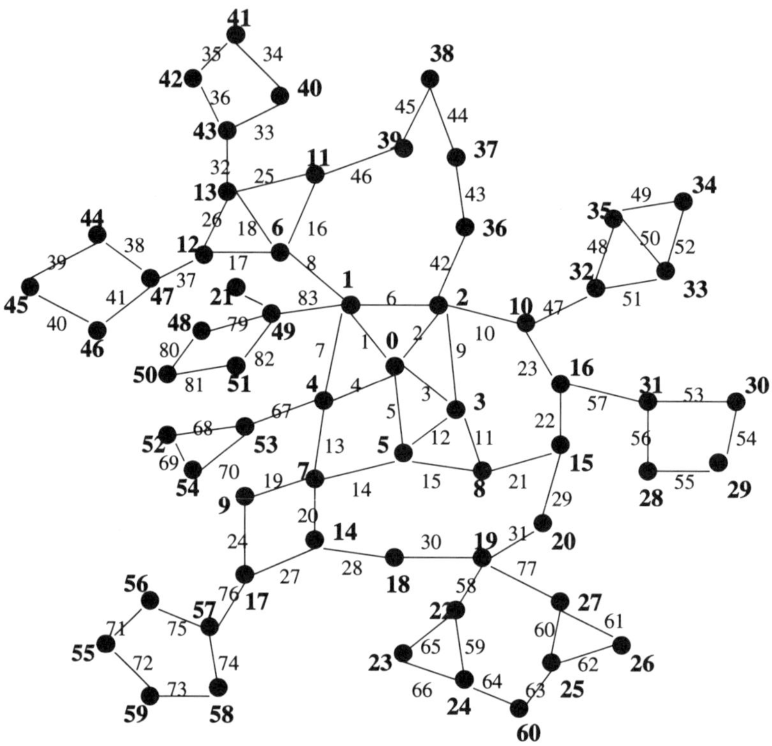

FIGURE 7. Example graph.

DIMACS Series in Discrete Mathematics
and Theoretical Computer Science
Volume **52**, 2000

The Design and Performance of Mobile TCP for Wireless Networks

Zygmunt J. Haas and Abhijit Warkhedi

{haas, warkhedi}@ee.cornell.edu

School of Electrical Engineering, Cornell University

Abstract – Wireless environments are characterized by lossy links and intermittent connectivity. In the past, considerable research has been conducted to improve the end-to-end performance of TCP over wireless networks. The focus of our research is on improving the end-to-end performance, while minimizing processing load of transport protocols on power-limited mobile devices. In this paper, we present the design and implementation of a transport protocol, called MTCP, for wireless networks. The primary goals of MTCP are as follows: a) alleviate the effects of wireless losses on end-to-end performance, b) minimize processing overhead on mobile devices, and c) allow seamless integration of mobile applications into existing internetworks. Our approach is based on the split-connection model, where the end-to-end connection between a mobile and a stationary host is partitioned into two connections: the connection between the mobile host and the mobile-gateway (or the basestation), and the connection between the mobile-gateway and the fixed host. The division allows us to develop MTCP, which is an optimized protocol emulating TCP functionality on the wireless link. The protocol is designed to be lightweight, in order to minimize and eliminate the unnecessary overhead placed on mobile devices. Our experiments with MTCP on different testbeds indicate a substantial improvement in throughput over the "regular" TCP under varied unreliable operational conditions of the wireless link. In addition, the results show that MTCP is much more efficient in utilizing CPU resources than TCP.

2000 *Mathematics Subject Classification.* Primary 68M12, 68M10.

1. Introduction

The vision behind mobile computing [1, 2] is to support ubiquitous access to various forms of information such as data, voice and (possibly) video. Integration of mobile devices into the existing internetwork consisting of stationary hosts has played a critical role in bringing the state-of-the-art a step closer to realizing this vision. At the same time, continued use of legacy technology in conjunction with wireless networks has given rise to certain problems due to the special characteristics of the wireless link and the requirements of the portable power-limited mobile devices.

One problem is the inability of existing end-to-end protocols such as TCP [3, 4] to effectively cope with highly unreliable wireless links. TCP is specifically designed to perform well in networks where losses occur mostly due to congestion problems. As a consequence, TCP fundamentally assumes that all packet losses in the network are caused due to congestion-related problems. This assumption, however, is incompatible with the properties of a wireless link where losses are due to handoffs and transmission errors caused by noise, interference, and channel fading.

A TCP sender identifies losses either by the arrival of three duplicate acknowledgments or by the absence of an acknowledgment within a timeout period. Upon identifying a loss, the TCP sender invokes various congestion control and avoidance mechanisms such as reducing the congestion window size and/or backing off the retransmission timer [3, 4]. These measures, however, cause an unnecessary reduction in the overall end-to-end throughput in networks consisting of links where packet losses are not due to congestion [5]. Several schemes have been proposed in an effort to alleviate the effects of non-congestion-related losses on TCP performance in wireless or similar high loss links [6, 7, 8].

Another problem posed by the existing communication protocols is the processing requirements placed on the mobile devices. As portable machines shrink in their physical parameters (weight and size), their computing and power capabilities are also reduced, especially due to the reduced capacity of the power storage devices. Thus, to maintain a constant level of performance of mobile applications, there is a need to reduce the processing load of communication protocols on mobile devices [9]. The focus of our research is to devise a new transport protocol that improves the end-to-end performance of TCP, while reducing the processing requirements on the mobile host. We base our approach on the split-connection model, where the transport-layer connection between a mobile and corresponding stationary host is partitioned into two connections: the connection between the mobile host and the mobile-gateway (also referred to as basestation), and the connection between the mobile-gateway and the corresponding host. The first connection is called the *wireless* segment, while the second portion is the *wired* segment. The division, by itself, is not a new idea and was already employed by the Indirect TCP (I-TCP) [10]. However, the contribution of this study is the design of a lightweight transport protocol, called MTCP, on the wireless segment that not only improves end-to-end performance, but also reduces the processing load on the mobile device, as well as the communication overhead on the wireless segment. Most of our protocol enhancement stems from the important fact that the connection between the mobile host and the mobile-gateway is a point-to-point link and that all the packets arrive in order. Consequently, as we will see in this paper, many functions can be either simplified or totally eliminated in the wireless segment of the transport-layer connection, leading to a lean protocol code on the mobile machine [9].

Following the split connection model employed by I-TCP, we divide the end-to-end logical connection into the MTCP connection communicating over the wireless segment and the "regular TCP" connection communicating over the wired segment. A transport-layer gateway, called the *Redirector*, is implemented on the mobile-gateway to seamlessly integrate both of these connection segments in such a way that the indirection is transparent to the mobile and the corresponding stationary hosts. This paper will primarily focus on the design and

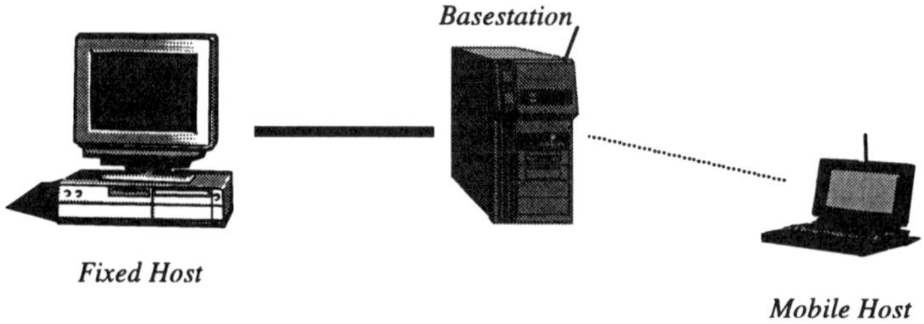

Figure 1: Wireless Network Model

implementation of the MTCP protocol and not on the details of the Redirector, as it has been well studied as part of the research on I-TCP [6].

In addition to designing the MTCP protocol, we present the results of experiments on our testbed under varied unreliable conditions. We evaluate the performance of our protocol using performance metrics such as throughput and CPU-usage. Our measurements, taken at different bit-error rates and raw data rates, indicate a substantial improvement in performance over Indirect TCP and end-to-end TCP.

The rest of this paper is organized as follows. Section 2 briefly describes and compares some previously proposed solutions to the problem of reliable transport protocols over wireless links. Section 3 presents the design of our MTCP protocol. In Section 4, we discuss some details of our protocol implementation. Section 5 presents the results and analysis of several experiments. A discussion on miscellaneous issues is included in section 6. We discuss our future plans in section 7 and conclude with a summary in section 8.

2. Related Work

Figure 1 shows a typical network model for wireless networks. There are fundamentally two different approaches to improving TCP performance in such networks. The first approach attempts to hide non-congestion-related losses from the TCP sender. The intuition behind this approach is that these losses are local to the wireless link and that the problem should be solved locally. The second approach explicitly notifies the TCP sender of non-congestion-related losses. This allows the TCP sender to differentiate between congestive and non-congestive losses and thereby to react accordingly. In general, we can broadly classify the above two approaches into three categories: a) link-layer scheme, b) split connection scheme, and c) end-to-end scheme. The link-layer and split connection methods belong to the first approach where the problem is solved locally, whereas the end-to-end scheme follows the second approach of explicitly notifying non-congestive losses back to the TCP source [11]. In the following, we provide a more in-depth look at these schemes.

2.1 Link-Layer Schemes

The goal of link-layer schemes is to perform link-layer retransmission on the wireless or similar lossy links. By doing so, most packet losses can be recovered and therefore shielded from the TCP sender. Two main classes of techniques are usually applied by these schemes: automatic retransmission requests (ARQ) and error correction, such as Forward Error Correction (FEC). The main advantage of this approach is that it fits well into the layered structure of network protocols. In addition, the link-layer approach operates independently of higher-layer protocols and does not maintain any per-connection state. The main concern, however, is that most of the link-layer mechanisms compete with reliable transport protocols, such as TCP, in order to recover from losses. This interaction can lead to degraded performance of higher-layer protocols. Furthermore, most link-layer schemes are poor in shielding the TCP sender on the fixed host from wireless losses, since the schemes do not attempt to either deliver retransmitted packets in-order or suppress duplicate acknowledgments sent by TCP on the mobile host [11]. This makes link-layer schemes especially ineffective in dealing with handoffs. The other concern is the additional processing burden that these schemes place on the mobile devices, which are already power and computing limited. Specifically, the ARQ mechanisms may use timeouts to perform retransmissions, while timeouts are also utilized at the transport layer. Link-layer schemes that use FEC to detect and recover from errors, duplicate the effort, since TCP already uses checksums to detect errors. These unnecessary processing demands can impose substantial burden on a mobile device depending on the capabilities of the device and the number of network applications running on it.

The TCP-aware link-layer solutions, such as the Snoop protocol [7], attempt to solve some of the issues just described. The Snoop protocol introduces a module, called the *snoop agent*, at the basestation. The agent monitors every packet that passes through each TCP connection and maintains a cache of TCP segments sent across the link that have not yet been acknowledged by the receiver. The state maintained at the Snoop agent for each connection is soft, and is not necessary for correctness. Snoop detects losses in the same way as TCP, i.e., by the arrival of several duplicate TCP acknowledgments or due to a local timeout. Upon detection of a packet loss, the Snoop module retransmits the lost packet from the local cache and suppresses any duplicate acknowledgments heading towards the TCP sender. The key feature of the Snoop protocol is that link-layer retransmissions are TCP-aware. Specifically, the Snoop protocol leverages off TCP acknowledgments arriving from the mobile host in order to detect packet losses. The duplicate acknowledgments are suppressed to shield the TCP sender from non-congestive-losses. Furthermore, Snoop uses timeout values that are smaller than TCP timeouts, so that it has an opportunity to recover from losses before TCP times out and invokes congestion avoidance mechanisms. The combination of these techniques reduces the overhead incurred due to link-layer retransmissions and also reduces the competing effects between the ARQ schemes of the transport and the link layers.

One disadvantage of the Snoop protocol is that, like other link-layer protocols, it could also suffer from not being able to completely shield the sender from wireless losses [11]. This can occur if the Snoop module fails to recover from packet losses before TCP times out. The problem can be minimized by using a finer timeout value at the Snoop module, however at the cost of incurring more processing load on the basestation. The second disadvantage is that the protocol violates the layering principle of protocol design due to its use of TCP acknowledgment for link-layer recovery. The third possible disadvantage of the Snoop approach is that it requires a slight modification in the TCP implementation of the mobile host, so that congestion and non-congestion losses can be differentiated. Specifically, the Snoop module is equipped to send negative acknowledgment (NACKS) to the TCP sender on the mobile host when missing packets are detected. As will be seen later, we will show that modification of the existing TCP implementations on *mobile hosts* is not a stumbling block in seamlessly integrating applications that use the TCP protocol, and therefore, not a major issue.

2.2 Split Connection Schemes

Split connection schemes split the end-to-end connection into two connections – one TCP connection between the sender and the basestation and another between the basestation and the receiver. The protocol on the wireless segment does not necessarily have to be TCP; it can be any other reliable transport protocol. The idea behind this approach is to separate loss recovery over the wireless segment from that across the entire end-to-end connection. This separation allows a reliable transport protocol on the wireless segment to quickly recover from packet losses, thereby shielding the TCP sender from the wireless link. Indirect TCP is an example of the split connection approach that uses "regular" TCP for its connection over the wireless segment. Experiments indicate that the choice of TCP on the wireless segment results in several performance problems. The TCP sender on the wireless segment still interprets all wireless losses as congestion problems and invokes various congestion control mechanisms leading to a sub-optimal performance. Another problem is that every packet incurs the additional overhead of traversing the transport layer twice at the basestation in order to hop over to the next connection. The problem can be solved by an efficient kernel implementation that avoids extra copies of packet buffers [11].

Category	Advantages	Disadvantages
Link-Layer (LL)	1. Fits well into layering structure. 2. Poor at shielding the sender from non-congestion losses. 3. Does not require connection state to be maintained at the basestation.	1. Competes with TCP and duplicates some effort. 2. Requires smaller timeout values: places more burden on mobiles. 3. Cannot solve the problem associated with handoffs.
Link-Layer-TCP Aware	1. More effective at shielding the sender from non-congestion losses than LL. 2. Depending on the implementation may not impose extra load on the mobile. 3. Only requires soft connection state to be maintained at the basestation.	1. Violates layering principles. 2. Possibly requires TCP on the mobile to be modified. 3. Requires extra link-layer messages for recovery.
Split-Connection	1. De-couples the error control on wireline segment from that of the wireless segment. 2. Allows development of a lightweight protocol on the wireless segment.	1. Requires connection state to be maintained at the basestation. 2. May incur more processing load at the basestation. 3. Acknowledgments are not end-to-end.
End-to-End	1. Explicit Loss Notification (ELN) can help sender to distinguish between congestive and non-congestive losses.	1. Slow to recover from losses. 2. Requires TCP implementations on fixed hosts to be modified.

Table 1: Comparison of the different approaches

One problem, particularly in the implementation of I-TCP, is that the end-to-end semantics of the TCP protocol are not preserved [11]. Proponents of the I-TCP approach, however, argue that most applications that use TCP also have some kind of support for application layer acknowledgment and error recovery. Such acknowledgments are often required, because TCP does not provide any notification to the sending application when the data is actually delivered to the peer application [6]. The other problem that is widely stated in the context of the split-connection scheme is that it requires substantial amount of state to be maintained at the basestation making handoff procedures complex and slow [11].

2.3 End-to-End Schemes

The two schemes discussed so far attempt to recover from non-congestion losses locally, thus shielding the TCP sender from the wireless link. On the other hand, most effective end-to-end solutions add an explicit loss notification (ELN) feature to TCP acknowledgments. When a packet is dropped on the wireless link, subsequent cumulative acknowledgments corresponding to the lost packet are marked by an agent on the basestation to indicate the occurrence of a non-congestion-related loss. The TCP sender uses this information to decide whether to invoke congestion control mechanisms or not. One problem with this approach is that end-to-end protocols are usually slow to recover from losses on the wireless link, since it takes more time for the TCP sender to detect problems [11]. The situation is exasperated in WAN environment, where the round-trip delays are much greater than in a LAN. The biggest problem with the end-to-end approach is the need to modify the existing TCP implementations on the *fixed hosts*, making the solution impractical for wide scale deployment.

2.4 Selective Acknowledgments

Typically, losses in the wireless environment consist of burst packet errors. The standard TCP protocol, which uses cumulative acknowledgments, does not provide the sender with sufficient information to recover from multiple packet losses. A selective acknowledgment (SACK) technique, called SMART Retransmission [12], contains the cumulative acknowledgment and the sequence number of the packet that caused the acknowledgment to be sent. The sender can use this information to construct a bitmap of the packets that have correctly reached the receiver. When the sender detects a gap in the bitmap, the corresponding packets are retransmitted without considering the possibility that the acknowledgments may be reordered or may simply be dropped. Researchers have studied SMART selective acknowledgment technique in conjunction with the above approaches (link-layer, split-connection and end-to-end) and compared them with each other [11]. The next section presents the performance results and the evaluation of these schemes.

2.5 Evaluation of Related Work

Table 1 summarizes the advantages and disadvantages associated with the different approaches. Various protocols in combination with SMART have been extensively studied and compared by a group of researchers in [11]. Here,

Name	Category	Description / Examples
E2E	End-to-end	Standard TCP-Reno
E2E-ELNRXMT	End-to-end	Standard TCP + ELN + retransmit on first duplicate ack.
E2E-SMART	End-to-end	Standard TCP + SMART selective acks
SPLIT	Split connection	Standard TCP on both segments
SPLIT-SMART	Split connection	Standard TCP + SMART on wireless segment
LL	Link-layer	Simple link-layer protocol
LL-TCP-AWARE	Link-layer	Snoop protocol
LL-SMART-TCP-AWARE	Link-layer	Snoop protocol + SMART

Table 2: Summary of various protocols

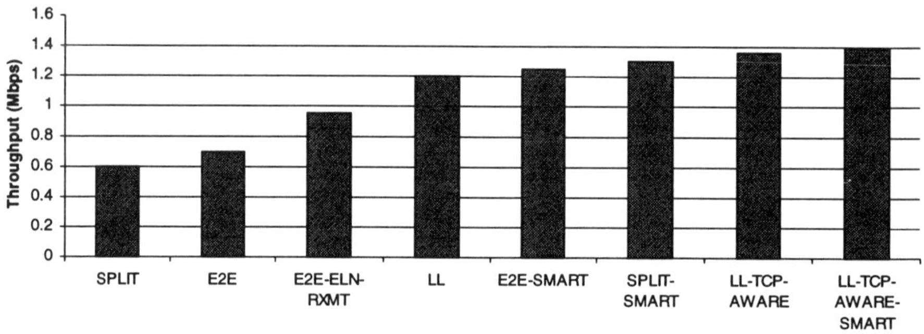

Figure 2: Performance of various protocols at bit-error rate of 1 error/65536 bytes[1]

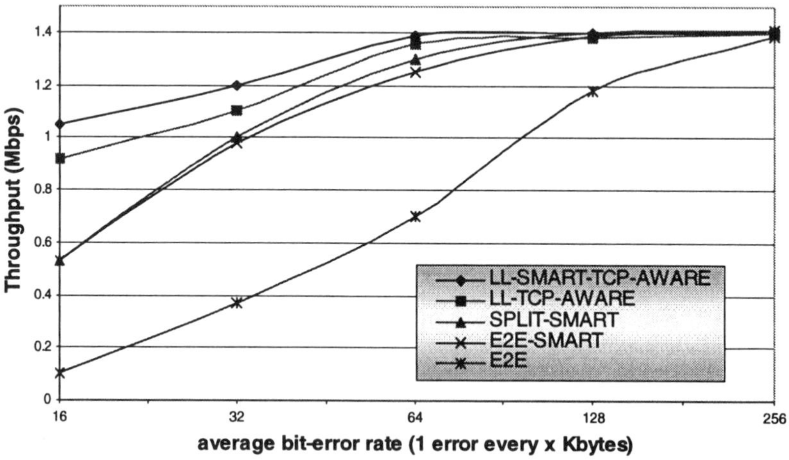

Figure 3: Comparison of throughput at various error rates, ranging from 1 error every 16 KB to 1 every 256 KB[1]

[1]*These results are obtained from [11].*

we briefly present some important elements of their findings. Their experimental testbed consisted of IBM ThinkPad laptops and Pentium-based PCs running BSD/OS 2.1 from BSDI. The machines were interconnected using 10Mbps Ethernet and AT&T Wavelan with a raw signaling bandwidth of 2Mbps. The peak throughput for TCP bulk transfers under ideal (no loss) conditions was 1.5Mbps in the local area testbed. Errors were generated on the lossy link using an exponentially distributed bit-error model and packet corruption was achieved by changing the TCP checksum. All measurements were taken on bulk transfers from TCP sender on the stationary host to the receiver on the mobile host using data packet size of 1400 bytes. Table 2 [11] enumerates some of the protocols that were evaluated. These are the protocols that we mainly focus on here.

The following observations can be made from the results in Figures 2 and 3:

a. SMART selective acknowledgments greatly improve the performance of each category of protocols. For example, the difference in throughput between SPLIT and SPLIT-SMART is very large, as is the difference in the case of the end-to-end protocols. Selective acknowledgments provide a mechanism to quickly recover from multiple packet losses and thereby reduce the occurrence of a timeout.

b. The SMART technique alone, however, does not perform very well under high loss rates. Both the LL-TCP-AWARE and the LL-SMART-TCP-AWARE protocols outperform SPLIT-SMART and E2E-SMART in heavy loss conditions. The key reason is that the TCP-aware link-layer protocols shield the TCP sender from wireless losses. Unlike the link-layer protocols, the SPLIT-SMART and the E2E-SMART protocols invoke congestion control mechanism upon experiencing a packet loss. In addition, as the congestion window size remains small, it significantly reduces the possibility for the TCP sender to fast retransmit lost packets.

c. Although, E2E performs slightly better than SPLIT under the LAN environment, results (not shown here) also indicate that SPLIT outperforms

E2E in the WAN environment [11]. This implies that local loss recovery is quicker and more effective than end-to-end recovery.

d. Minor performance differences between split-connection and other schemes under low loss condition indicates that the extra processing load on the basestation due to the connection split is negligible.

e. The LL protocol is not very effective in shielding the TCP sender from wireless losses. When a loss occurs, the LL protocol performs a local retransmission. However, enough packets from the TCP sender are typically in transit to create more than 3 duplicate acks. Since the LL protocol does not suppress any duplicates, they propagate to the sender and trigger a fast retransmission and the associated congestion control mechanisms [11]. As expected, the LL-TCP-AWARE protocol performs better than the LL protocol.

3. Mobile TCP (MTCP)

Based on the experience and the results of previously studied approaches, we conclude that the characteristics of an ideal solution must include the following:

- *local recovery from wireless losses*
- *capable of completely shielding the TCP sender from wireless (non-congestion) losses*
- *use of some form of selective acknowledgment scheme to recover quickly from losses*
- *no requirement for re-linking of applications*
- *minimization of processing load on mobile hosts*
- *effective in supporting mobility, i.e. handoffs*

Our approach in achieving many of these goals is to design Mobile TCP (MTCP), a lightweight transport protocol, for the wireless segment based on the split-connection model. The Mobile TCP protocol is based on our previous work in this area, as described in [9]. The key idea influencing our protocol design is the fact that Mobile-TCP operates over a single point-to-point connection. Thus, in many aspects, the design of MTCP resembles that of a link layer protocol [9].

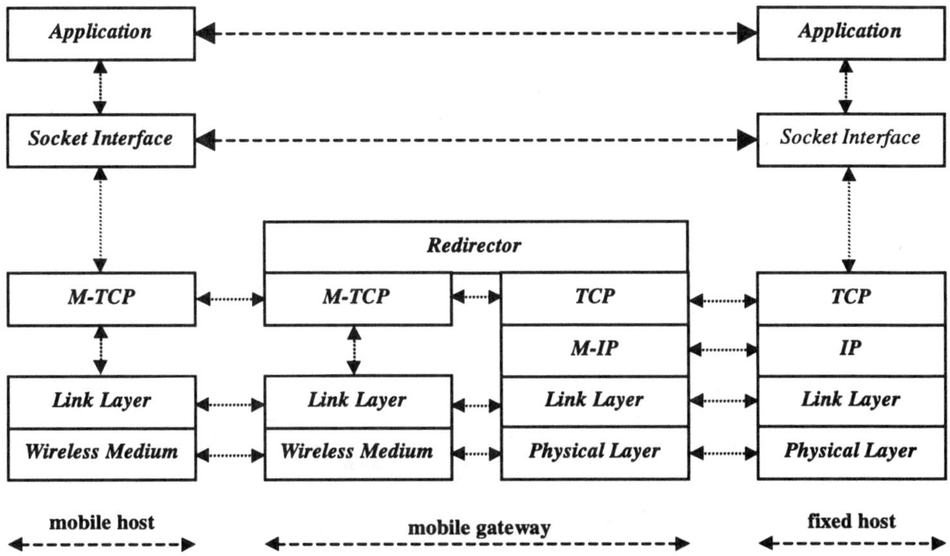

Figure 4: Mobile-TCP Protocol Stack

Furthermore, indirection provides the property that all packets arrive in order and all losses are non-congestion-related on the wireless segment. As a consequence, the design of MTCP can be greatly simplified and optimized, resulting in a significant reduction in the communication load over the wireless link.

The features of our MTCP implementation are as follows:

- elimination of the IP layer
- header compression
- no congestion control mechanisms and timer backoffs
- static window-based flow control
- selective acknowledgments optimized for burst packet losses
- explicit retransmission requests
- fast recovery from retransmission losses

Now, we proceed with the detailed description of the Mobile-TCP design. The next few sections deal with the design of connection control, flow control and error control mechanisms of the protocol.

0		16	32
0	*Code Bits*	*Connection ID*	
Sequence Number		*Acknowledgment Number*	
Reverse CID		*Checksum*	
Max Buffer Size		*Max Packet Size*	
Destination Port		*Source Port*	
Destination IP Address			
Source IP Address			

Figure 5a: Connection Control Packet

0		16	32
1	*Code Bits*	*Connection ID*	
Sequence Number		*Acknowledgment Number*	
Window Size		*Negative Ack Number*	
Data Length		*Checksum*	

Figure 5b: Data Packet

3.1 Connection Control

Connection control is concerned with setting up and tearing down connections. Under TCP, a host can initiate a connection with another host (called an active open) or can listen for and accept connections from other host (called a passive open). In the split-connection model, the mobile-gateway (MG) acts as an intermediary, which establishes independent connections with the mobile host (MH) and with the fixed host (FH). When the MH executes an active open, it communicates to the mobile-gateway the identity of the destination, as well as a set of parameters that will govern the operation of the connection between the MH and the mobile-gateway. The mobile-gateway then proceeds with setting up the connection with the fixed host using the destination IP address and port number specified by the mobile host. Similarly, when a MH executes a passive open, it informs the mobile-gateway that it is prepared to accept connections. This allows the mobile-gateway to accept or reject connection requests from FHs without communicating a priori with the MH. Similar technique is used at the mobile-gateway for connections initiated by the fixed hosts. Since the Mobile-TCP architecture, as shown in figure 4, completely eliminates the IP layer from the protocol stack, it subsumes the IP addressing functionality into its own packet headers.

In addition to eliminating some IP header information, we reduce the amount of wirelessly transmitted data by employing a technique similar to the Van Jacobson's header compression [13]. Since all the packets from the mobile host travel through the mobile-gateway, there is no need to communicate the full source and destination IP addresses (including the port numbers) in every packet. Instead, at the connection establishment stage, a connection ID (CID) is assigned for each direction and is used in subsequent exchanges of data over the wireless segment [9]. (CID_{MH} represents the CID from MH to MG and CID_{MG} is the CID in the reverse direction). The connection control and user data packet formats are shown in figures 5a and 5b. Note that the first bit is used to distinguish between the two formats. The MTCP protocol uses a three-way handshake similar to TCP to establish connections and exchange unique connection identifiers between the mobile host and the mobile-gateway. The bindings between CIDs and IP addresses (including port numbers) are cached at both ends of the connection. When sending a packet from the mobile to the network, the destination IP address is translated into the corresponding CID_{MH} at the mobile host. At the mobile-gateway, the CID_{MH} is expanded back into the destination IP address, which is used on the wireline segment of the connection. Similar operation is performed in the reverse direction. In practice, the Redirector, which sits above the transport-layer and is responsible for "routing" data from the source to the destination, only has to maintain a table of bindings between the MTCP and the TCP socket descriptors.

We follow similar implementation strategies as in I-TCP [6] to seamlessly establish an indirect connection such that full compatibility is ensured with the existing TCP/IP stack on the fixed host. The problem in case of I-TCP, however, is that it requires applications on the mobile host to use new socket calls for the purpose of connection establishment. We argue that this is unnecessary in our case, since this functionality is transparently embedded into the MTCP protocol. Later, we will address the issue of application re-linking on the mobile host. We direct our focus now on issues that have the most significant impact on the overall performance of the protocol.

3.2 Flow Control

As stated earlier, congestion control mechanisms in TCP affect the throughput by constricting the window size of the sender when losses are detected. Since all losses on the wireless segment of the connection are non-congestion-related, a better approach to flow control design is to de-couple it from error control strategies. In MTCP, flow control is achieved by a static sliding window protocol. This scheme is identical to the one used by TCP [4], except that congestion control mechanisms are not implemented. As a consequence, the sender window size is entirely controlled by the receiving end and the maximum size of the window is limited by the size of the MTCP receive buffer. An optimal buffer size should be selected to be at least as large as the bandwidth-delay product. A buffer size that is smaller than the bandwidth-delay product can cause the wireless link to be underutilized, while larger than optimal size can result in greater buffering requirements on the link-layer protocol drivers. In traditional networks, greater than optimal window sizes can also lead to severe congestion and traffic problems resulting in the loss of packets. However, this does not pose a problem in a point-to-point configuration, such as the one we are addressing here. In fact, in our experience, reasonably large window sizes significantly improve the chances of quickly detecting packet losses and allow for fast recovery, rather than relying on retransmission timeouts. Nevertheless, due to the various factors involved in choosing the receive buffer size, which are not necessarily in the realm of transport protocols, we have made it a configurable parameter, as is done in TCP. We would like to look at the issue of dynamically choosing the optimal buffer size in our future research.

3.3 Error Control

TCP implements error control functions in the sender as well as the receiver. Specifically, the TCP receiver's main objective is to perform error detection by the means of checksums and holes in received sequence numbers. When the receiver detects an error, it signals the sender of the loss using a duplicate acknowledgment. In turn, the sender identifies losses either by the arrival of three duplicate acknowledgments or by the absence of an acknowledgment within an estimated timeout period. Upon identifying a loss, the TCP sender invokes various congestion control mechanisms and then transmits the first unacknowledged packet followed by a number of outstanding (pending transmission) packets that are allowed by the window size. In essence, TCP takes a conservative approach to loss recovery by assuming that only a single packet is lost. The technique is well suited for networks that experience light congestion, but too slow to recover in environments with heavy losses.

MTCP adopts a strategy similar to TCP for error control with the exception of few key differences that make it far more robust in recovering from heavy loss conditions. These differences can be stated as the following:

1) Explicit retransmission request (RRQ) – Upon detecting an out of order packet, the receiver immediately requests the sender to quickly retransmit the lost packet(s) as opposed to waiting for three duplicate acknowledgments. This is a direct consequence of the fact that all packets on the wireless segment arrive in order, which is guaranteed by the split-connection model. An out of order packet thereby automatically signifies a loss on the wireless link.

2) Selective acknowledgments (SACKs) optimized for burst losses – MTCP adopts a technique for selective acknowledgment that is slightly different from the SMART scheme. Recall that in SMART every acknowledgment packet contains the sequence number of the packet that triggered the acknowledgment to be sent. This enables the sender to construct a bitmap of lost packets. The approach, however, has two problems: i) it is complex to implement, and ii) lost acknowledgments cause the sender to overestimate the magnitude of the loss. Instead, every MTCP acknowledgment contains a negative acknowledgment number, which conveys the size of the burst

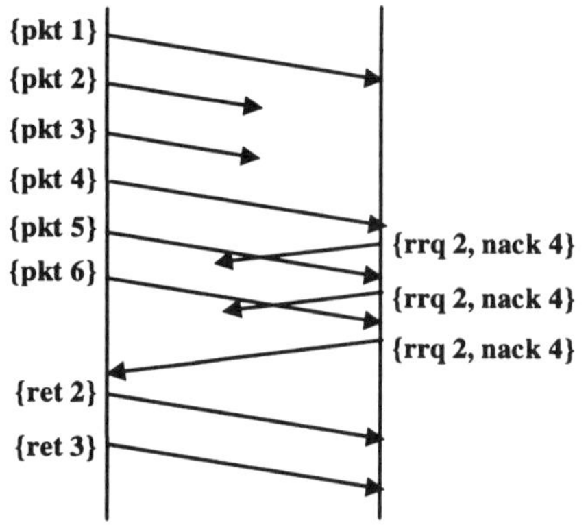

Figure 6a: MTCP SACK

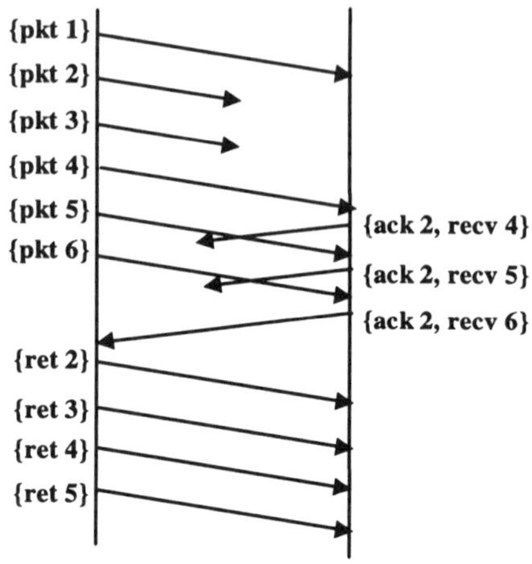

Figure 6b: SMART SACK

loss in bytes to the sender. Since this information is repeated in every acknowledgment, the MTCP sender obtains an accurate knowledge of burst losses even when some acknowledgment are lost. Figure 6 compares between the MTCP approach to SACKs versus the SMART technique. Figure 6a shows that the receiver requests three times the retransmission (RRQ) of packets 2 and 3. Two of these requests are corrupted or lost. Upon receiving the third request, the sender retransmits packets 2 and 3. On the other hand, SMART technique interprets the third retransmission request (ack 2, recv 6) as an indication of the loss of packets 2, 3, 4 and 5, causing it to retransmit in excess. Our scheme is comparatively much easier to implement than SMART and equally effective, if not more, in burst loss conditions. The limitation of the MTCP technique is that it cannot guess multiple sparse (non-burst) losses. We argue that most losses in wireless links are burst losses and therefore our technique is sufficient to handle this situation. Furthermore, intuition tells us that the MTCP window size is large enough (i.e. unrestricted by congestion control) to trigger fast retransmissions in most multiple sparse loss cases. Our performance results show that this scheme is indeed quite effective in combating losses.

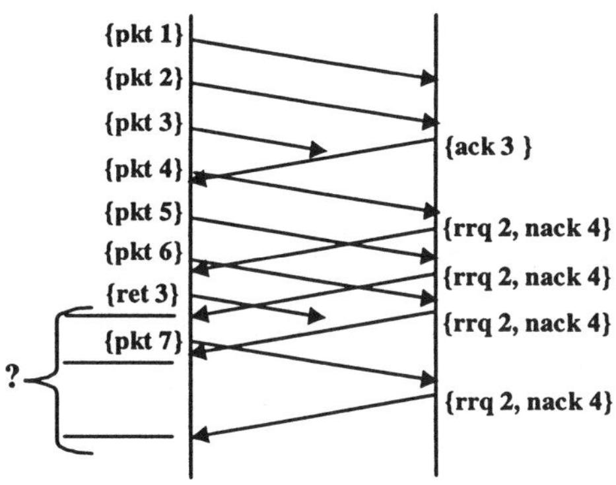

Figure 7a: Problem

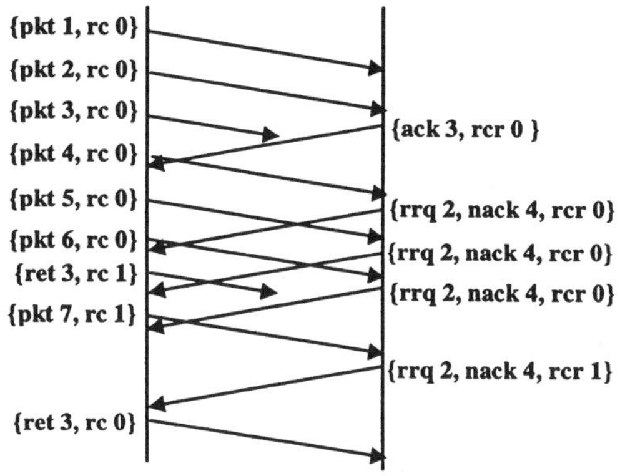

Figure 7b: Solution

3) Fast recovery from retransmission losses – Since losses are very common on the wireless segment, MTCP tries to take every step to quickly recover from errors. The same applies to packets lost during retransmission. When an MTCP sender receives a request for retransmission of a packet that was just retransmitted, the sender is unable to resolve whether the request is for the retransmitted packet or for the previously failed transmission. The situation is best illustrated in figure 7a. The key to differentiating between the two requests is for the sender to somehow know that the request was in response to the retransmission. In figure 7a, the retransmission requests (RRQs) triggered by packets 5 and 6 are indistinguishable from the RRQ triggered by packet 7. Ideally, the sender should respond with the retransmission of packet 3, when it gets the RRQ due to packet 7 and ignore the rest. However, the request packet does not contain adequate information for the sender to differentiate between the requests. One obvious possible solution is to use SMART-style sequence number of the received packet in the RRQ packet. We have opted for a more compact solution that can be achieved with just 2 bits (included in the CodeBits vs. an extra 16-bit field in the MTCP header). In our solution, every packet carries a retransmission-cycle number (RC) and every acknowledgment in the opposite direction contains a retransmission-cycle reply (RCR). When a receiver responds to the sender, it simply copies the previously received RC bit to the RCR bit of the acknowledgment packet.

Upon receiving a retransmission request, the sender checks to see if the RCR bit is equal to its present retransmission-cycle (PRC). If there is a match, the PRC bit is toggled and requested packets are retransmitted with the new cycle number. Otherwise, the retransmission request is simply ignored by the sender. This action prevents the sender from responding to duplicate retransmission requests. Figure 7b shows that with this scheme, the sender can now distinguish between the RRQs triggered by packets 5, 6 and the RRQ triggered by packet 7. Note that the technique works even for multiple retransmissions.

4. Implementation Details

We have implemented the MTCP protocol under the Microsoft Windows 95/NT platform. As the TCP source code for Windows was not available to us, we have designed and developed the entire MTCP protocol from scratch. The protocol is implemented as an intermediate protocol driver in the kernel mode and communicates with NDIS-compliant link-layer drivers [14] to send and receive packets on the network card. Thus, we achieve complete elimination of the IP layer from the processing path of the transport protocol. An important consequence of this is that buffer management is greatly simplified and optimized in MTCP. Furthermore, our implementation requires only a single memcpy per send or receive operation (apart from possible copies in the link-layer driver).

Most TCP implementations, such as BSD, employ two types of timers: fast timer (200ms) and slow timer (500ms) [4]. The TCP delayed acknowledgments are processed on fast timeouts and TCP retransmissions on slow timeouts. We would like timeout granularities to be on the order of the round-trip-time (RTT). However, in order to limit the overhead of processing timer interrupts, the MTCP protocol utilizes a single timer that goes off every 200ms to process both delayed acknowledgments and retransmissions. Interestingly, our experiments indicate that timeout values on the Microsoft TCP (MSTCP) implementation are more accurate than on the BSD implementation.

We have seen timeout values in the range of 200-500ms on MSTCP, while most timeouts on BSD are typically multiples of 500ms.

MTCP follows the standard BSD implementation to estimate the smoothed round trip time. Consequently, retransmission timeout interval in MTCP is computed as the sum of the smoothed RTT and four times its mean deviation [4].

5. Experimental Results

We performed various experiments to compare between the performance of MTCP, Indirect TCP and end-to-end TCP under wireless communication conditions. In the following sections, we present our experimental setup and discuss our results using various criteria for performance evaluation.

5.1 Experimental Setup

Our experimental testbed, shown in Figure 8, consists of Pentium-based personal computers running Windows 95 and Linux operating systems. The Win95 machine in the center of the figure is configured to act as a basestation using WinRoute software to route packets to and from the mobile host. As indicated in the figure, the basestation and the mobile host are connected to each other using either 10Mbps Ethernet or Proxim RANGELAN, a wireless LAN with a raw signaling bandwidth of 1.6Mbps. We chose to test the protocols under two different conditions to observe the effects of MTCP optimizations at different data rates. The peak throughput for TCP bulk transfers is measured to be approximately 8Mbps on the 10Mbps Ethernet and 0.70Mbps on the 1.6Mbps RANGELAN [15].

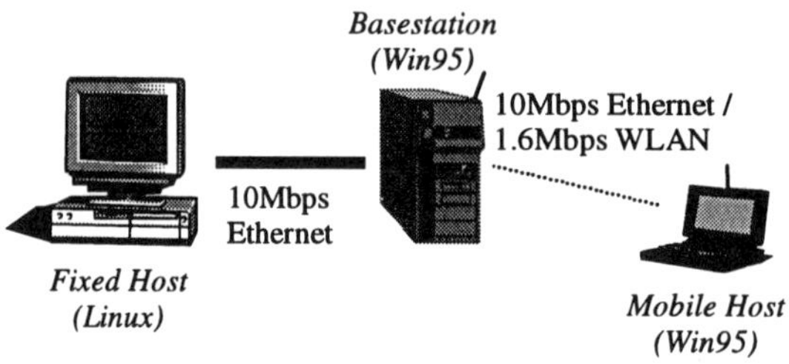

Figure 8: Network Configuration

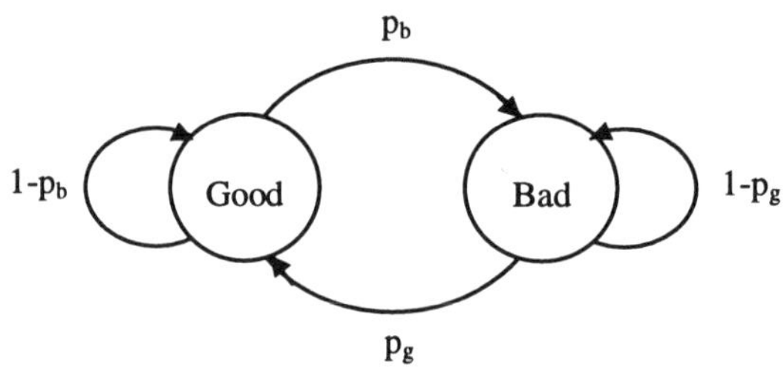

Figure 9: Burst-error model for lossy link

We simulate errors on the lossy link using a Markov chain burst-error model, shown in Figure 9. This model attempts to incorporate the fading effects experienced by the wireless channel. Specifically, the beginning of a fade corresponds to the transition from the **Good** state to the **Bad** state. Once in the bad state, the model continues to drop bits at a higher probability causing a burst loss of bits. Finally, the end of fade is marked by the transition from the **Bad** state to the **Good** state. The model is designed to be independent of the transmission rate of the channel, i.e. every bit is subject to the same drop probabilities regardless of its length in time. The properties of the burst-error model can be summarized by the following equations:

$$\text{[eq. 1]} \quad \text{Expected BER} = \frac{p_b}{p_b + p_g}$$

$$\text{[eq. 2]} \quad \text{Expected Burst Size in Bits} = \frac{1}{p_g}$$

$$\text{[eq. 3]} \quad \text{Expected Burst Size in Packets} = 1 + \frac{1}{Bp_g}$$

$$\text{[eq. 4]} \quad \text{PER}/\text{BER} = Bp_g + 1$$

where B is the average packet size in bits

Regardless of the error-model, a packet intruder driver modifies the checksum of outgoing packets in order to induce packet errors. The intruder specifies the burst-error parameters (p_b, p_g) and the size of the outgoing packet in bits to the Markov state machine shown in Figure 9, which thereby determines whether the packet will be corrupted or not. Losses are generated in both directions of the wireless link, so acknowledgments are dropped as well. For our simulations, the probability p_b is chosen between the range of 1E-7 and 120E-7 and p_g is fixed at 100E-6. This provides an approximate burst size of 10000 bits (or 1250 bytes) which corresponds to an expected burst loss length of 1.83 packets assuming that the packet size is 1514 bytes. The maximum packet size supported by our protocols and the physical media is 1500 bytes (excluding 14 bytes of the data link header). The maximum receive buffer size is set to 16 Kbytes for all the protocols.

5.2 Performance and Comparisons

We compared the performance of MTCP with Indirect TCP and end-to-end TCP using metrics such as normalized throughput, mean inter-buffer arrival time, mean CPU usage and mean energy consumption. Our experimental results indicate that MTCP performs significantly better than TCP depending on the packet error rate and the data rate of the link. In order to compare the results obtained for each protocol at different data rates, we based our evaluation of the performance on normalized throughput. We define normalized throughput as the percentage of the maximum throughput that is achieved under ideal (i.e., no loss) conditions. Figure 10 shows the performance of MTCP, I-TCP and end-to-end TCP for bulk transfer from the fixed host to the mobile host in Wireless (Fig. 10a) and Ethernet configurations (Fig. 10b). The results are shown for packet error rates ranging from 0.2% to 20% on a logarithmic scale. Note that the packet error rates shown in the figures correspond to errors occurring in the sender to receiver direction only. Since packets in the reverse direction only contain acknowledgments, the error rate in this direction is very small in comparison. The results indicate that MTCP is able to sustain a greater normalized throughput

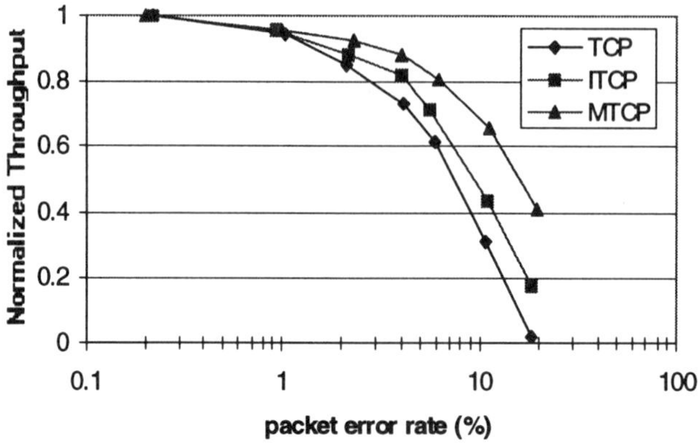

Figure 10a: Normalized Throughput vs. Packet error-rate for the wireless configuration.

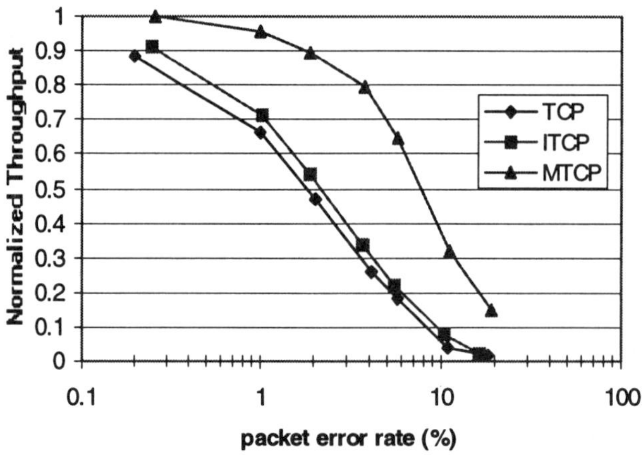

Figure 10b: Normalized Throughput vs. Packet error-rate for the ethernet configuration.

than both I-TCP and end-to-end TCP as packet error rates increase. The improvement is primarily due to: a) the elimination of congestion control mechanisms at the sender, b) the receiver-initiated retransmission requests (RRQs) that cause the sender to immediately retransmit lost packets, and c) the selective acknowledgment scheme for burst losses. The three techniques in combination improve the likelihood of fast re-transmitting lost packet(s), and thus reduce the possibility of an MTCP coarse timeout. The I-TCP performs worse than MTCP, since it continues to invoke congestion control mechanisms

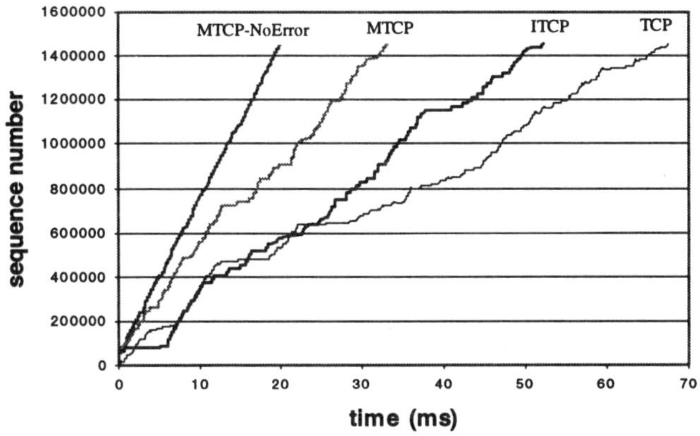

Figure 11: Protocol sequence trace at approximately 5% packet error rate in the wireless configuration.

upon packet loss on the wireless segment, causing an unnecessary reduction in the overall throughput. However, improved performance of I-TCP over end-to-end TCP can be attributed to local loss recovery, which reduces the time to recover from errors and also to the protocol differences that may exist between the Linux and the Windows implementation of the TCP/IP stack. Figure 11 graphically illustrates the progress of the data transfer from fixed host's standpoint in the Wireless configuration at approximately 5% packet error rate. It is apparent from this figure that both I-TCP and end-to-end TCP stall much more often in the presence of errors than MTCP and therefore, take longer time to complete the data transfer.

Figure 12a compares the results obtained for normalized throughput on Wireless and Ethernet configurations at various packet error rates. It can be clearly seen from the results that the protocols perform better in the Wireless configuration than in the Ethernet configuration by maintaining a higher normalized throughput as packet error rate increases. This can be explained by the fraction of the total transmission time spent idling in timeouts. As the transmission time decreases, the overhead due to coarse timeouts increases, thus resulting in lower normalized throughput in the Ethernet scenario. Since the packet error rate itself is dependent on the data rate of the connection, we also

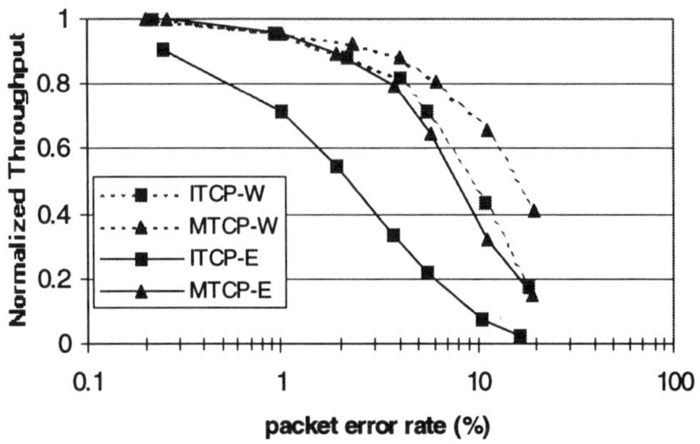

Figure 12a: Comparison of normalized throughput in wireless (denoted by suffix –W) and ethernet (-E) configurations at different packet error rates.

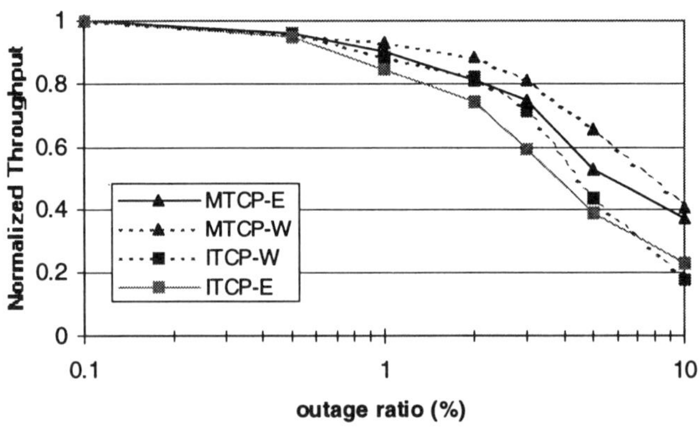

Figure 12b: Comparison of normalized throughput in wireless and ethernet configurations at different outage ratios.

compare the performance of the protocols at different outage ratios. The outage ratio, which is independent of the data rate, can be defined as the percentage of the time spent in the BAD state. The average length of an outage is kept constant at 10,000 Wireless-bit times, equivalent to the length of 10,000 Wireless bits in time. Since the ethernet data rate is 11 times faster than the Wireless data rate, the outage time corresponds to 110,000 Ethernet bits. The results of our experiments, shown in Figure 12b, indicate that the performance of the protocols in the Ethernet and the Wireless cases is closer than in Figure 12a. This can be attributed to the fact that the packet error rate decreases as the data rate increases.

On an average, an outage causes a loss of 10,000 Wireless bits and 110,000 Ethernet bits, which roughly corresponds to 2 packets and 11 packets in Wireless and Ethernet cases, respectively (assuming the packets size is 12,000 bits). Note that 2 packets are much more likely to be lost in the Wireless case than 1 packet because most outages will tend to overlap into the successive packet. Consequently, the ratio of the number of packets lost to the number of bits lost in the Wireless case is 2:10000, while it is only 1:10000 in the Ethernet case. Thus, the packet error rate in the Wireless case is approximately twice as much as in the Ethernet case at a particular outage ratio, which explains the results shown in Figure 12b.

We also studied the impact of different packet loss rates on the real-time constraints of an application. Due to the difficulty in computing the end-to-end delay accurately, as clocks on two machines need to be precisely synchronized, we introduce a metric known as inter-buffer arrival time. The inter-buffer arrival time can be defined as the time elapsed between the reception of two successive socket buffers at the application layer. Since our network topology consists of fixed connection routes, we can assume that the end-to-end delay without the presence of errors is fairly constant. It can be deduced that the additional end-to-end delay experienced in presence of errors is manifested in the extra inter-buffer arrival time seen at the application layer. Thus, inter-buffer arrival time can provide an insight into the changes in the end-to-end delay as well as the waiting time seen by the application in receiving data. In our simulation, the socket buffer size was chosen to be 1460 bytes. This particular choice of socket buffer size is important, since it is equal to the maximum data that can be sent in a single TCP packet. We can infer that the inter-buffer arrival time roughly corresponds to the inter-packet arrival time of packets arriving in order, since each buffer is placed in a separate packet. Figures 13a and 13b show the mean and the standard deviation of the inter-buffer arrival time measured over the lifetime of the connection at various packet error rates. We can see that MTCP performs better than I-TCP and end-to-end TCP in keeping a relatively low inter-buffer arrival time at higher packet loss rates. Interestingly, the standard

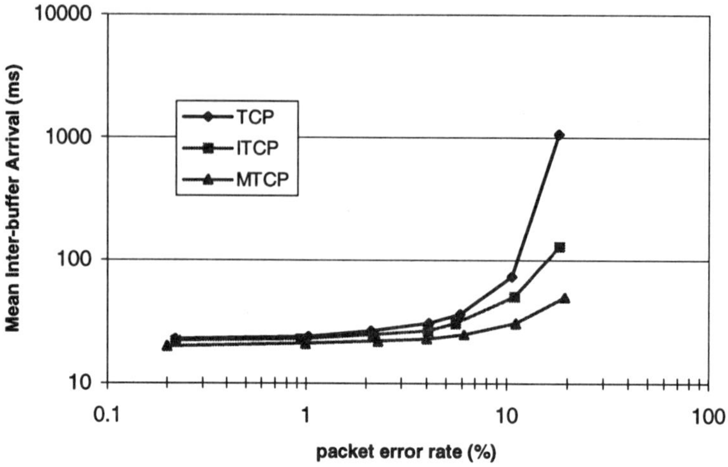

Figure 13a: Mean Inter-buffer Arrival Time vs. Packet Error Rate (Wireless Configuration)

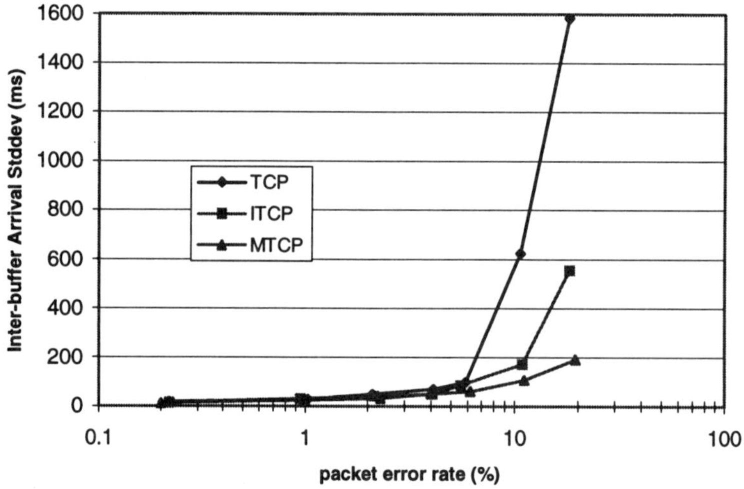

Figure 13b: Standard deviation of Inter-buffer Arrival Time vs. Packet Error Rate (Wireless Configuration)

deviation of the inter-buffer arrival time is also much lower in MTCP at higher packet loss rates, indicating that applications would be subject to less jitter and lower waiting time in receiving data. This is especially important for real-time applications, which are constrained by certain end-to-end delay and delay jitter requirements.

Protocol	Maximum throughput for bidirectional transfer (Kbytes/sec)	% CPU Usage
I-TCP	771.08	21
MTCP	866.06	12

Table 3: Throughput & CPU Usage for Bidirectional transfer under Ideal Conditions (for the ethernet configuration).

We now present various experimental results, which suggest an improvement in the utilization of the CPU resources by MTCP. As shown in Table 3, our experiments reported 12.3% difference in maximum throughput achieved by MTCP over I-TCP in the bi-directional data transfer tests for the Ethernet configuration. We claim that the bi-directional test is a more reliable indicator of processing limitations of a protocol, since it is not subject to various optimizations, as in the case of bulk-transfers. Specifically, factors such as the implementation of the Nagle's algorithm [16] and the timing of delayed acknowledgments can cause a significant impact on the throughput of unidirectional transfers. Higher throughput of MTCP under ideal conditions of the bi-directional tests suggests an improvement in the efficiency of protocol processing. Furthermore, the lower CPU usage by MTCP, as shown in Table 3, supports the implication that our protocol, in fact, reduces processing load on the end systems. We believe that majority of the contribution to this effect is due to the elimination of the IP layer. Removal of the IP layer from the TCP/IP stack provides benefits such as reducing protocol processing load, simplifying buffer management, decreasing the protocol header size, and possibly, reducing the number of memory copies. We also evaluated the impact of various packet error rates on the usage of CPU resources by each protocol according to the following metrics: Mean CPU Usage and CPU-Duration. Mean CPU Usage is simply the average % of the CPU used by the simple test application and the protocol during the lifetime of the connection. We subtract the background CPU usage (usage when no applications are running) from every CPU usage measurement in order to compute a more accurate estimate of the percentage of the CPU utilized in actually transferring the data. CPU-Duration approximates the area under the CPU usage curve as it varies over the duration of the connection. This measure is

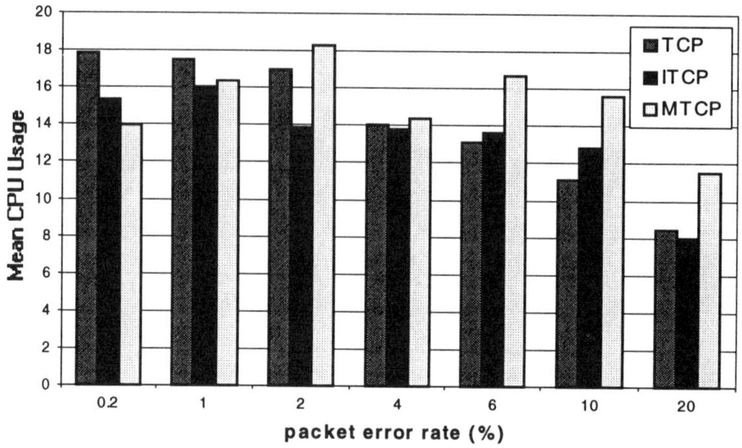

Figure 14a: Mean CPU Usage vs. Packet Error Rate (Wireless Configuration)

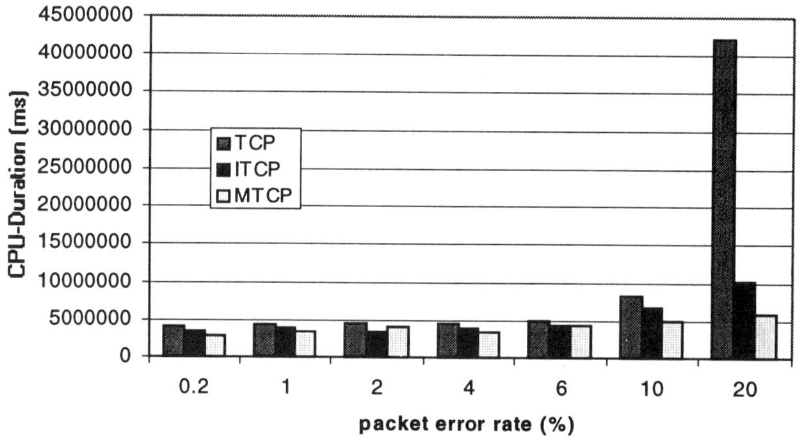

Figure 14b: CPU-Duration vs. Packet Error Rate for 1460*10000 byte transfer (Wireless Configuration)

estimated by taking the product of the average CPU usage and the time duration of the connection. The CPU-Duration signifies the total energy consumed by the application in order to transfer a certain number of bytes. The results of our experiments are shown in Figure 14a and 14b. In Figure 14a, we can observe that the CPU usage tends to reduce as the packet error rates increases and also that MTCP tends to maintain a higher utilization of the CPU than other protocols at higher packet error rates. This is explained due to the fact that there are more coarse timeouts in the case of I-TCP and end-to-end TCP, resulting in a less frequent use of the CPU. On the other hand, since MTCP times out less often, it completes the data transfer quicker than the other two protocols, thus offsetting the higher CPU usage. This is illustrated in Figure 14b, which shows that MTCP consumes less energy than I-TCP and TCP in order to transfer data over the network. These differences are more pronounced at higher packet error rates where timeouts are more frequent in I-TCP and TCP than in MTCP and unnecessary energy expended on tasks such as handling protocol timers is greater, while no data transfer takes place.

6. Other Issues

6.1 Application Re-linking

One of the criticisms of the split connection approach, and specifically Indirect TCP, is that applications running on the mobile host have to be re-linked with the I-TCP library in order to use special socket calls for the setup of the indirect connection [11]. Our approach in solving this issue is to subsume the mechanisms used to create an indirect connection into the MTCP protocol, making the connection establishment process completely transparent to the standard socket calls. Furthermore, the MTCP protocol driver can be designed to replace TCP drivers in the kernel mode, while interfacing with standard socket calls of the network operating system. These techniques can be used to seamlessly integrate any application into the rest of the Internet.

6.2 Handoff Support

In a typical wireless network, mobile hosts connect to the fixed network through mobile-gateways, also known as Point of Attachments (POAs), which act as routers between the wireless subnet and the rest of the network. As a mobile host roams, it may leave the coverage area of the current POA. To maintain connectivity, a handoff process is invoked such that the mobile host reconnects itself to a new POA. Typically, the handoff process causes the network layer to re-route all the packets to the new mobile-gateway via the old mobile-gateway [17].

The split-connection approach performs handoffs by migrating the state of all the indirect connections to the new mobile-gateway, while ensuring that the entire process is transparent to the connection endpoints on the mobile host and the fixed host. The details of the techniques involved in transferring the state are addressed in I-TCP [6]. We follow a similar strategy here, with the exception that a new *Connection ID* has to be reestablished for each socket connection on the mobile host. Specifically, as the *Connection ID* used on the old mobile-gateway might be in use on the new mobile-gateway, a new control packet is sent to the mobile host with the new value of the mobile-gateway's *Connection ID* on each connection [9].

The most commonly sited problem for Indirect-TCP is that it takes a longer time to complete the handoff process [11]. The main reason for this handoff latency is due to the complexity involved in migrating the state of all indirect connections associated with the mobile host. The state of a connection consists of the state of all TCP variables and also the data present in the TCP send and receive buffers at the time of the handoff. In order to reduce the handoff latency, we propose a scheme for setting up indirect connection, which optimizes the common case. Our scheme can be applied to Wireless Local Area Network (WLAN) technologies, such as those provided by Proxim or Wavelan. Figure 15 shows a typical WLAN configuration. In such configurations, mobile hosts connect to the wired network via devices known as Access Points (APs). An Access Point serves as a bridge between the Wireless Local Area Network and the Wired Local Area Network. Mobile hosts within the coverage area of the "nearest" AP can communicate with hosts on the wired network through the AP, which essentially acts as a basestation. When a mobile host moves into the coverage area of another AP with a stronger signal, it performs a handoff by registering itself with the new AP. Consequently, the new AP forwards all packets destined to the physical address (MAC address) of the handed off mobile host [15]. The main difference between a basestation and an Access Point is that

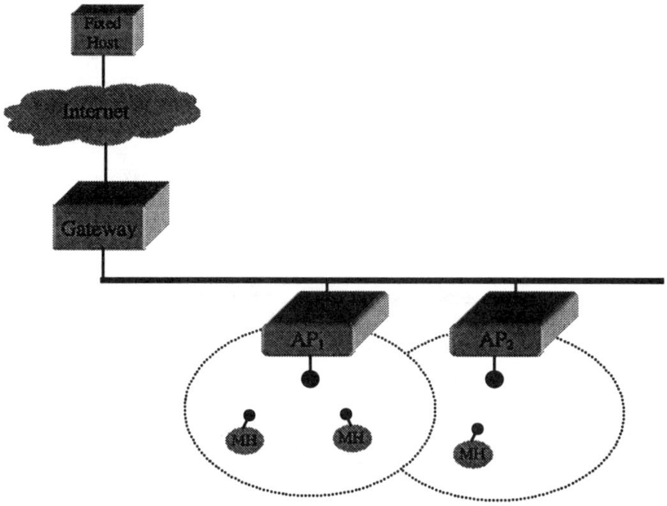

Figure 15: A Typical Wireless Local Area Network (WLAN)

the basestation is a router, while the Access Point is a bridge. Our scheme in the context of the present WLAN technologies is to split the end-to-end connection at the gateway and communicate with the mobile host over the AP using MTCP, as shown in Figure 16. This is possible since many WLAN technologies use Ethernet IEEE802.3 packet formats. The benefit of this scheme is that no transfer of state is required when mobile devices handoff within the same local area network, as handoffs are done on the MAC layer. In addition, we also get the benefit of using a streamlined protocol, such as MTCP, over the segment between the gateway and the mobile host. MTCP is particularly suitable for this segment, since the segment does not require any routing facilities in the protocol and, therefore, can be considered as a single hop link from the perspective of the transport layer. The packets that are lost during the handoff process can quickly be recovered, since these losses are handled separately by MTCP on the last segment. Essentially, losses occurring due to handoffs can be treated identical to the losses occurring on the wireless segment due to effects as interference and fading.

We propose the following technique for setting up the split connection in the WLAN configurations. Every mobile host registers itself with the MTCP-

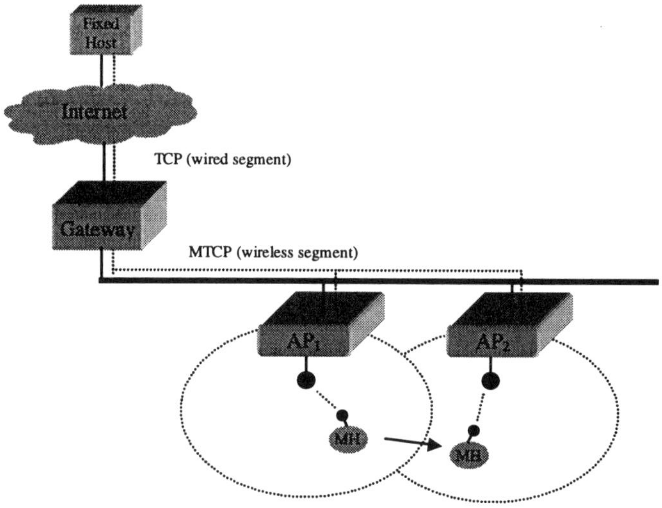

Figure 16: Split Connection using MTCP in WLAN

enabled gateway, which maintains a list of mobile devices in its local area network. The gateway acts like a transport-layer proxy, performing the function of setting up indirect MTCP connections on behalf of the registered mobile hosts and redirecting data from the wired segment to the "wireless" segment (the hop between the gateway and the mobile host) or vice versa. When a mobile host hands-off to an AP on another local area network, it un-registers itself from the previous gateway and registers itself with the new gateway. All packets destined to the mobile host are re-routed through the new gateway using Mobile-IP and the split connection-state is transferred from the old gateway to the new gateway much as in I-TCP. Note that the states of the split connections associated with a mobile still need to be transferred for an inter-local-area handoff. However, we argue that our scheme optimizes the common case of handoffs occurring within a LAN by eliminating much of the complexity.

7. Future Work

We are currently incorporating asymmetric protocol design techniques into MTCP to further offload processing from the mobile device. In an asymmetrically designed protocol, peer functions are implemented through algorithms and procedures that are of substantially different complexity, with the lower complexity used on the mobile device [9]. We are also working on improving the efficiency and accuracy of our timer implementation.

In this paper, a simple burst error model was used to study the behavior of TCP and MTCP under specific situations. We are looking into other error-models, which characterize a wider range of wireless link operational conditions.

One of the important advantages of the split approach is that it enables the development of lightweight transport protocols, such as MTCP. However, the advantages come at the cost of maintaining significant amount of state at the basestation, making handoffs relatively slow [11]. In this paper, we proposed one scheme for optimizing the split connection model in a Wireless Local Area Network configuration. We are currently investigating techniques in which the split-connection approach, and specifically MTCP, can be designed to reduce latency involving transfer of state from one basestation to another for the general case.

Finally, it is an interesting research area to investigate how TCP can be efficiently adapted into networks with multiple wireless hops, such as Mobile Ad-hoc Networks, or into hybrid networks consisting of a mixture of wireless and wired links.

8. Summary and Concluding Remarks

In this paper, we have presented the design of a lightweight transport protocol (MTCP) that improves the end-to-end performance of TCP in networks with wireless links and also reduces processing requirements on mobile devices. The protocol is based on the split-connection model, where the connection between the mobile host and the fixed host is partitioned into the wireless segment and the wired segment. The key idea behind MTCP is in the recognition that the wireless segment is, in fact, a single-hop connection. This allows elimination of some transport-layer functions and simplification of other functions, leading to a reduced communication load on the wireless segment and the mobile device.

The design of MTCP also incorporates various methods to efficiently recover from wireless losses. These techniques are simple to implement and our performance results indicate that they are very effective in sustaining good performance even under heavy packet losses. Our experiments have also highlighted the benefits of removing congestion control mechanisms from the wireless transport protocol. The elimination of such mechanisms enhances the effectiveness of the selective acknowledgment methods employed by MTCP. We have studied the behavior of MTCP and TCP at different data rates and different packet losses and our results indicate substantial improvements of MTCP over TCP in terms of the processing requirements and the overall throughput. It is important to point out that many of the techniques employed in MTCP, such as the elimination of the IP layer, enhancements in the error recovery mechanisms and handoff optimizations, can be applied to other approaches as well.

Finally, we have shown that MTCP can be designed to hide the process of establishing indirect connection from standard socket calls, such that applications do not need to be relinked. This addressed our goal of seamlessly integrating mobile applications into the Internet.

References

[1] G.H Forman and J. Zahorjan, *The Challenges of Mobile Computing*, IEEE Computer, April 1994.

[2] T. Imielinski and B. R. Badrinath, *Mobile Wireless Computing: Solutions and Challenges in Data Management*, Rutgers University, Department of Computer Science, Technical Report

[3] J.B. Postel. *Transmission Control Protocol*. RFC, Information Sciences Institute, Marina del Rey, CA, September 1981. RFC-793.

[4] W. R. Stevens. *TCP/IP Illustrated, Volume 1*. Addison-Wesley, Reading, MA, Nov 1994.

[5] R. Caceres and L. Iftode. *Improving the Performance of Reliable Transport Protocols in Mobile Computing Environments*. IEEE JSAC, 13(5), June 1994

[6] A. Bakre and B. R. Badrinath, *I-TCP: Indirect TCP for mobile hosts*, in Proc. 15th International Conference on Distributed Computing Systems, Vancouver, Canada, June 1995, pp. 136-143.

[7] H. Balakrishnan, S. Seshan, and R. H. Katz. *Improving Reliable Transport and Handoff Performance in Cellular Wireless Networks*, Wireless Networks, Vol. 1, No. 4, pp. 469-481, Dec 1995.

[8] R. Yavatkar and N. Bhagwat. *Improving End-to-End Performance of TCP over Mobile Internetworks*. In Mobile 94 Workshop on Mobile Computing Systems and Applications, December 1994.

[9] Z.J. Haas, *Mobile-TCP: An Asymmetric Transport Protocol Design for Mobile Systems*, ICC'97, Montreal, Canada, Jun 8-12, 1997.

[10] B. R. Badrinath, A. Bakre, T. Imielinski, and R. Marantz, *Handling Mobile Clients: A case for indirect interaction*, in Proc. 4th Workshop on Workstation Operating Systems, October 1993.

[11] H. Balakrishnan, V. Padmanabhan, S. Seshan, R. H. Katz., *A Comparison of Mechanisms for Improving TCP Performance over Wireless Links*, In Proceedings of ACM SIGCOMM 96.

[12] S. Keshav and S. Morgan. *SMART Retransmission: Performance with Overload and Random Losses*. In Proc. Infocom '97, 1997.

[13] V. Jacobson, *Compressing TCP/IP Headers for Low-Speed Serial Links*, RFC 1144.

[14] Microsoft Developer Network (MSDN) Library, Jan 1998.

[15] Proxim RANGELAN2 7100: User's Guide, 1995.

[16] Nagle, J. 1984. *Congestion Control in IP/TCP Internetworks*, RFC 896, 9 pages (Jan.)

[17] C. Perkins, editor, *IP Mobility Support*, Internet Engineering Task Force, Internet Draft, draft-ietf-mobileip-protocol-15.txt, 9 Feb. 1996.

DIMACS Series in Discrete Mathematics
and Theoretical Computer Science
Volume **52**, 2000

Supporting Transaction Service Handoff in Mobile Environments

Abdelsalam (Sumi) Helal, Jin Jing, and Ahmed Elmagarmid

ABSTRACT. In a mobile computing environment, a user carrying a portable computer may need to execute a *mobile transaction* over multiple overlay networks. To efficiently support the execution of mobile transactions, a transaction service handoff is needed to reduce the overhead of message transmission across network boundaries. As a result of user mobility and service handoff, the operations of transaction could be submitted to different data servers in multiple overlay networks. The distribution of operations implies that the transmission of messages (such as those involved in a two phase commit protocol) may be required among these data servers in order to coordinate the execution of these operations. In this paper, we will address transaction service handoff issues that occur in multiple overlay networks during the execution of transactions. Specifically, we will examine the application of *partitioned data* for distributed data services and management to support the transaction service handoff. We show that, for operations pertaining to resource allocation, the message overhead (e.g., for a 2PC protocol) introduced by the distribution of operations is undesirable and unnecessary. We introduce a new algorithm, the *Reservation Algorithm* (RA), that does not necessitate the incurring of message overheads for the commitment of mobile transactions. We address two issues related to the RA algorithm: a *termination protocol* and a *protocol for non-partition-commutative operations.* We perform a comparison between the proposed RA algorithm and existing solutions that use a 2PC protocol.

1. Introduction

Advances in wireless networking technology have engendered a new computing paradigm, called *mobile computing*, in which users carrying portable devices have access to a shared infrastructure independent of their physical location. In a mobile environment that consists of multiple overlay networks, service level handoff will be very important for the following reasons. First, different networks might exhibit tremendous differences in network characteristics to applications and services. The differences might range from network latency and bandwidth to network topology and structure (e.g., circuit switch based or packet switch based). As a result, the applications or services in each network must have been customized and optimized for the characteristics of network. The service level handoff will allow mobile users to utilize the optimized applications and services provided in the local application and service servers. Second, the overhead of message transmission and transform

2000 *Mathematics Subject Classification.* Primary 68U35.

could be significant across network boundaries. The overhead could greatly affect the performance of application services after mobile users move to a different network environment. Therefore, the service level handoff is an interesting issue in supporting the mobility of users over multiple overlay networks. In this paper, we are particularly interested in the issues of transaction service handoff and the application of partitioned data for the handoff.

Following the concepts and terms introduced in [**9, 7, 5**], a mobile computing environment consists of two distinct sets of entities: *mobile hosts* and *fixed hosts.* Some fixed hosts, called *Mobile Support Stations* (MSSs), are augmented with a wireless interface to communicate with mobile hosts. A mobile host can move from one *cell* (or radio coverage area) to another or from one network to another while retaining its network connections. As a result of user mobility and service handoff, the operations of transaction could be submitted to different data servers in multiple overlay networks. Mobility of hosts engenders a new kind of locality that migrates as hosts move. A user carrying a portable computer can submit the operations of a transaction to distributed data servers from different locations or networks. The distribution of operations implies that the transmission of messages (such as those involved in a two phase commit protocol) may be required among these data servers in order to coordinate the execution of these operations. In this paper, we will address transaction service handoff issues that occur in multiple overlay networks during the execution of transactions. Specifically, we will examine the application of *partitioned data* for distributed data services and management to support the transaction service handoff.

1.1. The Problem. Conventional methods for replicated data management are expensive because more than one site may be required to form the quorum necessary to run an update transaction. To overcome this restriction, some approaches reported in the literature have taken into account the semantics of applications to improve the response time and throughput of update transactions and to increase system resiliency.

One of the application classes that has recently been extensively studied and has been used to improve response time involves the problem of *resource allocation.* Consider an application where a data item represents the number of tickets to be sold. If the item is replicated, more than one site may be required to form the quorum necessary to perform an update. If the item resides in a central site, requests for tickets originating at that site can be satisfied locally, while all other sites in the system must exchange a series of messages with the central site. An alternative to either of these approaches is to *partition* the "tickets" data item among all the sites. [1] Each site is allocated a fraction of the tickets and will use them to process transactions as long as enough tickets are locally available. As a result, the overhead associated with communications is avoided for most transactions. Therefore, by partitioning data among server sites, transactions with resource allocation operations can be performed in a single site if the allocation updates do not violate local resource constraints.

Problems involving resource allocation can also be found in such mobile application domains as mobile sales and inventory applications [**15**] and mobile shopping

[1]Some proposed approaches, including site escrow [**14, 16**], demarcation protocol [**6, 1**], and Data Value Partitioning protocol [**19**], can be used for dynamically partitioning data among different server sites.

applications [**4**]. In a mobile environment, a mobile host can query or update a database, which is distributed in multiple data servers over a fixed network, from different locations. A mobile host is also likely to incur long disconnection periods due to the limitations of battery energy and the mobility of hosts. This long-disconnection characteristic may cause mobile transactions that access data from servers to be long-lived.

By exploiting the semantics of applications, it is possible to partition data items into geographically distributed servers and allow mobile clients to perform updates in local or nearby partitioned copies. The partition approach, therefore, improves not only the performance and scalability, but also the availability of transaction processing for update operations. However, traditional partitioned data management may incur extra message costs due to the mobility of hosts. The costs run counter to the motivation for the use of partitioned data and negate some of the advantages of this approach. The following example that uses the site escrow approach proposed in [**14, 16**] illustrates the problem:

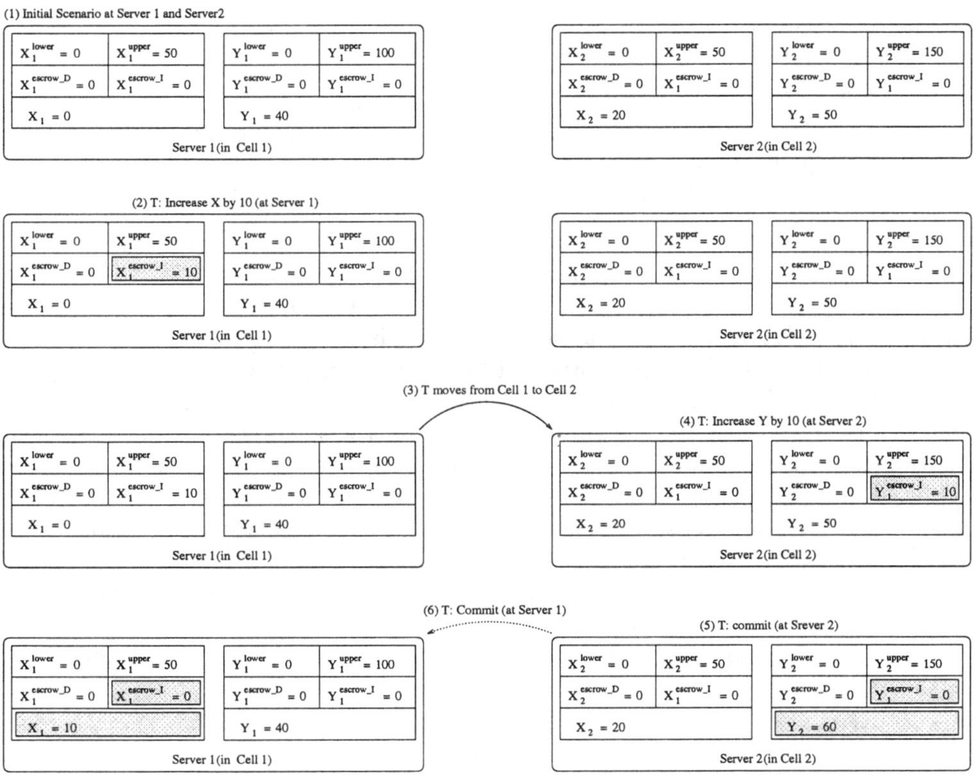

FIGURE 1. A mobile transaction example

EXAMPLE 1.1. Consider a mobile database system where X and Y are numeric objects with the resource constraints $X_{min}(=0) \leq X \leq X_{max}(=100)$ and $Y_{min}(=0) \leq Y \leq Y_{max}(=250)$, respectively. Initially, the value of X is partitioned into local numeric objects X_1 and X_2 in servers 1 and 2, respectively, with $X_1+X_2=X$.

Similarly, Y, X_{min}, X_{max}, Y_{min}, or Y_{max} are also partitioned into servers 1 and 2 such that $Y_1 + Y_2 = Y$, $X_1^{lower} + X_2^{lower} = X_{min}$, $X_1^{upper} + X_2^{upper} = X_{max}$, $Y_1^{lower} + Y_2^{lower} = Y_{min}$, and $Y_1^{upper} + Y_2^{upper} = Y_{max}$.

Assume that a mobile host submits a resource allocation transaction T with the following operations: [Increase X by 10], followed by [Increase Y by 10], and commit. Figure 1 shows the execution procedure of the transaction in the mobile database system. The mobile transaction host for T submits [Increase X by 10] to Server 1 from Cell 1. Server 1 checks the escrow variable $X_1^{escrow\_I}$ and makes a worst-case decision to determine whether the operation can be executed. If $X_1^{escrow\_I} + X_1 + 10 \leq X_1^{upper}$, then $X_1^{escrow\_I}$ is increased by 10. [2] The transaction host then moves to Cell 2 to submit [Increase Y by 10] to Server 2. Similarly, Server 2 increases the escrow variable $Y_2^{escrow\_I}$ by 10.

As a result of this mobility, the two operations are actually executed in two different servers. At commit time, a two phase commit protocol must be used to shift the increases from the escrow variables to the local numeric objects. That is, $X_1 = X_1 + 10$, $Y_2 = Y_2 + 10$, $X_1^{escrow\_I} = X_1^{escrow\_I} - 10$, and $Y_2^{escrow\_I} = Y_2^{escrow\_I} - 10$.

The distribution of operations in the above example presents two problems which arise with the site escrow approach:

- the message transmissions involved in a 2PC protocol increase the traffic over the fixed networks; and
- the use of a 2PC protocol will reduce site autonomy.

These problems obviously run counter to the motivation for the use of partitioned data and negate some of the advantages of this approach. Of course, if the transaction host in Example 1.1 remains fixed, transaction T can be executed in either server without resulting in message transmissions.

To avoid the use of a 2PC protocol at commit time, it was suggested in [**15**] that the move of a transaction host to a new cell should be accompanied by the transfer of the escrow log for the transaction to the local server under the cell. At the end of transaction, a commit operation can be executed at the local server without communication with other servers. However, the transfer procedure itself requires the use of a 2PC protocol and therefore still generate high traffic over the fixed network. When the host repeatedly moves between two cells during the execution of a transaction, the repeated log transfers between the two servers cause particularly heavy message overhead.

In a mobile computing system, the mobility factor is of the utmost importance in the design of a distributed algorithm. Because the physical distance between two points does not necessarily reflect the network distance, the communication path can grow disproportionately to actual movement. For example, a small movement which crosses network administrative boundaries can result in a much longer path. In a longer network path, communications traverse more intermediaries and consume more network capacity. The mobility of hosts can cause that even a short transaction to involve a long communication transmission.

---

[2] Actually, an uncommitted operation on object X_1 may be logged in an *escrow log*. For simplicity, we shall use the escrow variables $X_1^{escrow\_I}$ (= a) and $X_1^{escrow\_D}$ (= a) to represent the log information for operations [Increase X by a] and [Decrease X by a], respectively.

1.2. Contributions of the paper. In this paper, we present an approach that avoids both heavy message transmissions and the use of a 2PC protocol. A low message overhead among servers for each operation (including commit and abort) will improve the response time of an operation requested by a mobile host. One benefit of fast response time is that the mobile host will not need to expend precious battery resources while waiting for the acknowledgement of requested operation.

The approach we propose in this paper is called *Reservation Algorithm* (RA). In the site escrow approach, an escrow log is used for both commitment/recovery and constraint-maintaining purposes for uncommitted transactions. In contrast, this algorithm ensures resource constraints for the operations of uncommitted transactions by simply modifying bound (lower or upper) variables at local server sites. For commit and recovery purposes, the algorithm stores the operations in a *reservation log*. For example, for the operation [Increase X by 10] in Example 1.1, this algorithm needs only to decrease the bound variable X_1^{upper} by 10 and to store the operation in a reservation log at Server 1. The results of operations are returned to the mobile host along with acknowledgement messages. The mobile host stores the returned results in its reservation log for the transaction. Conceptually, the reservation log in the mobile host is a logical copy of logs maintained in server sites. At the commit time (after the host moves to Cell 2), the mobile host sends its logical reservation log, along with a commit request, to the local server 2 in the current cell. Server 2 will use the log information to perform the actual resource allocations, i.e., increasing both X_2^{upper} and X_2 by 10. This algorithm will ensure that the resource constraint, $X_{min} \leq \sum_{i=1}^{2} X_i^{lower} \leq X \leq \sum_{i=1}^{2} X_i^{upper} \leq X_{max}$, is continualy maintained.

The RA approach allows the *resource reservations* for the operations of uncommitted transactions and the actual *resource allocations* to be executed at different servers without the need for communication. The resource reservations involve the modification of bound variables and the update of reservation logs. Modifications of bound variables ensure the maintenance of resource constraints for the operations of uncommitted transactions. The process of resource allocation will restore modified bound variables and allocate resources at any partitioned data site (which may be different from the site where the reservations were performed).

Although the overall framework of the RA approach is straightforward, two interesting issues related to this approach merit deeper investigation. The first issue is the design of a *termination protocol.* In a mobile environment, an active mobile transaction may be aborted unilaterally by a data server. Such a *unilateral abortion* may be triggered by an extended long disconnection by or a total failure (destruction or loss) of the mobile host. In this case, the server may decide to abort the transaction to release reserved resources. Unfortunately, the server can not make this decision only on the basis of the information in its local reservation log, since the mobile host can make a commit decision without communication with the server. The purpose of a termination protocol is to guarantee that the commit decision of a mobile host will not contradict with the unilateral abort decision of a data server.

The second issue is the development of a protocol for non-partition-commutative operations on partitioned data. Assume that a data item X is partitioned among N sites such that $X = \sum_{i=1}^{N} X_i$ where X_i is the partitioned copy of X in site i. We say an operation O is a *partition-commutative* operation (pc-operation) if

$O(X) = O(X_j) + \sum_{i=1,i\neq j}^{N} X_i$ for any j $(1 \leq j \leq N)$; otherwise, it is a *non-partition-commutative* operation (npc-operation). An example of an npc-operation can be found in a banking application. In this application, both withdrawal and deposit operations are pc-operations, while an interest-posting operation is an npc-operation. It is obvious that an npc-operation on partitioned data can not be performed in any single site if the data is partitioned over more than one site. A protocol is therefore needed to coordinate the execution of such operations.

In this paper, we explore the following problems related to our proposed reservation approach:

1. Development of termination protocols that can be included in the reservation approach. These protocols should ensure that a unilateral abort by a data server and a commit by a mobile host would not be made simultaneously for a mobile transaction;
2. Determination of the effect of the proposed reservation approach on the npc-operations on partitioned data and of a protocol to permit these operations to accommodate the reservation approach.
3. Comparison of the message cost of a reservation algorithm that includes the required protocols for termination and npc-operations with that of existing site escrow or escrow log transfer algorithms.

The remainder of this paper is organized as follows. Section 2 introduces the system model and relevant terminology. In Section 3, we describe the basic reservation algorithm and the required termination protocols. Section 4 discuss a protocol for npc-operations. Section 5 presents a performance evaluation of the proposed algorithm and the traditional 2PC protocol in terms of message costs in the fixed network. Related research is discussed in Section 6, and concluding remarks are offered in Section 7.

2. The Mobile Transaction Model

Figure 2 presents a general mobile database system model similar to those described in [**9, 7, 5**] for mobile computing systems. In this model, both a database *server* and a database are attached to each fixed host. A database server is intended to support basic transaction operations such as resource allocation, commit, and abort.

Each MSS has a *coordinator* which receives transaction operations from mobile hosts and monitors their execution in database servers within the fixed networks. Transaction operations are submitted by a mobile host to the coordinator in its MSS, which in turn sends them to the distributed database servers within the fixed networks for execution. For example, the coordinator will send a resource allocation operation to a local server if a partitioned copy is in the local site.

A mobile host may submit transactions in one of two ways:

1. An entire transaction may be submitted in a single request message; the whole transaction thus becomes one submission unit. The mobile host also delivers execution control to its coordinator and awaits the return of the results of the transaction execution.
2. In contrast, the operations of a transaction may be submitted in multiple request messages. A submission unit thus consists of one operation or a group of operations; the mobile host interactively submits the operations of a transaction to its coordinator. A subsequent operation can be submitted

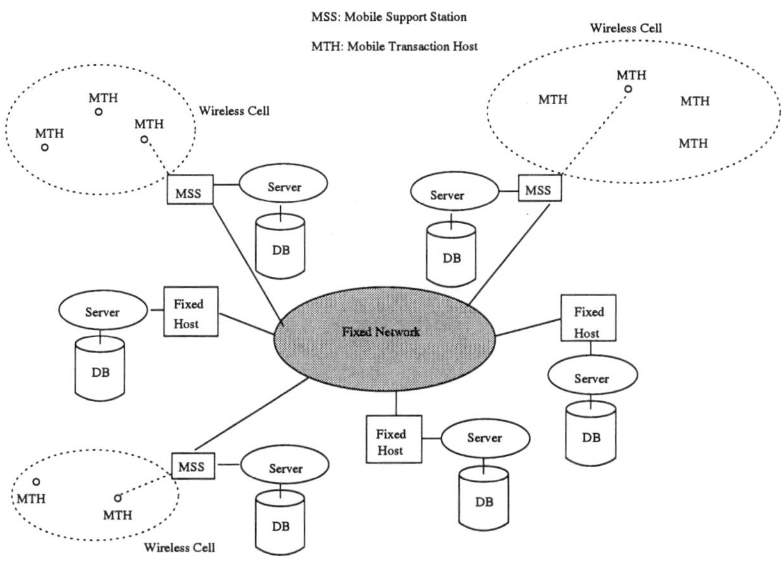

FIGURE 2. Mobile database system model

only after those previous have been executed and the results returned from the coordinator.

While the first approach involves a single coordinator for all the operations of a transaction, the second approach may involve multiple coordinators because of the mobility of the host. For example, a mobile host may move into a new cell after it obtains the results of previously submitted operations. In the new cell, it will submit the remainder of the transaction operations to the coordinator in the appropriate new MSS. The first approach is described in [**20**] and related issues regarding the interface between the mobile host and the coordinator are discussed. Our proposed model employs the second approach to transaction submissions. This approach supports the interactive execution of transactions and therefore offers increased flexibility in transaction computations.

We assume that a mobile host may move at any time. It may move away from its current cell after submitting an operation and before receiving a reply from the coordinator. The new coordinator will determine whether the host needs to obtain acknowledgement messages from previous coordinator after registering in the new cell. In this case, additional procedures are needed to locate the mobile host and convey to it the results of submitted operations. For simplicity, we assume that each service area supported be a server covers only a single cell. In reality, one service area may support more than one cell [**10, 11**].

We also assume that only one transaction may be initialized by a mobile host at any time. That is, a mobile host can initialize a transaction only after the previous transaction has finished. The transaction submitted from the mobile host is termed a *mobile transaction* and the host is called a *mobile transaction host.* A mobile transaction consists of a set of pc-operations and npc-operations which are bracketed by a $BEGIN\_TRANSACTION$ statement and an $END\_TRANSACTION$ statement.

3. Reservation Algorithm For Mobile Transactions

3.1. Basic Structure of the Reservation Algorithm. Suppose that the value of X is partitioned into local numeric objects X_i at server i $(1 \leq i \leq N)$ such that $X = \sum_{i=1}^{N} X_i$. Similarly, the bound value X_{min} (or X_{max}) is initially partitioned into X_i^{lower} (X_i^{upper}) in every server i $(1 \leq i \leq N)$ such that $X_{min} = \sum_{i=1}^{N} X_i^{lower}$ (or $X_{max} = \sum_{i=1}^{N} X_i^{upper}$). A *reservation action* for [Increase X by a] (or [Decrease X by a]) at server i involves the operation $X_i^{upper} = X_i^{upper} - a$ when $X_i + a \leq X_i^{upper}$ (or $X_i^{lower} = X_i^{lower} + a$ when $X_i^{lower} \leq X_i - a$). A *release action* for [Increase X by a] (or [Decrease X by a]) at server i involves the operation $X_i^{upper} = X_i^{upper} + a$ (or $X_i^{lower} = X_i^{lower} - a$). A *allocation action* for [Increase X by a] (or [Decrease X by a]) at server i performs the operations $X_i = X_i + a$ and $X_i^{upper} = X_i^{upper} + a$ (or $X_i = X_i - a$ and $X_i^{lower} = X_i^{lower} - a$). An operation O at server i is *safe* if $X_i^{lower} \leq O(X_i) \leq X_i^{upper}$. An operation O at server i is *unsafe* but *resolvable* if $X_i^{lower} \leq O(X_i) \leq X_i^{upper}$ does not hold but $X^{min} \leq O(X) \leq X^{max}$ holds.

Each reservation action (release, or allocation action) should be implemented as an atomic unit. Conventional database techniques can be used at each server to ensure that the actions that change the bound and resource variables will be atomic and persistent. When an action is completed, any locks on bound and resource variables will be released. Each server will record all the executions of these actions in a reservation log.

Assuming that no abortion is invoked by the servers, the reservation algorithm follows this general format:

1. The mobile host sends each pc-operation of a mobile transaction to the coordinator in the current cell, which will forward it to a local or nearby server where a partitioned data copy resides.
2. If an pc-operation at a server is safe, the server then executes a local reservation action for the pc-operation. Otherwise, the server invokes a resource repartition procedure (such as the point-to-point demarcation protocol [**6**] or a dynamic quorum-based protocol [**14**]) to requisition additional partitioned data resources from other servers. Upon the successful completion of the resource repartition procedure, the reservation action can be executed at the local server. The result of the reservation action is returned to the mobile host that submitted this pc-operation through its coordinator. If the operation is neither safe nor resolvable, a failure message will be returned.
3. The mobile host records the results of the reservation action of each pc-operation and the pc-operation itself in a reservation log. If all pc-operations of a mobile transaction succeed from the execution of reservation actions, the mobile host sends a COMMIT message along with the reservation log of the transaction to the coordinator in the current cell. Otherwise, it sends an ABORT message along with the reservation log to the coordinator. The coordinator then submits an allocation action (for COMMIT) or release action (for ABORT) for each pc-operation in the log to the local or nearby server.

Note that, due to mobility, the server at which the allocation actions are executed may not be the same as that from which the mobile host reserved these resources for the transaction (see Figure 3).

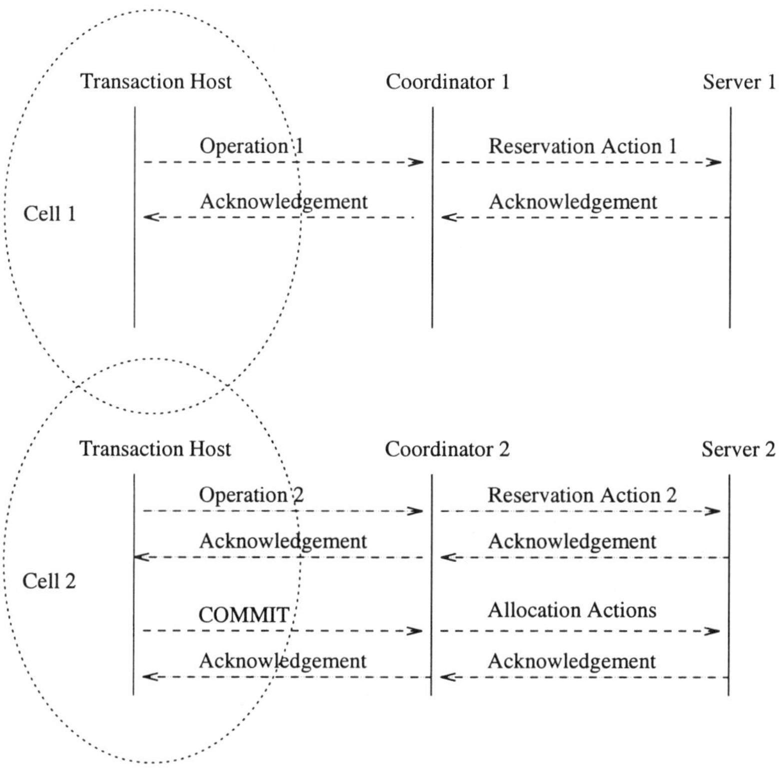

FIGURE 3. An example of basic reservation algorithm execution

In the escrow approach, the escrow log serves both to check local resource constraints and to commit or recover transactions. In contrast, the reservation log in the reservation approach serves only the latter purposes. In the escrow approach, for example, when an uncommitted transaction attempts to perform an pc-operation such as [Increase X by a], the server will use the escrow variable X^{escrow} in the escrow log to ascertain whether a given local resource constraint will be satisfied for the pc-operation; i.e., whether $X + X^{escrow} + a \leq X^{upper}$ holds. In the reservation algorithm, on the other hand, the maintenance of the constraint will not involve the use of the escrow variable X^{escrow}. Any effect of uncommitted transactions on resource constraints has been addressed by modifying the values of the bound variables by the reservation actions.

A mobile host may move from one cell to another at any time. In this paper, we assume that each service area supported by a server covers only a single cell. If the host has left a cell before receiving the acknowledgement message for last submitted pc-operation, this message will be forwarded by the coordinator in the new cell. After the host registers in the new cell, the coordinator will determine whether any acknowledgement messages are still outstanding from previous coordinator. To handle host mobility, each coordinator runs the following handoff algorithm:

1. If a recently arrived host has received acknowledgement messages for all submitted pc-operations, the coordinator in the new cell will require only

a request of pc-operations from the host. Otherwise, before accepting new pc-operations, it will send an acknowledgement request to the coordinator in the previous cell and then forward these acknowledgements (if any) to the host.

2. If a coordinator receives an acknowledgement request from another coordinator for the pc-operations submitted by a host that has left its cell, it will forward the acknowledgement message to the requesting coordinator.

It is clear that such an approach will maintain resource integrity constraints, provided that no data server is allowed to abort a mobile transaction. Any reservation action for a pc-operation can be performed only if the pc-operation is safe locally or resolvable globally. This safety or resolvability property implies that the bound variable updates by the reservation action always maintain resource constraints. A release action is actually the reversal of a reservation action and is invoked only if the reservation action has been executed. An allocation action will also reverse the bound variable change made by a reservation action and will update the corresponding resource variable. The increased or decreased value for the resource variable always matches that for the bound variable. Therefore, the basic reservation algorithm maintains resource constraints. The algorithm also ensures a serializable execution of committed transactions with pc-operations without requiring locks to be held until the commit time (a lock may be needed during the execution of each action), since all pc-operations are mutually commutative.

In a mobile environment, however, a mobile host may be unreliable or may suffer a total failure such as destruction or loss. In this case, a server may take an abort action for an uncommitted mobile transaction. We will address this issue and related problems in the rest of this section.

3.2. Unilateral Abortion Anomaly. In a mobile database environment, a server may decide to abort a transaction if the mobile transaction host has disconnected from any server for an extended period. Such an abortion allows the system to release resources reserved by the host. A mobile host may be somewhat unreliable, and an unexpected long disconnection period may be caused by a total failure of the device through destruction or loss. Abortion of a transaction avoids the indefinite holdings of reserved resources.

An abortion of this sort may result in an anomaly, called a *Unilateral Abortion Anomaly*, if the mobile host attempts to commit a mobile transaction through its current coordinator without following certain protocols.

EXAMPLE 3.1. Consider a mobile database system consisting of servers 1 and 2 with a resource constraint $X \leq 20$. Suppose that a mobile host had submitted the operation [increase X by 10] for transaction T to server 1 before it was disconnected from the system. The host then moves and establishes a connection to server 2 after server 1 has decided to abort the transaction during its disconnect period. Because server 1 made the abort decision without following certain protocols, the mobile host, which has no knowledge of the abort, may commit the transaction at server 2. While the abort will actually release reserved resources by increasing X_1^{upper} by 10, the commit will consume the reserved resources by increasing X_2 and X_2^{upper} by 10 in server 2. As a consequence, the total upper bound of X becomes 30 rather than 20; i.e. $X_1^{upper} + X_2^{upper} = 30 > 20 = X_{max}$, which violates the resource constraint $\sum_{i=1}^{2} X_i^{upper} \leq X_{max}$.

This anomaly arises because both the commit and abort actions are performed independently at different servers without any coordination. Recall that the coordinator allocates reserved resources at any local server site without communication with other servers from which these resources were gathered.

It is also possible that both a server and a mobile host may simultaneously decide to abort a transaction. Recall that a release action on behalf of a mobile host can be executed at any partitioned copy server. The protocols must guarantee that these reserved resources will not be released redundantly by both abortions. That is, each reserved resource can be released only once, even though these releases invoked by a mobile host can be performed at servers other than those where these resources were originally reserved.

Therefore, a protocol to avoid the unilateral abortion anomaly should ensure the following two conditions:

1. a transaction cannot be simultaneously committed by a mobile host and aborted by a data server; and
2. each reserved resource in a transaction can be released only once if a given transaction is aborted by a server and a mobile host.

3.3. Termination Protocols. In the proposed reservation algorithm, the allocation actions for the commitment of a transaction do not take the responsibility of writing the commit status into logs in other servers where reservation actions were executed. A *termination protocol* should therefore be applied so that an abort decision made by a server will not coincide with a commit decision made by the mobile host.

We assume that, when a server makes an abort decision, it can release the resources on x only if they were reserved by the transaction from the server. In other words, the server cannot release the resources on y if they were reserved from other servers. Two termination protocols which are candidates for inclusion in our reservation algorithm are:

1. All-Copy Voting Protocol: Let $N(x)$ be the set of partitioned copy sites for a data item x and $D(t)$ be the set of data items that transaction t has reserved. We assume that each item in $D(t)$ is only reserved once by transaction t. The server can abort an uncommitted transaction t and release the resources reserved on x in the server only if it receives an abort vote from each site in $N(x)$ for data item x in $D(t)$.

   A two phase protocol should be used to ensure that an abort decision and released resources will be recorded in reservation logs at all sites which have voted for the abort. In the first phase, the server sends an abort request to all the sites in $N(x)$ for reserved item x in $D(t)$.

   After all the sites return their votes, the server enters the second phase. If all the sites vote for the abort and the reserved item has not been released in any of these sites (by a mobile host), the server sends the abort decision to those sites and locally releases the resources reserved on x. Once a site voting for the abort receives the abort decision, the abort status for the transaction t will be recorded in its reservation log along with the names of released items. If any site has voted for commit or the reserved item has been released, the server will do nothing except recording the status into its local reservation log.

2. Transaction-Proxy Voting protocol: When a mobile transaction is initialized, the system can specify a server as the proxy for the transaction. If a server wishes to abort the uncommitted transaction and release resources reserved in the server, it must obtain a vote from the transaction proxy.

   The proxy will vote for abort only if it has not voted for commit or the reserved item has not been released. Once the transaction proxy votes for the abort, the abort status will be recorded in its reservation log along with the names of released items.

To integrate the all-copy voting protocol into our proposed reservation algorithm, each allocation action for a partitioned data item x should involve a determination of whether any partitioned copy server has voted for an abort decision. If not, the action can be executed at any local or nearby server and a commit flag (vote) can be recorded in the reservation log of the server without communication with other servers.

When a mobile host requests a commit operation for a transaction, its coordinator should execute all the allocation actions for the transaction in an atomic unit. If all the allocation actions can be executed at a single site (i.e., there is a local partitioned copy for each reserved resource), the commit operation can be performed locally. Otherwise, a 2PC protocol is needed to ensure the atomic execution of these actions. The procedure is required because each server may unilaterally abort a transaction and release reserved resources at any time. If the commit operation is not executed in a atomic unit, a server may unilaterally abort the transaction and execute a release action between two allocation actions, resulting in an undesired inconsistent termination decision. When a mobile host requests an abort operation for a transaction, the coordinator can execute the release actions for the transaction individually, without requiring atomic execution. However, each release action must ensure that the resource in question has not been previously released by an unilateral abort action by another server. This can be determined by examining the log information at the local server.

To incorporate the second termination protocol discussed above, the reservation algorithm should be modified in the following manner. Before a coordinator executes any allocation action, the server should get a commit vote from the transaction proxy. Once it obtains this vote, the allocation actions for a transaction can be executed individually; For a release action, the server should obtain an abort vote from the proxy. Furthermore, each server should inform the proxy which reserved resources to be allocated or released. Whenever other servers require a vote from the proxy, this information regarding released items should be supplied to them to prevent the redundant release of a reserved item by different servers.

A reservation algorithm that integrates either an all-copy voting or transaction-proxy voting termination protocol will be free from the unilateral abortion anomaly. Either protocol will require that the sites (or copies) voting for an abort always intersect with the sites (or copies) voting for commitment. By recording the released resources in the logs of voting sites, any reserved resource will not be released more than once by different servers.

3.4. Discussion. The two termination protocols described above could be subject to blocking even in the case of site failure. In the all-copy voting protocol, when a site which holds a copy of a reserved item fails, other servers can not execute the termination protocol to release the reserved item. In the transaction-proxy voting protocol, the potential for blocking is even higher. If the transaction proxy fails, no server can execute the termination protocol until it recovers.

Counterbalancing these blocking problems, the all-copy voting protocol offers low message overhead and supports a high degree of site autonomy for the commitment of transactions because all allocation actions can be executed locally or at nearby sites. The transaction-proxy voting protocol also offers low message overhead for the commitment or abortion of all transactions but does not support high site autonomy because both commit and abort decisions depend on the vote of a server designated prior to the execution of the transaction.

It has been generally held that non-blocking termination and efficient commitment are two incompatible goals in a distributed system. Our reservation algorithm illustrates the validity of this observation. While a reservation algorithm which incorporates either termination protocol permits a low-cost and efficient commitment of transactions, it imposes some restrictions on the execution of the termination protocol. The 3PC protocol, in contrast, involves no blocking in the event of site failure but has high message overheads for the commitment of transactions.

Finally, we note that, when a transaction is committed, this decision is not broadcast immediately to all log sites where reservation actions for the transaction were executed. In this case, some logs will still contain the *pending reservation* information about the transaction, potentially resulting the invocation of an abort request by the server. Although this will not create inconsistencies if all servers follow one of the termination protocols, it cause some unnecessary messages to be sent over the network. The pending reservation and other log information for a committed transaction can be removed if the system can periodically circulate the commit decision to other servers or piggyback the decision on other messages sent to servers.

4. Protocol for Non-Partition-Commutative Operations

In this section, we will examine the effect of our proposed reservation algorithm on npc-operations with partitioned data and discuss the design of a protocol to accommodate the reservation algorithm with such operations. We assume that serializability is used as the correctness criterion for the execution of transactions. We first review the execution of an npc-operation in a traditional distributed environment in which the host is fixed during the execution of a transaction and each reservation action and its allocation action (or release action) are performed in the same site.

4.1. The Problem. Assume that a data item X is partitioned among n servers such that $\sum_{i=1}^{n} X_i = X$. An npc-operation *npcO* on the partitioned data X can be performed by the coordinator of a transaction in two different ways. In the first approach, the coordinator collects all the values of partitioned copies from all n servers and executes the operation over the sum of these values; i.e., $npcO(\sum_{i=1}^{n} X_i)$. At commit time, the coordinator will repartition the result of the operation and write these repartitioned copies back to the n servers. In the second method, the coordinator sends the operation directly to all n partitioned copy servers. Each server i will perform the operation over the value of the partitioned copy; i.e., $npcO(X_i)$. An operation *npcO* on X is successful (i.e., each server can write the results back to a database) if and only if the operation $npcO(X_i)$ succeeds at every server i $(1 \leq i \leq n)$. The discussions in the rest of this section will be suitable to either approach.

To ensure a serializable execution, a lock protocol could be followed to coordinate the execution of pc-operations and npc-operations. If a pc-operation is to be executed on a partitioned copy X_i, a *Partition-Commutative Lock* (PC_LOCK) must be obtained from the partitioned copy X_i. If an npc-operation is to be executed, a *Non-Partition-Commutative Lock* (NPC_LOCK) should be obtained from all partitioned copies of X. A PC_LOCK is compatible with other PC_LOCKs but conflicts with another NPC_LOCK. Two NPC_LOCKs conflict with each other.

In a traditional distributed environment where the host is fixed during the execution of a transaction, all the actions of a pc-operation on X will be executed in a single partitioned copy server. A PC_LOCK should be set on the partitioned copy before these actions can executed. For an npc-operation on X, the coordinator of the transaction will send an NPC_LOCK request to all partitioned copy servers. Each server then determines whether the NPC_LOCK can be granted. If no pending PC_LOCK is set on the copy, the server grants the request and sends a reply message to the coordinator. Otherwise, the request is blocked at the server. The coordinator collects reply messages from these sites. Once all sites reply to the request, it concludes that there is no pending PC_LOCK at any server and the NPC_LOCK is granted. Thus, only one round of message exchanges is needed between the coordinator and any partitioned copy server for an NPC_LOCK request. At commit or abort time, these PC_LOCKs and NPC_LOCKs will be released.

If the transaction host may move among different cells during the execution, a reservation action and a allocation action for a pc-operation in the reservation algorithm may be executed at different servers. In this case, an interesting question is how a PC_LOCK that was set when the reservation action was requested can be released after the allocation action completes.

At the simplest level, a server executing the allocation action can immediately forward a PC_UNLOCK to the server where the PC_LOCK was set. If the transaction host moves frequently, however, this method will generate heavy message traffic. If there are m such pc-operations with reservation actions and allocation (or release) actions on different servers, m messages will be needed. The method obviously runs counter to the motivation of the use of the reservation algorithm.

A second method is to delay the forwarding of the PC_UNLOCKs until an NPC_LOCK request arrives. Each server then sends to other servers a single message containing a batch of PC_UNLOCKs that were executed since last npc-operation on the copy was completed. The method will release a PC_LOCK on a copy until an NPC_LOCK request arrives on the copy. If each partitioned copy server for a partitioned data item must forward the PC_UNLOCK messages to all other servers, then, in the worst case, the message overhead will be approximately n^2, where n is the number of partitioned copies.

In this paper, we suggest a method that requires each copy server to send all granted PC_LOCKs and delayed PC_UNLOCKs to the coordinator which is requesting a NPC_LOCK on that copy. The coordinator collects these PC_LOCKs and PC_UNLOCKs and then attempts to match a pending granted PC_LOCK with a delayed PC_UNLOCK. The message overhead will be $2n$. If the m PC_UNLOCKs for granted PC_LOCK between two npc-operations are uniformly distributed and m is greatly larger than the number n of the partitioned copies (i.e., $m >> n$), this method will prove to be more efficient than the first approach described above. If $n >> 2$, then the method is also superior to the second approach above because $n^2 > 2n$.

4.2. The Protocol. For a pc-operation, the *PC_LOCK* and *PC_UNLOCK* operations can be executed at different servers. A *PC_LOCK* is always executed at the server at which a reservation action is executed, while a *PC_UNLOCK* is always executed at the server at which an allocation action is executed. The granted *PC_LOCK* will not actually be released until an *NPC_LOCK* request arrives.

To request an *NPC_LOCK* for an npc-operation, the coordinator can execute a protocol with two rounds of message exchanges. In the first phase, the coordinator collects the *PC_LOCK/PC_UNLOCK* information from all copy servers. The coordinator cannot enter into the second phase until all servers reply and each *PC_LOCK* is matched by one *PC_UNLOCK* on the item to be accessed by the npc-operation. In the second phase, the coordinator sends a confirmation message to every copy server and the server releases the *PC_LOCK* and sets the *NPC_LOCK* on the partitioned copy.

To guarantee that no other *PC_LOCK* will be set after the first phase of the protocol, a new lock mode, called *NPC_INTEND*, must be used. This lock mode locks the copy at all the partitioned copy servers before these servers reply to the *NPC_LOCK* request in the first phase. If this step were bypassed, the copy server would be unable to force the setting of the *NPC_LOCK* in the second phase if other *PC_LOCK*s are granted on the copy after the first phase. A requested *NPC_INTEND* is compatible with a granted *PC_LOCK* but not with a granted *NPC_INTEND* or *NPC_LOCK*. A granted *NPC_INTEND* is, however, incompatible with a requested *PC_LOCK* or a requested *NPC_INTEND* or *NPC_LOCK*.

When the first phase of an *NPC_LOCK* request arrives at a copy server, an *NPC_INTEND* is set on the copy if no other *NPC_INTEND* or *NPC_LOCK* applies to that copy. Otherwise, the first phase of the *NPC_LOCK* request is blocked at the server. This scenario implies that two different npc-operations are requesting the *NPC_LOCK*. If there is a *PC_LOCK* on the copy, the first phase of the *NPC_LOCK* request will immediately set the *NPC_INTEND* on the copy. On the other hand, a granted *NPC_INTEND* will prevent any other new requested *PC_LOCK* or *NPC_LOCK*.

If the *PC_LOCK*s and *PC_UNLOCK*s collected in the first phase are not matched, it can be concluded that some partitioned copies are locked by other pc-operations. In this case, the coordinator will wait for further *PC_UNLOCK* messages from partitioned copy servers. A partitioned copy server will forward any newly executed *PC_UNLOCK* to a coordinator if an *NPC_INTEND* requested by the coordinator has been set on the copy. Once the coordinator collects matched *PC_UNLOCK*s, an *NPC_LOCK* confirmation message is sent to all the partitioned copy servers.

After the copy server receives a confirmation message from the coordinator in the second phase, all *PC_LOCK*s that have been matched by *PC_UNLOCK*s at the coordinator are removed and the *NPC_INTEND* lock is upgraded to an *NPC_LOCK*. The lock compatibility matrix appears in Figure 4.

Tj(lock) / Ti(request)	NPC_INTEND	PC_LOCK	NPC_LOCK
NPC_INTEND	No	Yes	No
PC_LOCK	No	Yes	No
NPC_LOCK	No	No	No

FIGURE 4. Lock compatibility matrix

We have shown that the granted PC_LOCK can be removed when another transaction prepares to perform an npc-operation on the copy. The pending period of a PC_LOCK may last until the first NPC_LOCK by another transaction. It is obvious that the pending PC_LOCK does not block either other PC_LOCK requests or NPC_LOCK requests. Thus, no blocking is incurred by a pending PC_LOCK on a copy of an item until the copy is accessed by an npc-operation.

Merging the locking protocol into the reservation algorithm is a straightforward procedure. A coordinator can send a PC_LOCK request with the corresponding reservation action to a partitioned copy server. The server starts to execute the reservation action only if the PC_LOCK is set on the copy. The PC_UNLOCK is set at any copy server after the completion of the allocation or release action. For an npc-operation, a coordinator sends an NPC_LOCK request to all partitioned copy servers. The coordinator can send any action of the npc-operation to copy servers only after the first phase of the locking protocol ends. Therefore, the npc-operation (or related actions) can be sent to copy servers with the NPC_LOCK confirmation message. At commit time, these NPC_LOCKs are released; note that a 2PC protocol is required to commit the npc-operation transaction.

5. Comparison of RES, ELT, and 2SE

As discussed in the introduction, one advantage offered by the reservation algorithm is the increased autonomy made possible by avoiding the use of a 2PC protocol. The algorithm also results in improved performance during normal execution, as no communication between server sites is needed for the commitment of some transactions. In this section, we present a comparative analysis of message costs incurred by the execution of the reservation algorithm (RES), the Escrow Log Transfering algorithm (ELT), and the 2PC-Site-Escrow algorithm (2SE) over fixed networks. The 2SE algorithm is the direct application of the site escrow method presented in [**14, 16**] in a mobile environment (as shown in Example 1). The ELT algorithm is a modification of a site escrow method which always transfers the escrow log for a mobile transaction to the local server in the current cell of the transaction host. This algorithm was described in [**15**]. Through this comparison, we wish to demonstrate the effect of the parameters of mobility and data partition on the message costs of these algorithms and to discover those circumstances in which the RES algorithm offers lower message costs than others.

5.1. An Analytical Model. We shall first describe a general equation that models the message costs incurred by the execution of various algorithms in accessing partitioned data. Let C^{alg} be the average number of messages per second required by the execution of a given algorithm alg. Then, C^{alg} can be expressed as:

$$C^{alg} = C^{alg}_{pc} + C^{alg}_{npc} + C^{alg}_{rpp} + C^{alg}_{com} + C^{alg}_{hd}$$

where C^{alg}_{pc} is the expected number of messages per second for the execution of pc-operations; C^{alg}_{npc} is the expected number of messages per second for the execution of npc-operations; C^{alg}_{rpp} is the expected number of messages per second for the execution of the repartition protocol (e.g., the site escrow protocol or the demarcation protocol); C^{alg}_{com} is the expected number of messages per second for the execution of commit operations; and C^{alg}_{hd} is the expected number of messages per second for the execution of the handoff protocol.

C^{alg} is obviously a function of such parameters as data partition, transaction host mobility, transaction rate, and transaction access pattern. To present a detail message cost equation for each algorithm, we shall now define these parameters and specify some assumptions.

Without loss of generality, we assume that there are N data servers, with each server attached to an MSS. In fact, some servers may be attached to fixed hosts which have no wireless communication interface. Our model can be generalized to include this case by assigning each of these servers to its closest MSS. *Partition-commutative transactions* (pc-transactions), which contain only pc-operations, and *non-partition-commutative transactions* (npc-transactions), which contain only npc-operations, arrive in Poisson distributions with an average arrival rate of λ_{pc} and λ_{npc}, respectively. The data are randomly accessed by a transaction. The average number of partitioned data items accessed by a pc-transaction is n_{pc}. The average number of partitioned data items accessed by an npc-transaction is n_{npc}.

The average number of partitioned copies per partitioned data item is represented as p (≥ 2). The probability that a pc-operation hits a partitioned copy at the local server is p/N. Let N_{npc} be the average number of server sites where the npc-operations of an npc-transaction are executed. The number will be between p and N.

We assume that a mobile transaction host can move away from the current cell (or server) only after each operation request submitted from the cell has been acknowledged by the coordinator in the same cell. In other words, we ignore the case in which the host moves to a new cell before it receives an acknowledgement for previously submitted operations. The probability of the mobility of each transaction host is m.

We now derive the basic expressions that describe the message costs for the RES, 2SE, and ELT algorithms. In these expressions, we ignore the message costs for aborted transactions and assume that no conflict exists between an npc-transaction and a pc-operation at any server. We also assume that a transaction host always submits a commit or abort operation to the local server which has all partitioned copies for the execution of the allocation or release actions. Our analytical model will not treat the message costs between mobile hosts and MSSs and consider only the message costs among data servers over MSSs.

C_{pc}^{alg}:: For each pc-operation in each of the three algorithms, if there is a partitioned copy at local site, then no communication is needed. Otherwise, the operation will be sent to a nearby partitioned copy site. The expected number of messages per second for pc-operations in these algorithms is:

$$C_{pc}^{RES} = C_{pc}^{ELT} = C_{pc}^{2SE} = 2\lambda_{pc}n_{pc}(1 - p/N)$$

C_{npc}^{alg}:: For each npc-transaction in the ELT and 2SE algorithms, only one round of messages is needed to obtain NPC_LOCKs, while, in the RES algorithm, two rounds of messages are needed. The operation is piggybacked along with the NPC_LOCK request messages. Because we assume that no conflict exists between an npc-transaction and a pc-operation at any server, the exact two rounds of messages will be sufficient for an NPC_LOCK request in the RES algorithm. The expected number of messages per second for pc-operations in these algorithms is:

$$C_{npc}^{RES} = 4\lambda_{npc}n_{npc}p, \quad and \quad C_{npc}^{ELT} = C_{pc}^{2SE} = 2\lambda_{npc}n_{npc}p$$

C_{com}^{alg}:: When the commit operation (the $END\_TRANSACTION$ operation) is requested, a 2PC protocol will be executed for all npc-transactions in each algorithm. The number of messages involved in the 2PC protocol, which is dependent upon the number of update transactions and the average number of servers updated by each transaction, can be expressed as $4\lambda_{npc}N_{npc}$. For the 2SE algorithm, a 2PC protocol will be executed for all distributed pc-transactions. Let N_{pc} be the average number of server sites where the pc-operations of a pc-transaction are executed and θ be a parameter such that $N_{pc} = n_{pc}m\theta$. Because $n_{pc}m$ is the average number of movements per pc-transaction among different servers, it is obvious that $N_{pc} < n_{pc}m$ and $0 < \theta \leq 1$. The number of messages per second involved in the commitment of distributed pc-transactions can therefore be expressed as $4\lambda_{pc}n_{pc}m\theta$. Thus, we arrive at the expected number of messages per second for the commit operation for each algorithm as follows:

$$C_{com}^{RES} = C_{com}^{ELT} = 4\lambda_{npc}N_{npc}, \quad and \quad C_{com}^{2SE} = 4\lambda_{npc}N_{npc} + 4\lambda_{pc}n_{pc}m\theta$$

C_{rpp}^{alg}:: For any of the RES, ELT, and 2SE algorithms, a repartition protocol RP (e.g., the site escrow protocol or the demarcation protocol) shall be used to dynamically reallocate or repartition resources so that each pc-operation is safe. The message overhead for the execution of the repartition protocol are in general independent of the algorithm used. Instead, the overhead is a function F_{RP} of database parameters, transaction parameters, and repartition protocol parameters. The expected number of messages per second for the execution of partition protocol can be expressed as:

$$C_{rpp}^{RES} = C_{rpp}^{ELT} = C_{rpp}^{2SE} = F_{RP}$$

C_{hd}^{alg}:: We assume that the mobile host moves to a new cell only after it has received acknowledgements for all operations submitted from the old cell. Therefore, in both the RES and 2SE algorithms, the current server does not need to contact the remote server for acknowledgement messages after the host moves to a new cell. However, in the ELS algorithm, a handoff protocol must be executed to transfer the context information from the previous server to the current server. The expected number of messages per second for the execution of the handoff protocol can be expressed as:

$$C_{hd}^{RES} = C_{hd}^{2SE} = 0, \quad and \quad C_{hd}^{ELT} = 4\lambda_{pc}n_{pc}m$$

Totally, the expected number of messages transmitted per second for each algorithm can be given by the expressions:

$$\begin{aligned} C^{RES} &= 2\lambda_{pc}n_{pc}(1-p/N) + 4\lambda_{npc}n_{npc}p + 4\lambda_{npc}N_{npc} + F_{RP} \\ &= C_0 + 2\lambda_{npc}n_{npc}p \end{aligned} \tag{5.1}$$

$$\begin{aligned} C^{ELT} &= 2\lambda_{pc}n_{pc}(1-p/N) + 2\lambda_{npc}n_{npc}p + 4\lambda_{npc}N_{npc} + 4\lambda_{pc}n_{pc}m + F_{RP} \\ &= C_0 + 4\lambda_{pc}n_{pc}m \end{aligned} \tag{5.2}$$

$$\begin{aligned} C^{2SE} &= 2\lambda_{pc}n_{pc}(1-p/N) + 2\lambda_{npc}n_{npc}p + 4\lambda_{npc}N_{npc} + 4\lambda_{pc}n_{pc}m\theta + F_{RP} \\ &= C_0 + 4\lambda_{pc}n_{pc}m\theta \end{aligned} \tag{5.3}$$

where $C_0 = 2\lambda_{pc}n_{pc}(1-p/N) + 2\lambda_{npc}n_{npc}p + 4\lambda_{npc}N_{npc} + F_{RP}$.

5.2. The Comparison. C^{2SE} is never larger than C^{ELT}, because $0 < \theta \leq 1$. From equations 5.1 through 5.3, we observe that, to make C^{RES} smaller than both C^{2SE} and C^{ELT} (i.e., $C^{RES} < C^{2SE}$ and $C^{RES} < C^{ELT}$), the following inequality should apply:

$$\begin{aligned} & C^{RES} < C^{2SE} < C^{ELT} \\ \Rightarrow\ & 2\lambda_{npc}n_{npc}p < 4\lambda_{pc}n_{pc}m\theta < 4\lambda_{pc}n_{pc}m \\ \Rightarrow\ & \xi < 2m\theta/p \leq 2m/p \end{aligned} \tag{5.4}$$

where $\xi = \lambda_{npc}n_{npc}/\lambda_{pc}n_{pc}$.

The inequality in 5.4 above illustrates the following relationships among the RES, ELT, and 2SE algorithms:

1. When $m = 0$, message costs for the RES algorithm are no better than for either the ELT and 2SE algorithms. In other words, when all transaction hosts do not move, the RES algorithm offers no advantage over the ELT and 2SE algorithms. In fact, in this case, no additional messages are needed for pc-operations if there is a partitioned copy at the local site, while more messages are required by the RES algorithm than by the ELT and 2SE algorithms for npc-operations.
2. When $\xi = 0$, the RES algorithm always performs at least as well as the ELT and 2SE algorithms in terms of message cost. In this case, no npc-operations are to be executed.

3. When $\xi > 1$, message costs for the RES algorithm are no better than for either the ELT and 2SE algorithms. In fact, when data has been partitioned, p is always equal to or larger than 2 and $2m\theta/p$ or $2m/p$ is no larger than 1. The observation indicates that the number of npc-transactions should not be larger than the number of pc-operations. Note that, when $\xi > 1$, no partitioned-data algorithm offers lower message costs than does the algorithm for non-partitioned data. In this instance, the message costs of npc-transactions (in two or more partitioned copy sites) may offset the message savings made by pc-operations in a partitioned copy site. Non-partitioned data may actually require fewer message exchanges, because any npc-transaction or pc-operation needs at most one round of messages between the remote no-partitioned data server and the transaction server.

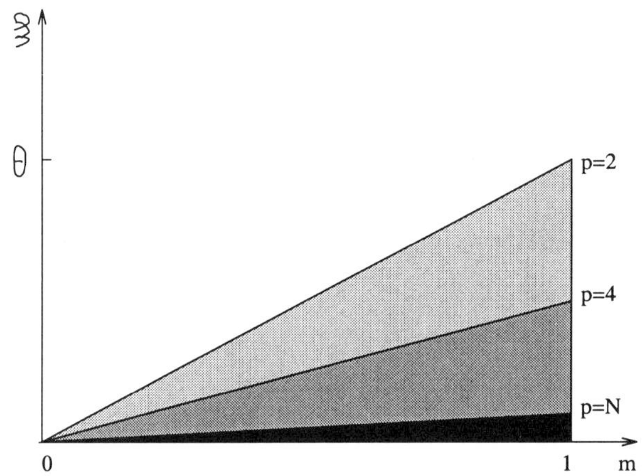

FIGURE 5. The relation of parameters ξ, m, and p

Figure 5 shows the relation among the parameters m, p, and ξ expressed in inequality 5.4 as the mobility parameter rises from 0 to 1. The shaded area in the figure indicates all the possible values of parameters ξ and m for a given p for which the the RES algorithm performs better than both ELT and 2SE algorithms in terms of message cost.

5.3. Satisfiability. Assume that C^{CEN} is a message cost function for non-partitioned and centralized algorithm in which no data is partitioned and all pc-operations or npc-operations are sent to a central site that stores data copies. The cost equation can be expressed as:

$$\begin{aligned} C^{CEN} &= C^{CEN}_{pc} + C^{CEN}_{npc} + C^{CEN}_{com} \\ &= 2\lambda_{pc}n_{pc}((N-1)/N) + 2\lambda_{npc}n_{npc}((N-1)/N) \\ &\quad + 2(\lambda_{pc} + \lambda_{npc})((N-1)/N) \end{aligned} \tag{5.5}$$

where $2\lambda_{pc}n_{pc}((N-1)/N)$ is the expected number of messages per second for pc-operations, $2\lambda_{npc}n_{npc}((N-1)/N)$ for npc-operations and $2(\lambda_{pc}+\lambda_{npc})((N-1)/N)$ for commit operations.

We now examine the *satisfiability* of C^{RES} with respect to C^{CEN} under the condition of inequality 5.4. Let C^X be a cost function for the algorithm X with parameter vector V. We say C^A is *satisfiable* with respect to C^B if there is a vector value v' in the domain of the parameter V such that $C^A < C^B$ with the vector value v' for the parameter V.

The satisfiability of C^{RES} with respect to C^{CEN} implies that in some situations the RES algorithm will involve lower message overhead than the CEN algorithm. However, if C^{RES} is NOT satisfiable with respect to C^{CEN} under inequality 5.4, the algorithm RES may not be valuable because it will involve higher message overhead than the CEN algorithm. In other words, in this case, $C^{RES} < C^{CEN}$ can not be satisfied even though $C^{RES} < C^{2SE} < C^{ELT}$ with $\xi < 2m\theta/p \leq 2m/p$. Our expectation would be that the RES algorithm would not only offer lower message costs than the ELT or 2SE algorithms but would also prove superior to the CEN algorithm in non-partitioned data environments.

Let now examine a scenario in which $C^{RES} < C^{CEN}$ under condition (4). Consider a database consisting of a fully partitioned data item X with no resource constraint over two servers (i.e., the constraint can be expressed as $-\infty < X < \infty$, $p = N = 2$, and each server has a partitioned copy). Assume that each pc-transaction has two pc-operations, i.e., $n_{pc} = 2$, and each npc-transaction has only one npc-operation, i.e. $n_{npc} = 1$. In this database, for any transaction accessing data item X, the message cost for repartition protocol is zero; i.e., $C^{RES}_{rpp} = 0$, as any pc-operation is always safe. C^{RES}_{pc} is also equal to zero, as $p/N = 1$. When $\xi = 0$ (i.e., there is no npc-operation), $C^{RES} = 0$ but $C^{CEN} \neq 0$. That is, $C^{RES} < C^{CEN}$. Assume $\xi > 0$. We compute the inequality (i.e., $C^{RES} < C^{CEN}$) from equations 5.1 and 5.5 as follows:

$$\begin{aligned}
& 2\lambda_{pc}n_{pc}(1 - p/N) \; + \; 4\lambda_{npc}n_{npc}p \; + \; 4\lambda_{npc}N_{npc} < \\
& \quad 2\lambda_{pc}n_{pc}((N-1)/N) \; + \; 2\lambda_{npc}n_{npc}((N-1)/N) \\
& \quad + \; 2(\lambda_{pc} + \lambda_{npc})((N-1)/N) \\
\Rightarrow \; & 4\xi p \; + \; 4\xi N_{npc}/n_{npc} \; < \; 1 \; + \; \xi \; + \; 1/n_{pc} \; + \; \xi/n_{npc} \\
\Rightarrow \; & \xi < 3/28
\end{aligned}$$

From this computation, we have the inequality $\xi < 3/28$. That is, when $0 < \xi < 3/28$, $C^{RES} < C^{CEN}$ can be satisfied. So, when $0 < \xi < \min(3/28, 2m\theta/p)$, $C^{RES} < C^{RES}$ and $C^{RES} < C^{2SE} < C^{ELT}$. Therefore, we have the following theorem:

THEOREM 5.1. *C^{RES} is satisfiable with respect to C^{CEN} with either $\xi = 0$ or $0 < \xi < 2m\theta/p$.*

6. Related Work

Some of the problems involved in supporting transaction services and distributed data management in a mobile environment have been identified recently in [**9, 3**]. The management of distributed data has been identified in [**9**] as a research area on which the mobility of host has a large impact. In [**3**], it is predicted that future applications of mobile computing will demand various transactional and transaction-like services.

A prototype of transaction service for mobile hosts is currently being implemented on the Coda file system [**13, 18**] to support continued services in a disconnection mode. This prototype uses the *optimistic concurrency control* method presented in [**17**] to enforce the serializable execution of transactions submitted from mobile hosts. The optimistic concurrency control method is generally suitable for applications, such as those in a file system environment, of low data contention. The prototype, however, did not address the issue of the mobility of transaction hosts and its effect on the management of distributed data.

The impact of mobility on distributed algorithms has recently been investigated in [**5**]. This research also emphasizes the reduction of the message costs in networks in which a mobile host involved in the execution of distributed algorithms moves across different cells. Unlike the work presented here, that research did not utilize the semantics of data to minimize the message overhead caused by the mobility of hosts.

As stated previously, the notion of using partitioned data to reduce message overhead and increase system throughput in distributed database environments has been investigated in the literature [**2, 6, 14, 16, 19, 1**]. These efforts address principally the efficient repartition or reconfiguration of a partitioned data item among different sites so that an operation on the data item can be performed at a local site. The research presented here, in contrast, utilizes the partitioned data to efficiently deal with the distribution of operations caused by the mobility of a transaction host. This problem did not arise in a traditional distributed environment with a fixed location of transaction host.

Some commonalities are present between previous work on repartition protocols and our efforts toward a reservation algorithm. The execution of a mobile transaction in the reservation algorithm can be thought of as involving a series of repartition procedures. A reservation action is a repartition procedure that moves a portion of the partitioned data from a local server to a mobile host, and the allocation (or release) action involves a repartition procedure that moves a portion of partitioned data from a mobile host to a local server. Specifically, like the demarcation protocol presented in [**6**], these repartition procedures are performed by updating the bound variables of a partitioned data item. However, the requirements for reliability procedures are quite different. In the previous approach, a repartition procedure is performed only among relatively reliable distributed servers in fixed networks. The procedure can be executed as a atomic unit. In our work, a repartition procedure is performed between a data server and a mobile host, and the series of repartition procedures for a mobile transaction is executed as an atomic unit. Guaranteeing the atomicity of the series of repartition procedures therefore poses an additional issue, particularly in the development of an atomic protocol which can handle the problems introduced by the failure and extended long disconnections of the mobile host.

In [**15**], the problem of using the site escrow method (2SE) for mobile transactions was discussed. To avoid the use of a 2PC protocol at commit time, the authors in [**15**] suggested the Escrow Log Transfering (ELT) method. The method executes a handoff protocol to move the escrow log of a transaction from the server in previous cell to the new server in the current cell before the transaction can continue its execution. This method carries with it a heavy message overhead when a mobile host moves frequently across cells.

Our approach to a pending partition-commutative lock is similar to that explored in [**12**] for a pending read lock in a replicated database for mobile transactions. In [**12**], the commutative semantics of read locks are utilized to reduce message overhead for the distributed read operations of a mobile transaction. The partition-commutative lock is applied to prevent a conflicting non-partition-commutative operation from accessing a partitioned copy, while a read lock is used to prevent a write operation from updating a replicated copy.

The issue of termination for a 2PC or 3PC protocol has been well studied (see [**8**] for details). A server will execute a termination protocol only after it enters the first phase of a 2PC or 3PC protocol but before it receives the commit decision from the coordinator. In our algorithm, a server is always required to execute a termination protocol before it can make an abort decision. The termination protocol in our algorithm, like a 2PC protocol, may be blocked if there is a link failure or a site failure.

7. Conclusions

In a mobile computing environment, a user carrying a portable computer can execute a *mobile transaction* over multiple overlay networks. To efficiently support the execution of mobile transactions, a transaction service handoff is needed to reduce the overhead of message transmission across network boundaries. In fact, because the physical distance between two points does not necessarily reflect the network distance, the communication path can grow disproportionately to actual movement. A small movement which crosses network administrative boundaries can result in a much longer path. In a longer network path, communications traverse more intermediaries and consume more network capacity. The mobility of hosts can cause even a short transaction to involve a long communication transmission. Moreover, application services have been designed optimally for the characteristics of network. The service level handoff will mobile users to utilize the optimized applications and services provided in the local application and service servers.

As a result of user mobility and service handoff, the operations of transaction could be submitted to different data servers in multiple overlay networks. The distribution of operations implies that the transmission of messages (such as those involved in a two phase commit protocol) may be required among these data servers in order to coordinate the execution of mobile transactions. In this paper, we address transaction service handoff issues that might occur in multiple overlay networks during the execution of transactions.

Specifically, we have examined the application of *partitioned data* for distributed data services and management to support the transaction service handoff. We have shown that, for operations pertaining to resource allocation, the message overhead (e.g., for a 2PC protocol) introduced by the distribution of operations is undesirable and unnecessary.

We have introduced a new algorithm, the *Reservation Algorithm* (RA), that does not necessitate the incurring of message overhead for the commitment of mobile transactions. We have discussed two issues related to the RA algorithm: *termination protocol* and *protocol for non-partition-commutative operations.* The algorithm ensures a serializable execution of transactions. We have performed a comparison between the proposed RA algorithm and existing solutions that use a 2PC protocol.

The algorithm proposed in this paper requires the transmission of reservation log information from a mobile host to its current coordinator through a wireless channel when the host decides to commit a transaction. These transmissions do not usually involve additional message exchanges, as they are piggybacked on the commit request message of the transaction. These transmissions can be structured to consume only minimal bandwidth on wireless channels by representing the reservation log by logical operations rather than physical pages.

Although the algorithm discussed in this paper applies directly only to operations over partitioned data, it can be merged into other locking algorithms to support operations on non-partitioned data. For example, in [**12**], we have shown that, if non-partitioned data are replicated among different servers, then a read unlock for an non-partitioned data item can be executed at any copy site, including sites other than that on which the read lock is set. The locking schema utilizes the replicated copies of data items to reduce the message costs incurred by the mobility of the transaction host. Therefore, the reservation algorithm can be augmented with the locking schema to support operations on both partitioned and non-partitioned data. Such an augmented algorithm can improve the efficiency of concurrency control protocols in a mobile environment if the number of read operations on non-partitioned data and pc-operations on partitioned data dominates that of write operations on non-partitioned data and npc-operations on partitioned data.

Finally, we note that the algorithm presented in this paper is a pessimistic concurrency control protocol. Each pc-operation obtains reserved resources when it is invoked by a mobile host. The message overhead which arises from the mobility of hosts can obviously be avoided through a more optimistic approach which defers reservation actions until commit time. Increasingly optimistic approaches, however, carry with them increasingly high transaction abort rates, offering a tradeoff between the message overhead and the abort rate. The choice of a pessimistic or an optimistic approach therefore depends on the requirements of the specific applications and the parameters of the system environment.

References

[1] G. Alonso and A. E. Abbadi. Partitioned data objects in distributed databases. *the International Journal on Distributed and Parallel Databases*, 3(1), 1995.

[2] R. Alonso, D. Barbará, and H. Garcia-Molina. Data caching issues in an information retrieval system. *ACM Transactions on Database Systems*, 15(3):359–384, September 1990.

[3] R. Alonso and H. Korth. Database issues in nomadic computing. In *Proceedings of the ACM SIGMOD Conference on Management of Data*, pages 388–392, 1993.

[4] A. Asthana, M. Cravatts, and P. Krzyzanowski. An indoor wireless system for personalized shopping assistance. In *1994 Workshop on Mobile Computing Systems and Applications*, 1994.

[5] B. R. Badrinath, A. Acharya, and T. Imielinski. Structuring distributed algorithms for mobile hosts. In *Proc. of the 14th International Conference on Distributed Computing Systems*, Poznan, Poland, June 1994.

[6] D. Barbará and H. Garcia-Molina. The demarcation protocol: A technique for maintaining constraints in distributed database systems. *The VLDB Journal*, 3(3):325–354, 1994.

[7] D. Barbará and T. Imielinski. Sleepers and workaholics: Caching strategies for mobile environments. In *Proceedings of the ACM SIGMOD Conference on Management of Data*, pages 1–12, 1994.

[8] P. Bernstein, V. Hadzilacos, and N. Goodman. *Concurrency Control and Recovery in Databases Systems*. Addison-Wesley Publishing Co., 1987.

[9] T. Imielinski and B. R. Badrinath. Wireless mobile computing : Challenges in data management. *Communication of ACM*, 37(10), 1994.

[10] R. Jain and N. Krishnakumar. Network support for personal information services to pcs users. In *Proceedings of the IEEE conference on Networks for Personal Communications (NPC '94)*, Long Branch, NJ, Mar. 1994.

[11] R. Jain and N. Krishnakumar. Service handoffs and virtual mobility for delivery of personal information services to mobile users. Technical Report TM-24696, Bellcore, Dec. 1994.

[12] J. Jing, O. Bukhres, and A. Elmagarmid. Distributed lock management for mobile transactions. In *Proc. of the 15th International Conference on Distributed Computing Systems*, Vancouver, Canada, June 1995.

[13] J. Kistler and M. Satyanaranyanan. Disconnected Operation in the Coda File System. *ACM Transactions on Computer Systems*, 10(1), February 1992.

[14] N. Krishnakumar and A. J. Bernstein. High throughput escrow algorithm for replicated databases. In *Proceedings of the Eighteenth International Conference on Very Large Data Bases*, 1992.

[15] N. Krishnakumar and R. Jain. Protocols for maintaining inventory databases and user service profiles in mobile sales applications. In *Proceedings of the Mobidata Workshop*, Rutgers University, Nov. 1994.

[16] A. Kumar and M. Stonebraker. Semantics-based transaction management techniques for replicated data. In *Proceedings of the ACM SIGMOD Conference on Management of Data*, 1988.

[17] H. Kung and J. Robinson. On optimistic methods for concurrency control. *ACM Trans. Database Syst.*, 6(2):213–226, June 1981.

[18] Q. Lu and M. Satyanaranyanan. Isolation-only transactions for mobile computing. *ACM Operating Systems Review*, 28(3), July 1994.

[19] N. Soparkar and A. Siberschatz. Data-value partitioning and virtual messages. In *Proceedings of the Conference on Principles of Database Systems*, 1990.

[20] L. Yeo and A. Zaslavsky. Submission of transactions from mobile workstations in a cooperative multidatabase processing environment. In *Proc. of the 14th International Conference on Distributed Computing Systems*, Poznan, Poland, June 1994.

CISE Department, University of Florida, Gainesville, FL 32611
E-mail address: helal@cise.ufl.edu

GTE Laboratories, 40 Sylvan Road, Waltham, MA 02454
E-mail address: jjing@gte.com

Computer Sciences, Purdue University, W. Lafayette, IN 47907
E-mail address: ake@cs.purdue.edu

DIMACS Series in Discrete Mathematics
and Theoretical Computer Science
Volume **52**, 2000

How to Combine a Column and Row Generation Method with a Column or Row Elimination Procedure - Application to a Channel Assignment Problem

Brigitte Jaumard, Christophe Meyer, and Tsevi Vovor

ABSTRACT. Solving exactly large instances of the channel assignment problem often requires to solve large linear programs with respect to the number of columns or rows. Considering the models proposed by Jaumard *et al.* (1998) which make use of column generation methods, it is necessary to combine those methods with some row generation procedures in order to handle the huge number of constraints. However, in order to keep a reasonable size of the linear programs, it is often interesting to eliminate some rows and columns from time to time. In this paper, we explore strategies which aims at combining column and row generation and elimination procedures, while providing a guarantee to reach the optimal solution. Computational experience are presented on some data provided by Bell Mobility for an urban cellular network.

1. Introduction

Linear programs resulting from combinatorial optimization problems often have a large number of variables and/or constraints. In many cases, these problems are so large that they usually cannot be solved directly by the basic simplex algorithm, and if it is possible such a solution is very expensive in terms of memory and computation time requirements.

When the constraint matrix can be implicitly defined, a *column generation* method can be used. A linear program with a small subset of the variables (i.e., the columns of the constraint matrix) is first solved. The variables which are not in this subset are then implicitly inspected through the solution of an auxiliary problem - often called *pricing problem* - in order to find one with a negative reduced cost (assuming it is a minimization problem). If such a variable is found, it is added to the subset, the resulting augmented linear program is solved and we iterate. If no variable with a negative reduced cost is found, the current solution is optimal. The column generation method was introduced by Gilmore and Gomory [**GG61**] to

---

1991 *Mathematics Subject Classification.* Primary: 90C06, 90B80.

Work of the first author has been supported by FCAR (Fonds pour la Formation de Chercheurs et l'Aide à la Recherche) grant 98ER1048, by a NSERC (Natural Sciences and Engineering Research Council of Canada) grant GP0036426 and by a NSERC Research Collaborative grant CRD209996 with Bell Mobility.

The postdoctoral fellowship of the second author was supported by NSERC network grant NET0200815 to $\mathsf{nc}m_2$ (Network for Computing and Mathematical Modeling).

Work of the third author has been supported in part by Bell Mobility.

solve the continuous relaxation of the cutting stock problem. Since then, it has been successively used to solve continuous relaxation of various combinatorial problems such as clustering [**MT98**], vertex [**MT96**] or edge [**NP91**] coloring, routing and scheduling [**DDSS95**] to cite only a few. See also the recent survey of Barnhart *et al.* [**BJN+98**].

When considering the solution of linear integer problems, cutting-plane methods can be viewed as row generation ones. First, the continuous relaxation is solved: if the solution satisfies the integrality constraints, then it is optimal. Otherwise one or more valid constraints, violated by the fractional solution are generated and added to the linear program, and the process is iterated. For most combinatorial problems, large classes of valid inequalities are known, which allows to find violated constraints by solving auxiliary problems (often called *separation problems*). One of the first combinatorial problems to be attacked in this way was the *Traveling Salesman Problem* (Dantzig, Fulkerson and Johnson [**DFJ54**], Padberg and Rinaldi [**PR91**]). See [**HP85**][**NW88**] for surveys and other applications. Observe that row generation methods can also be used for weaker relaxations than the linear programming one (when there is a huge number of constraints) with the aim of solving the linear relaxation. This is one of the points developed in this paper.

In order to stay with a reasonable memory size, it can be necessary to remove from time to time variables [**BJN+98**] or constraints [**Mil78**]. As long as only variables (respectively constraints) or constraints are generated, and that their elimination is performed only when a strict improvement (respectively deterioration) of the objective value has been obtained, we still have the convergence to the optimal solution of the linear program corresponding to the whole set of variables and constraints.

Variable and constraint generation methods have been rarely used together. This can be explained by the fact that adding new constraints can destroy the structure of the pricing problems. In particular, the reduced cost of a variable depends on the coefficients of this variable in the additional constraints which are generated and which may complicate significantly the pricing problem. A solution is to introduce in the pricing problem a $0-1$ variable for each new constraint, which takes the value 1 or 0 depending on whether the new column is present in the corresponding constraint or not (Vanderbeck and Wolsey [**VW96**]). However it is not always possible to assume that the new variables have a 0 coefficient in the new constraints, since the constraints may become invalid (see, e.g., Mehrotra [**Meh92**]). These difficulties disappear if we divide the variables into groups and only consider new constraints in which all variables of a given group have the same coefficient (see Desrosiers *et al.* [**JDD84**] where however the number of new constraints is at most 2).

In Nemhauser and Park [**NP91**], a combination of column and row generation is used to solve the edge coloring problem. The problem is formulated as a set partitioning problem in which the columns correspond to matchings of the graph. To strengthen the linear relaxation, odd-circuit constraints are added. The pricing problem is then a matching problem with an additional variable and constraint for each odd circuit constraint. This is a difficult problem, which is solved by branch-and-cut. In [**PR91**], Padberg and Rinaldi propose an algorithm to solve the symmetric traveling salesman problem. Their formulation have an exponential number of constraints and a polynomial, but large, number of variables. To solve the continuous relaxation, they combine variable and constraint generation, but allow constraint eliminations only. The *ABACUS* software [**Thi95**][**JT98**] allows variable or constraint generation as well as removal, but no indication is provided on how to proceed with the combination of row and column removals in order to guarantee the convergence to an optimal solution.

When variables and constraints are generated together, the simultaneous removing of variables and constraints without threatening the convergence of the algorithm is not trivial. This problem seems not to have been considered yet. In this paper, we propose a convergent algorithm that combines variable and constraint generation, and in which both variables and constraints can be removed. We use the channel assignment problem (see, e.g., Jaumard *et al.* [**JMMV98**]) to illustrate this algorithm.

In Section 2, we first recall the formulation of the channel assignment problem as a 0-1 linear integer program. This $0-1$ linear program can be reformulated in different ways, depending on how its variables are grouped: the reformulation, which involves a huge number of variables is described in Section 3. In Section 4 we explain how to solve this reformulation. We first present the variable generation procedure (Section 4.1) and the constraint generation one (Section 4.2). We next describe the variable-constraint generation-removal algorithm, together with a convergence proof in Section 4.3. Finally in Section 5, we provide some numerical results for an instance of the channel assignment problem corresponding to a Bell Mobility urban cellular network.

2. The channel assignment problem

A mobile cellular network can be viewed as a collection of cells C_i $(i \in I)$, each of them having H_i antennas located at a base station. A required number d_i of channels must be assigned to each cell C_i in order to satisfy the demand of users covered by the antennas of the cell. A tolerance is often associated with this demand ($\underline{d}_i \le d_i \le \overline{d}_i, i \in I$) in order to ensure the existence of a feasible channel assignment.

Channels are taken from a subset of the frequency channel management chart (see, e.g., Lee [**Lee95**]), which can be identified to a set $\mathcal{F} \subseteq \mathbb{N}$ of natural numbers. Channels are often assigned using block: we will denote by $B_k, k \in K$ the set of available blocks.

The assignment of channel blocks to cells must satisfy several interference constraints. The most important of them are the *cochannel interference constraints* and the *adjacent channel interference constraints.* The first express that a given channel cannot be assigned simultaneously to two cells that are too close from each other (the proximity is defined in terms of the signal transmission quality). The second express that for cells that are too close from each other we cannot assign on one of them a channel f, and on the other a channel f' such that $|f - f'| = 1$. Denote by J^{co} the pair of cells for which there is a cochannel interference constraint, and by J^{adj} the pair of cells for which there is an adjacent channel interference constraint. We have generally $J^{adj} \subseteq J^{co}$. Another class of interferences corresponds to the *antenna channel spacing interference*: two channels assigned to the same antenna of a given cell C_i must differ by at least spa_i (spa_i is typically between 15 and 18).

A classical objective is to minimize the unsatisfied demand (see, e.g., Jaumard *et al.* [**JMMV98**]). Introducing the variables y_{ik} which take the value 1 if block

B_k is assigned to cell C_i, and 0 otherwise, the channel assignment problem can be formulated as:

$$\max \sum_{i \in I} \sum_{k \in K} |B_k| y_{ik}$$

$$\text{s.t.} \begin{cases} \sum_{k \in K} |B_k| y_{ik} \geq \underline{d}_i, & i \in I \quad (2.1) \\ \sum_{k \in K} |B_k| y_{ik} \leq \overline{d}_i, & i \in I \quad (2.2) \\ y_{ik} + y_{jk} \leq 1, & (i,j) \in J^{co}; \quad k \in K \quad (2.3) \\ y_{ik} + y_{j\ell} \leq 1, & (i,j) \in J^{adj}; \\ & k, \ell \in K : B_k \text{ and } B_\ell \text{ adjacent} \quad (2.4) \\ \sum_{f=\underline{f}}^{\underline{f}+spa_i-1} \sum_{k : f \in B_k} y_{ik} \leq H_i, & \underline{f} \in \mathcal{F}, \quad i \in I \quad (2.5) \\ y_{ik} \in \{0,1\}, & i \in I; \quad k \in K \end{cases}$$

where 2 channel blocks B_k and B_ℓ are said adjacent if there exists $f \in B_k$ and $f' \in B_\ell$ such that $|f - f'| = 1$. Constraints (2.1) and (2.2) express that the number of channels assigned on each cell must be in a given range, (2.3) and (2.4) define the cochannel and adjacent channel interference constraints, and (2.5) expresses the channel spacing constraints (under some assumptions on the definition of the blocks; see Jaumard *et al.* [**JMMV99**]).

Depending on how we group the variables y_{ik}, this problem has several reformulations which fit in the following mathematical program expression:

$$(2.6) \quad \min \left\{ \sum_{g=1}^{\overline{g}} c^g x^g : \sum_{g=1}^{\overline{g}} A^g x^g \leq b; \quad x^g \in S^g \subseteq \{0,1\}^{n_g} \text{ for } g = 1, \ldots, \overline{g} \right\}$$

where $\overline{g}$ is the number of groups of variables, n_g is the number of variables in group g, A^g a matrix of size $m \times n_g$, c^g a vector of size n_g, $g = 1, \ldots, \overline{g}$ and b a vector of size m.

We present below two possible group definitions of the variables for the channel assignment problem.

- In the first one, each group of variables refers to a channel block. Therefore $\overline{g}$ is the number of blocks $|K|$ and n_k is equal to the number of cells $|I|$ for all k (we use the index k rather than g to denote a group in this paragraph). The set S^k corresponding to the group of variables associated with the block B_k is defined by

$$S^k = \left\{ (y_{1k}, y_{2k}, \ldots, y_{|I|k}) \in \{0,1\}^{|I|} \ : \ y_{ik} + y_{jk} \leq 1, \ (i,j) \in J^{co} \right\}.$$

  The vector c^k is equal to $(-|B_k|, -|B_k|, \ldots, -|B_k|)$. The matrix A^k is $\begin{pmatrix} -|B_k| \, I_{n_k} \\ |B_k| \, I_{n_k} \\ M \\ M' \end{pmatrix}$ where I_{n_k} is the identity matrix of dimension n_k corresponding to the demand constraints and M and M' are $0-1$ matrices whose lines correspond respectively to the adjacent channel interference constraints and to the channel spacing constraints.

  The vector b is a vector of size $m = 2|I| + |J^{adj}|$ (number of pairs of adjacent blocks) $+|\mathcal{F}|\,|I|$, equal to $(-\underline{d}_1, -\underline{d}_2, \ldots, -\underline{d}_{|I|}, \overline{d}_1, \overline{d}_2, \ldots, \overline{d}_{|I|}, 1, 1, \ldots, 1, H_1,$

$H_1, \ldots, H_1, H_2, H_2, \ldots, H_2, \ldots\ldots, H_{|I|}, H_{|I|}, \ldots, H_{|I|})$ (in practice some constraints can be aggregated, which reduces the size of m: see, e.g., Jaumard *et al.* [**JMMV98**][**JMMV99**]).

- In the second formulation, each group of variables y_{ik} refers to a cell, hence $\overline{g} = |I|$ and $n_i = |K|$ for all i. The feasible set S^i of the group of variables associated with cell C_i is defined by the constraints:

$$\sum_{k \in K} |B_k| y_{ik} \geq \underline{d}_i$$

$$\sum_{k \in K} |B_k| y_{ik} \leq \overline{d}_i$$

$$y_{ik} + y_{i\ell} \leq 1, \qquad i : (i,i) \in J^{adj}; \quad k, \ell \in K : B_k \text{ and } B_\ell \text{ adjacent}$$

$$\sum_{f=\underline{f}}^{\underline{f}+spa_i-1} \sum_{k : f \in B_k} y_{ik} \leq H_i, \qquad \underline{f} \in \mathcal{F}$$

$$(y_{i1}, y_{i2}, \ldots, y_{i|K|}) \in \{0,1\}^{|K|}.$$

The vector c^i is equal to $(-|B_1|, -|B_2|, \ldots, -|B_{|K|}|)$. The matrix A^i is a $0-1$ matrix whose lines correspond to the cochannel interference constraints, and to the adjacent channel interference constraints involving different cells. The vector b is the vector with all components equal to 1, of size $m = |J^{co}| \times |K| + |\{(i,j) \in J^{adj}, i \neq j\}| \times$ (number of pairs of adjacent blocks).

3. A reformulation of the $0-1$ linear program

We now present a reformulation of the general $0-1$ linear program (2.6). Replacing x^g by $\sum_{t=1}^{\overline{t}_g} \lambda_t^g x^{gt}$ with $\lambda^g \in \{0,1\}^{\overline{t}_g}$ where $x^{gt}, t = 1, \ldots, \overline{t}_g$ are the points of S^g, we obtain the following equivalent formulation:

$$(BLP) \quad \min \quad \sum_{g=1}^{\overline{g}} \sum_{t=1}^{\overline{t}_g} \lambda_t^g (c^g x^{gt})$$

$$\text{s.t.} \quad \begin{cases} \displaystyle\sum_{g=1}^{\overline{g}} \sum_{t=1}^{\overline{t}_g} \lambda_t^g (A^g x^{gt}) \leq b \\ \displaystyle\sum_{t=1}^{\overline{t}_g} \lambda_t^g = 1, & g = 1, \ldots, \overline{g} \\ \lambda_t^g \in \{0,1\}, & t = 1, \ldots, \overline{t}_g, \quad g = 1, \ldots, \overline{g} \end{cases}$$

(where the equality constraints are redundant).

The continuous relaxation of this problem is:

$$(RBLP) \quad \min \quad \sum_{g=1}^{\overline{g}} \sum_{t=1}^{\overline{t}_g} \lambda_t^g (c^g x^{gt})$$

$$\text{s.t.} \quad \begin{cases} \sum_{g=1}^{\overline{g}} \sum_{t=1}^{\overline{t}_g} \lambda_t^g (A^g x^{gt}) \leq b & \\ \sum_{t=1}^{\overline{t}_g} \lambda_t^g = 1 & g = 1, \ldots, \overline{g} \\ \lambda_t^g \geq 0, & t = 1, \ldots, \overline{t}_g, \quad g = 1, \ldots, \overline{g}. \end{cases}$$

This problem has $n' = \sum_{g=1}^{\overline{g}} |S^g|$ variables and $m' = m + \overline{g}$ constraints. Since S^g is a subset of $\{0,1\}^{n_g}$ whose size is 2^{n_g}, the number of variables n' can be very large, possibly exponential. It is therefore not realistic to take into account all variables simultaneously, so we resort to a column generation method. On the other hand, the number of constraints m' is roughly the same as for the original formulation and, depending on the definition of the sets J^{co} and J^{adj}, can also be very large. Often only a small part of them are binding at the optimum. So again it can be interesting to consider a subset of them at the beginning, and to generate those that are needed during the optimization. In the next section, we present an algorithm to solve problem $(RBLP)$ that dynamically generates variables and constraints, and that also removes variables and constraints when they are no longer needed.

4. Solution of the continuous relaxation

At iteration ℓ, we assume that only a subset of the variables and of the constraints has been considered in $(RBLP)$. We denote by $T^\ell = (T^{\ell 1}, \ldots, T^{\ell \overline{g}})$ and C^ℓ respectively the set of variables and constraints, and by $RBLP(T^\ell, C^\ell)$ the resulting linear program. We will assume that the convexity constraints on the λ_t^g variables are always included in C, which implies, together with the positivity of the variables, that the restricted linear programs are bounded. We will also assume that they are feasible: this can be ensured by adding an additional positive variable in all inequality constraints and penalizing it in the objective function (note that equality constraints can either be expressed as 2 inequality constraints, or can be used to eliminate a variable). These assumptions imply that problem $(RBLP)(T, C)$ has a finite optimal solution for all $T \subseteq \overline{T}$ and all $C \subseteq \overline{C}$. Denote by λ^* an optimal solution, and by π^* an optimal solution of its dual.

4.1. Variable generation procedure. Let λ^* be an optimal vector of problem $RBLP(T^\ell, C^\ell)$, π^* be the optimal dual values associated with the inequality constraints of the form $Ax \leq b$ and $\pi_g^{0*}, g = 1, \ldots, \overline{g}$ the optimal dual values associated with the equality constraints. The variable generation phase consists in finding variables with a negative reduced cost, or proving that there exists none. The reduced cost of variable λ_t^g is given by

$$r_t^g = c^g x^{gt} - \pi^* A^g x^{gt} - \pi_g^{0*}.$$

The variable of group g with the smallest reduced cost is obtained by solving the following *auxiliary problem*

$$\min\left\{(c^g - \pi^* A^g)x - \pi_g^{0*} : x \in S^g\right\}.$$

If the optimal value is negative, the variable λ^g corresponding to the optimal solution x^{*g} is generated. Otherwise no variable is generated.

4.2. Constraint generation. Assume again that $\lambda^* = (\lambda^{*1}, \ldots, \lambda^{*\overline{g}})$ is an optimal solution of $RBLP(T^\ell, C^\ell)$, and compute $x^* = (x^{*1}, \ldots, x^{*\overline{g}})$ with

$$x^{*g} = \sum_{t \in T^{\ell g}} \lambda_t^{*g} x^{gt}, \qquad g = 1, \ldots, \overline{g}.$$

We next examine which constraints of problem (2.6), not included in problem $RBLP(T^\ell, C^\ell)$, are violated and add them or a subset of them to the current problem $RBLP$. If the constraints can be parameterized, the most violated constraint can also be obtained by solving an auxiliary problem in a similar way than for variable generation. In order to reinforce the linear relaxation ($RBLP$), we can also add valid inequalities. For instance, in the channel assignment, we can consider additional constraints that are valid for the node packing polytope $conv(\{y \in \{0,1\}^{|I||K|} \mid y_{ik} + y_{i\ell} \le 1,\ (i,j) \in J^{adj}, k, \ell \in K : B_k \text{ and } B_\ell \text{ adjacent}\})$, such as maximal clique inequalities (see, e.g., Nemhauser and Wolsey [**NW88**]). The number of these constraints can be very large, so violated constraints must be generated through the solution of a separation problem.

4.3. Algorithm GENCOLROW**.** The following notations are used in the description of the algorithm:

$separ_\ell$	:	indicate if the constraint or the variable separation procedure ($separ_\ell$ = VARIABLE or $separ_\ell$ = CONSTRAINT) was called at iteration ℓ,
$nbgen_\ell$	:	number of constraints or variables generated at iteration ℓ,
z_ℓ	:	value of the relaxation at the end of iteration ℓ,
z^{best}	:	value of the current best known solution.

The function *selectSeparation()* chooses between variable or constraint separation for the current iteration. Functions *variableSeparation()* and *constraintSeparation()* return respectively the number of variables and constraints generated. If the *constraintSeparation()* procedure has been called at the previous iteration and if no constraint has been generated, then the *variableSeparation()* procedure is called at the current iteration, and conversely.

Algorithm GENCOLROW:

Step 1 (Initialization):

$z^{best} = +\infty$; $nbgen_0 = nbgen_1 = +\infty$; $separ_1$ = VARIABLE; $\ell = 1$.
Initialize C^1 with initial constraints and T^1 with initial variables. Solve $RBLP(T^1, C^1)$, obtaining an optimal solution λ^{*1} with value z_1.

Step 2 (Decision):

$\ell \leftarrow \ell + 1$; $T^\ell = T^{\ell-1}$; $C^\ell = C^{\ell-1}$.
If ($nbgen_{\ell-1} = nbgen_{\ell-2} = 0$) stop: $\lambda^{*\ell-1}$ is optimal.
If ($separ_{\ell-1}$ = CONSTRAINT and $nbgen_{\ell-1} = 0$) go to Step 3.
If ($separ_{\ell-1}$ = VARIABLE and $nbgen_{\ell-1} = 0$) go to Step 4.
If (*selectSeparation()* = CONSTRAINT) go to Step 4.

Step 3 (Variable generation):
$separ_\ell$ = VARIABLE.
$nbgen_\ell$ = *variableSeparation()*.
If ($nbgen_\ell > 0$) add the new variables to T^ℓ and go to Step 5; otherwise set $z_\ell = z_{\ell-1}$ and return to Step 2.
Step 4 (Constraint generation):
$separ_\ell$ = CONSTRAINT.
$nbgen_\ell$ = *constraintSeparation()*.
If ($nbgen_\ell > 0$) add the new constraints to C^ℓ and go to Step 5.
Step 4b (Feasibility):
If ($z_{\ell-1} < z^{best}$) update z^{best} and remove from C^ℓ a subset of the non-binding constraints.
Set $z_\ell = z_{\ell-1}$ and return to Step 2.
Step 5 (Solving the LP):
Solve $RBLP(T^\ell, C^\ell)$, obtaining an optimal solution $\lambda^{*\ell}$ with value z_ℓ.
Step 6 (Elimination):
If ($separ_\ell$ = VARIABLE and $z_\ell < z_{\ell-1}$) remove from T^ℓ a subset of the variables with strictly positive reduced cost.
Return to Step 2.

THEOREM 4.1. *Assuming that the algorithm used to solve the linear programs of Step 5 cannot cycle, the above algorithm finds an optimal solution of problem (RBLP) in a finite number of iterations.*

PROOF. A first point to show is that when the algorithm stops at Step 2 of iteration ℓ, $\lambda^{*\ell-1}$ (with components $\lambda_t^{*\ell-1}, t \notin T^{\ell-1}$ equal to 0) is indeed an optimal solution of problem $RBLP$. This is the case because $\lambda^{*\ell-1}$ is feasible for ($RBLP$) as an optimal solution of $RBLP(T^{\ell-1}, C^{\ell-1})$ and since no violated constraints could be generated at the last or at the last but one iteration. The objective function value of $\lambda^{*\ell-1}$ is equal to the optimal value of $RBLP(T^{\ell-1}, C^{\ell-1})$, i.e., $z_{\ell-1}$. Now let $\pi^{*\ell-1}$ be the optimal dual values of $RBLP(T^{\ell-1}, C^{\ell-1})$ (with $\pi_i^{*\ell-1} = 0$ for $i \notin C^{\ell-1}$): $\pi^{*\ell-1}$ is dual feasible to ($RBLP$) as an optimal dual solution of $RBLP(T^{\ell-1}, C^{\ell-1})$ and as no variable with negative reduced cost could be found. Moreover its value is equal to the optimal value of the dual of $RBLP(T^{\ell-1}, C^{\ell-1})$, i.e., $z_{\ell-1}$. It follows that $\lambda^{*\ell-1}$ is an optimal solution of problem ($RBLP$).

It remains to show now that the algorithm cannot cycle. First note that the eliminations of constraints in Step 4b can occur only a finite number of times. Indeed, this happens only when we improve on the best feasible value z^{best}. But z^{best} is non-increasing and can take only a finite number of distinct values, as these values correspond to the optimal values of a finite number of possible linear programs. Hence it suffices to consider the algorithm from the last passage through Step 4b. Note that after the last passage through Step 4b, constraints are added but are never removed. Since we cannot add a constraint that is already in the LP, there is a finite number of such addition. Consider the algorithm starting from the last time we add a constraint. Since we only add variables and that the removed variables are not in the current basis, z_ℓ is non-increasing and can take only a finite number of distinct values. Consider the algorithm after the last improvement of z_ℓ: z_ℓ is constant, variables are added and no one is removed. Since we cannot add a variable that is already in the LP (because we only generate variables with strictly negative reduced cost, and that the variables already in the LP have a reduced cost positive or null), the algorithm will stop after a finite number of iterations. □

Remarks:

- We have implicitly assumed when describing the algorithm that the linear programming solver in Step 5 has no limit on the number of rows and columns that it can handle. In practice it could happen that despite the elimination procedures in Step 4b and Step 6, the number of rows and columns is still too large for the LP solver (for example in presence of degeneracy, when the objective value does not change). In order to be able to remove additional columns and rows while maintaining the convergence, we would have to add an anticycling procedure similar to those used in linear programming, for example the Bland rule (see, e.g, Bazaraa and Jarvis [**BJ77**]).
- The algorithm can be thought of as consisting in major and minor iterations. A major iteration corresponds to the improvement of the best known solution. At the end of each major iteration, constraints can be removed. During minor iterations, only variables can be removed.
- The proof extends easily to the following variants of the algorithm:

  Variant 1: : in Step 4, each time a constraint is generated, variables with strictly positive reduced cost can be removed. Idem in Step 4b if there is an improvement of the best known value. The proof remains unchanged.

  Variant 2: : Step 4b and Step 6 are respectively replaced by

  Step 4b' (Feasibility):
  If ($z_{\ell-1} < z^{best}$) update z^{best} and eliminate variables with strictly positive reduced cost from T^{ℓ}.
  Set $z_\ell = z_{\ell-1}$ and return to Step 2.

  Step 6' (Elimination):
  If ($separ_\ell =$ CONSTRAINT and $z_\ell > z_{\ell-1}$) eliminate non-binding constraints from C^{ℓ}.
  Return to Step 2.

  The role of variable elimination and constraint elimination must be inverted in the proof (i.e., during minor iterations we remove constraints instead of variables).

5. Computational results

We have tested 5 variable/constraint generation/deletion strategies for the solution of an instance of the channel assignment problem derived from an urban network of Bell Mobility for which we used the formulation with the first grouping of variables described in Section 2. A strategy is noted $xy|z$ where x and y refer to the constraints while z refers to the variables. Parameter x is equal to 1 if constraints are dynamically generated; 0 otherwise (i.e. all the constraints are included in the first $RBLP$ problem). Parameter y is equal to 1 if constraints are dynamically removed; 0 otherwise (i.e. constraints are never removed once they have been generated). Finally, parameter z is equal to 1 if variables are dynamically removed; 0 otherwise (since variables are in exponential number, they are always dynamically generated).

In order to reduce the computational burden, we have omitted the channel spacing constraints from the formulation. We start the algorithm GENCOLROW with the demand constraints (except of course when $x = 0$ in which case all constraints are included), and with the slack variable (see the beginning of Section 4 for its definition) as the only variable. The maximum number of variables generated by iteration varies: while the slack is non null (i.e., while we are not feasible), or when we had to solve exactly the pricing problem, we can generate up to $|K|$ variables (one per block). Otherwise, the maximum number of variables generated is set to 5 (with still at most one variable per block). At any time the maximum number of constraints generated per iteration is set to 100. There is no limit on the number of variables or constraints that we can remove per iteration.

However the removing of constraints is possible only if, since the last removing at least 500 constraints were generated. Similarly the removing of variables is only tried if, since the last removing at least 100 variables were generated.

For the instance considered, the number of cells in the network is 85. There are 85 demand constraints, 63 so called convexity constraints and 34112 adjacency constraints.

The program was run on an SS20/514MP workstation (dantzig). We measured the total CPU time *CPU* (in seconds), the part of the time devoted to respectively the solution of the restricted linear program (Step 5 of the algorithm) *CPU%LP*, the generation of variables *CPU%pricing* and the generation of constraints *CPU%cutting*. We also counted the number of LP *#LP*, the number of constraints generated $\#con^+$ and removed $\#con^-$ and the number of generated variables $\#var^+$ and removed $\#var^-$.

Results are reported in Table 1. We observe that for the pure column generation strategy (00|0), the number of linear programs is the smallest among the 5 strategies, but their solution is costly. This is due to the large number of constraints (34260). Note that the instance considered is small compared to the problems of the real life which easily have more than 500 cells. It is clearly not possible to take into account all constraints simultaneously.
When generating constraints only when needed (strategy 10|0), the CPU time is divided by more than a factor 7 while the number of LPs is multiplied by 4. Note that only 4502 among the 34112 adjacency constraints were needed, and that the generation of these constraints took less than 1% of the total CPU time.

Strategy	00|0	10|0	11|0	10|1	11|1
CPU	171h23mn	22h45mn	14h27mn	16h49mn	11h58mn
CPU%LP	98.97	92.65	87.07	79.62	67.14
CPU%pricing	0.95	6.11	10.24	17.36	26.96
CPU%cutting	0.00	0.21	0.45	0.58	1.11
#LP	483	1438	2037	2878	4155
$\#con^+$	0	4502	25685	4525	31232
$\#con^-$	0	0	24629	0	30327
$\#var^+$	5904	4814	4516	12637	12149
$\#var^-$	0	0	0	11610	11054

TABLE 1. Numerical comparison of several generation/deletion strategies

If we allow the deletion of constraints (strategy 11|0), we obtain an additional 36% decrease of the CPU time. The number of *LP* increases, but they are more

easy to solve. Moreover, the deleted constraints are easily regenerated if needed (less than 0.5% of the total CPU time is spent to generate constraints).
Strategy 10|1 is the counterpart of strategy 11|0: deletion of variables replaces deletion of constraints. We still observe that while the number of LPs increases, the time spent to solve them decreases (803mn compared to 1264mn in strategy 10|0). However the cost of regenerating deleted variables is greater than that of regenerating constraints in strategy 11|0, which explains the relative bad performance of this strategy.
When both deletion of variables and constraints are allowed (strategy 11|1), we observe the best performance in terms of CPU. The CPU time is decreased by 17% and more with respect to a strategy that allows the elimination of only variable or only constraint, and by 47% with respect to a strategy that remove none of them.

In conclusion, these results show that removing both variables and constraints can help to solve the linear relaxation of the channel assignment more efficiently. Further improvements could be obtained by increasing the efficiency of the variable generating procedure, and/or by considering more sophisticated deletion strategies, such as removing only constraints or variables that were inactive during the p previous iterations, where p is some parameter.

References

[BJ77] M. S. Bazaraa and J. J. Jarvis, *Linear programming and network flows*, Wiley, New York, 1977.

[BJN+98] Cynthia Barnhart, Ellis L. Johnson, George L. Nemhauser, Martin W. P. Savelsbergh, and P. H. Vance, *Branch-and-price: column generation for solving huge integer programs*, Operations Research **46** (1998), no. 3, 316–329.

[DDSS95] J. Desrosiers, Y. Dumas, M. M. Solomon, and F. Soumis, *Time constrained routing and scheduling*, Handbooks in Operations Research and Management Science: Network Routing (M. O. Ball *et al.*, ed.), vol. 8, North-Holland, 1995, pp. 35–139.

[DFJ54] G. B. Dantzig, D. R. Fulkerson, and S. M. Johnson, *Solution of a large-scale traveling salesman problem*, Operations Research **2** (1954), 393–410.

[GG61] P. C. Gilmore and R. E. Gomory, *A linear programming approach to the cutting stock problem*, Operations Research **9** (1961), 849–859.

[HP85] K. Hoffman and M. Padberg, *LP-based combinatorial problem solving*, Annals of Operations Research **4** (1985), 145–194.

[JDD84] F. Soumis J. Desrosiers and M. Desrochers, *Routing with time windows by column generation*, Networks (1984), 545–565.

[JMMV98] B. Jaumard, O. Marcotte, C. Meyer, and T. Vovor, *Comparison of column generation models for channel assignment in cellular networks*, Les Cahiers du GERAD G-98-49, GERAD, September 1998.

[JMMV99] Brigitte Jaumard, Odile Marcotte, Christophe Meyer, and Tsevi Vovor, *How to include antenna spacing requirement in the channel assignment problem*, Unpublished Manuscript, 1999.

[JT98] Michael Jünger and Stefan Thienel, *Introduction to ABACUS - a branch-and-cut system*, Operations Research Letters **22** (1998), 83–95.

[Lee95] W.C.Y. Lee, *Mobile Cellular Telecommunications: Analog and Digital Systems*, 2nd ed., McGraw-Hill, 1995.

[Meh92] A. Mehrotra, *Constrained graph partitioning: Decomposition, polyhedral structure and algorithms*, Ph.D. thesis, Georgia Institute of Technology, Atlanta, California, 1992.

[Mil78] P. Miliotis, *Using cutting planes to solve the symmetric traveling salesman problem*, Mathematical Programming **15** (1978), 177–188.

[MT96] Anuj Mehrotra and Michael A. Trick, *A column generation approach for graph coloring*, INFORMS Journal on Computing **8** (1996), no. 4, 344–354.

[MT98] Anuj Mehrotra and Michael A. Trick, *Cliques and clustering: A combinatorial approach*, Operations Research Letters **22** (1998), 1–12.

[NP91] George L. Nemhauser and Sungsoo Park, *A polyhedral approach to edge coloring*, Operations Research Letters **10** (1991), 315–322.

[NW88] George L. Nemhauser and Laurence A. Wolsey, *Integer and combinatorial optimization*, Wiley, New York, 1988.

[PR91] Manfred Padberg and Giovanni Rinaldi, *A branch-and-cut algorithm for the resolution of large-scale symmetric traveling salesman problems*, SIAM Review **33** (1991), no. 1, 60–100.

[Thi95] S. Thienel, *ABACUS - A Branch-And-CUt System*, Ph.D. thesis, Universität zu Köln, 1995.

[VW96] François Vanderbeck and Laurence A. Wolsey, *An exact algorithm for IP column generation*, Operations Research Letters **19** (1996), 151–159.

GERAD & ncm$_2$ & MITACS, Département de mathématiques et de génie industriel, Succursale Centre-ville, Case Postale 6079, Montréal (Québec) Canada H3C 3A7

E-mail address: `brigitt@crt.umontreal.ca`

Ecole Polytechnique de Montréal, Département de mathématiques et de génie industriel, Succursale Centre-ville, Case Postale 6079, Montréal (Québec) Canada H3C 3A7

E-mail address: `christop@crt.umontreal.ca`

Ecole Polytechnique de Montréal, Département de mathématiques et de génie industriel, Succursale Centre-ville, Case Postale 6079, Montréal (Québec) Canada H3C 3A7

E-mail address: `tsevi@crt.umontreal.ca`

DIMACS Series in Discrete Mathematics
and Theoretical Computer Science
Volume **52**, 2000

On Mobility and Agents

Anupam Joshi

ABSTRACT. The ability to access networked resources, both computational and information, from mobile clients will become increasingly importance in the emerging pervasive computing scenario. There is significant ongoing work in the research community in this area. The objective of this paper is to argue that an agent oriented approach can be used to create systems to enable the goal of ubiquitous access. We show examples from existing work, including our own, where an agent abstraction has been used. We also outline our ongoing work to use a multiagent approach to enable the next generation, ad-hoc network based mobile applications.

1. Introduction

The trend of computing today is clearly towards an environment where computers are pervasive. This is enabled by two trends which are poised to accelerate in the near future. One is the shrinking form factor and cost, and the increasing MIPS of computers. This has allowed for affordable palmtop computers and desktop computers that have the power of supercomputers from less than a decade ago. It will lead to both high end petaflops machines as well as embedded computers in most engineered artifacts. These artifacts span the scale from roads and bridges to cars to even items of clothing. The second is the advent of high speed optical networking and RF wireless networking. The former permits extremely low cost – extremely high bandwidth network backbones and residential connections, the latter provides for universal connectivity. Related to these two are possible developments of micro sensors and actuators (MEMS).

Concurrently, with the advent of dynamic content (e.g. server parsed HTML, CGI) and executable content (e.g. Java), integration of security mechanisms, and emerging metadata standards, it is clear that the Web is transforming into the basis of a globally distributed computing and information access system. For example "Road Warriors" (business users with palmtop/laptop type devices) constitute a large and growing segment of users. Their work typically involves accessing and modifying corporate information repositories (with multimedia data) over low bandwidth networks (wireless, phonelines). More often than not, the applications they use are web enabled and use the browser as a "thin" client. Given the emergence of the Web as the "kernel" for distributed information and computation access, a

---

2000 *Mathematics Subject Classification.* Primary 68P20, 68N99, 68U35, 68T99.

This work was supported in part by an IBM faculty development award and NSF award IIS 9801711.

system to permit mobile web access can allow a variety of networked resources to be accessed from mobile clients.

This paper aims to discuss how agents can be used to solve the challenging problems in middleware and data management for mobile access. In particular, we will describe how agents have been used to support web access in the past, and outline our ongoing effort to create a framework that would support applications for next generation ad-hoc networks.

It is well understood in the mobile research community that the web (more specifically, the HTTP protocol) is designed to work in wired, high bandwidth environments, and does not operate particularly well when the access point is a mobile host[**JWM+96, JWH97, Kat94**]. The reasons behind this are quite well known (e.g. low/variable bandwidth, disconnections, etc.[**JWH97, FZ94**]). Further, the mobile host is typically resource poor. Even though the very high end laptop machines can now deliver performance comparable to low-end to moderate desktops, most "thin and light" subnotebooks and PDA type devices are constrained in terms of CPU, power, memory, disk, display capabilities etc.

An argument can be made that these bandwidth and resource related problems are transient, and eventually wireless network speeds as well as the resources available on mobile platforms will increase. That may well be the case, yet the speeds on wired networks and the resources (memory, CPU speed, etc.) on static hosts will also increase. As has been amply demonstrated over the last decade, software catches up and uses all available hardware resources – Existing (and novel) applications will evolve and become dependent on these enhanced capabilities. Thus while the absolute performance measures of mobile systems will undoubtedly improve, what we call the *bandwidth gap* and *resource gap* will remain.

2. Agents to Enable Mobile Web Access

The problems in web browsing caused by the low/variable bandwidth of wireless networks, and the mobile client's limited resources are compounded by the current model of web information access, where the user has to navigate the information space to find information that s/he needs. Bandwidth, a precious resource in the mobile scenario, is thus wasted by transmitting a lot of useless data across the wireless channel in the browsing process. This problem relates to *personalization* of the web space, *i.e.* locating the information needed by a user and possibly filtering it. Given the synchronous nature of the HTTP protocol, a continuous connection is also required throughout the information retrieval process. Disconnections typically require that the information be fetched again. More generally, the problem is a lack of *asynchronous operation* – *i.e.* not allowing the mobile user to disconnect in the process of information retrieval. Further, web servers have no knowledge about the client's resources and assume that it can handle the data sent by them. For example, a user may follow a hyper-link to a VRML world on a machine that does not have the capability of handling a VRML plugin. The client will retrieve the data from the server, and then not be able to use it. This results in consumption of available bandwidth and wastage of time while transferring data (and potentially even the plugin) across the wireless network – a *capability mismatch* in terms of handling content. Moreover, the mobile client expends battery resources in receiving this data, which it cannot use. With the ever increasing use of different types of multimedia content on the web, this phenomenon represents a growing problem.

2.1. Proxy Agents. There exists a large body of work, including our own, which handles the problem of *capability mismatch* for multimedia content in wireless web access. The solution, typically, has been to use a client-proxy-server model[**BJA98, FB96, BMMM96, NSN$^+$, JdT$^+$95**]. The proxy transcodes multimedia formats according to some predefined rules, usually in a manner that trades quality for bandwidth. This proxy has been viewed as an intelligent agent in our prior work[**JWH97**], as well as in the work of Brewer et. al.[**BKC$^+$98**]. The basic idea is that the proxy, given certain (often implicit) rules, monitors (or is otherwise provided) network QoS and knows about client side resources. It then autonomously transforms the data stream being sent to the client so as to best match the network QoS and client capability. Thus the proxy represents an agent embodying the weak sense of agency.

However, as PDA or thin-light notebooks become more popular, the proxy's functionality (and hence the computational power it needs) will increase. For example, some proxies now seek to deal with videos as well as images. This includes work done at Berkeley, as well as our own recent work[**BJA98**]. Our work has also examined other questions that relate especially to PDA type mobile clients, such as what to do with active content and HTML fonts / styles which the PDAs typically cannot handle. This adds functionality to the proxy and thus increases the computational load it places on the hardware. There are also proxy approaches that even render the HTML in a format appropriate for the PDA and then transmit that[**FGG$^+$98**], clearly adding to the computational burden at the proxy machine.

We feel that a purely proxy based solution will become increasingly non scalable, especially with the number of users connecting wirelessly expected to grow. The role of proxies has been recently questioned elsewhere as well – there is some debate as to whether proxy based solutions are really needed to provide networking services to mobile clients (cf. a panel on this theme at MobiCom98). Further, in ad-hoc networks which will be used in conjunction with Bluetooth/HomeRF enabled devices, it is not clear whether the client-proxy-server architecture is even applicable.

2.2. Agents to Negotiate End-End Performance. The alternative to proxy based approaches are end-end systems, where the server adapts the data stream it sent back based on some "negotiation" with the client. Such approaches are well known in networking and systems literature.

In the web context, dual versions (graphics heavy vs. text only) of web pages, with the user explicitly choosing between them, represent a simple form of the end-end approach . To the best of our knowledge, it was Seshan *et al.*[**SSK97**] who first presented the notion that the Web clients could use network performance parameters to download documents from a server at different "fidelities", and explicitly mentioned that this was something beyond text only pages. Implicit in their paper was the idea that the server would indeed have different fidelities to present to the client. Such a scenario goes against the basic assumption underlying proxy based transcoding, namely that the servers will have a document at a fixed "resolution" of a given medium.

Recent developments in multimedia coding also support an end to end approach. For example, new formats (e.g. wavelet based codes) that allow for efficient progressive transmittal of images and videos, and/or reconstruction of data at arbitrary resolutions[**VYZ$^+$99**] from the same bitstream are being developed.

Moreover, with disk storage increasingly inexpensive, it is not altogether unrealistic to expect that content providers will react to the market and make available multimedia content in multiple (fixed) resolutions. A recent example is the mapblast driving directions site (http://www.mapblast.com/), which provides driving directions for the in special formats for palmpilot and WinCE devices, in addition to the regular HTMLized format.

With an end-end approach, the server can also carry out more complex transformations (such as summarizing text, converting text to audio etc.) based on the client device needs. Of course, these may also be computed offline and stored. The end-end approach is predicated upon the client being able to "identify" itself to the server. There has been a very recent initiative (CC/PP from W3C, http://www.w3.org/TR/NOTE-CCPPexchange) to use RDF based HTTP extensions as a standard mechanism for this. The WAP forum (http://www.wapform.org) Wireless Application Environment also provides for mechanisms where state information allows a server to send content optimized for narrowband delivery to a client. Alternatively, the client must have a way to express its preferences for particular resolutions to the server. This is similar in concept to expressing preferences for languages or formats as envisaged in HTTP/1.1 . In our recent work[**BJA98, Jos99**], we have shown how HTTP/1.1's content negotiation mechanism can be leveraged to negotiate multimedia fidelity between the web browser and the server.

In more general terms, the end–end approach represents an instance where the client (agent) and the server (agent) negotiate to get the right content to the user. Unlike the one-shot approach of our existing work, we can envision a more complex negotiation process. The server could present the client with choices for the resource requested in terms of access cost, accuracy or fidelity of information, alternative resources in case the server does not have the requested resource etc. For example, consider the case of a laptop client interacting with a VOD server (clearly, this hypothetical scenario must wait for the bandwidths promised in 3G or 4G wireless networks). Suppose the user requests *www.blockbuster.com/Shakespeare_in_love.* The server could respond with a choice to either provide MPEG2 content (broadcast quality) at a certain price, or QCIF MPEG1 content at a lower price. If this film was not presently available (due to too many requests already being serviced), the server could suggest the movie *Elizabeth* as an alternative. The client could make a counter offer (willingness to pay a premium to get the movie now, requesting a reservation for a future time etc. etc.). In other words, agent based end-end approaches can facilitate a more comprehensive interaction between the mobile client and the appropriate servers. Given the disconnection related problems, we envision that the negotiation on behalf of the mobile client would be done by a proxy on the wired side.

2.3. Agent Based Information Personalization. The functionality of information personalization has, to the best of our knowledge, not been much explored in the context of supporting mobile web access. We have shown[**JWH97, KKW$^+$96**] how this can be used to minimize information flow on the wireless network and implemented it in a proxy agent called W^3IQ *(Web Intelligent Query).*

Broadly, its operation can be described as follows: The user connects to the proxy agent, makes a request for information, and disconnects. The proxy then tries to assemble a set of URLs that would satisfy the request, and returns them to the user. Should the user be disconnected, the results of his prior request(s) are made available at reconnection. These results are URLs that the system feels will contain the information the user requested. Users then prioritize the URLs provided by the system by rating them, depending on their perception of the relevance of the URL. The user feedback allows the system to infer a "semantic" match between the user's query terms and the information in the document. This helps the system create a user's profile which is used to control the information s/he would receive in response to future queries. The ranking information, the URLs, and the corresponding query are all stored by the W^3IQ proxy as metadata. Over time, then, the proxy builds a repository of ranked URLs which can be used to answer future queries by this user or other users. The system allows a proxy to pass the query to other W^3IQ proxies in the network, to garner any information related to the query in their local metadata. This allows all proxies to share information (specifically, ratings in regard to particular query terms) about URLs obtained by one proxy. When no W^3IQ agent has the needed information, the query is passed on to Lycos.

However, this functionality can also be taken away from proxies and passed on to servers. The idea that web sites (more generally, information sources) should "adapt" based on the user has been discussed in the last few years. Besides our own work on web mining[**JK98, NKJ99**], there is also Etzioni's description of adaptive web sites[**PE98**], and the Webminer project of Han *et al.*[**ZXH98**]. For instance, we have created a system that mines a server's access logs for patterns of traversal. Specifically, we determine browsing "sessions" (URLs from the site accessed by the same host within a defined timespan). We define a similarity metric between two sessions based on how much overlap there is between the URLs visited. When comparing URLs, this metric takes into account the document hierarchy within the server. Given the sessions and distance between them, we use relational fuzzy algorithms to cluster them. This provides associations between URLs visited – in other words URLs belong to the same cluster when they are visited in the same session. Given such information, the server can then use cookie based mechanisms to associate a user with an access pattern, and tailor the information sent back. Once a user is identified with particular information access patterns, the server can tailor the information returned in response to their requests. The same result can be achieved by involving the user directly in the process (for example, by providing rankings or other feedback), although clearly this is more burdensome for the user. An even more primitive form of this idea can be seen in most "personalized" search engines, where users are asked to explicitly provide keywords that match their interest.

3. Agents to create next generation mobile applications

In this section we outline our vision of how agents will be used to create new type of applications that will take advantage of the next generation wireless systems. There is an emerging paradigm shift in the way we thing about designing and building complex systems in general, and the way we approach the design of mobile aware applications in particular. We are moving away from the traditional discipline of developing a single, overarching design for a complex system in which all of the parts are carefully engineered to fit together and toward a new approach in which the individual atomic components are designed to be autonomous ("active"), self-describing ("articulate"), highly interactive ("social"), and adaptive ("intelligent").

These components can be viewed as agents, and a multiagent systems approach can provide the basis for creating a robust environment in which independent devices(from computers, PDAs, routers, printers to even home appliances and cars etc.) existing in a location can and will discover, interoperate, and cooperate with the other devices in their vicinity in an as-needed and as-desired basis. The underlying communication technology is presently being developed by consortia such as Bluetooth (http://www.bluetooth.com/) and HomeRF (http://www.homerf.org/). Note that while we propose these ideas in the context of wireless ad-hoc networks, they will also be applicable to wired networks.

An important enabling component of this vision is the work in distributed object-oriented systems in general, and Java/Jini in particular, which our agent oriented abstraction will build upon. Specifically, systems such as Jini which enable the formation of a community of devices at the syntactic level make it convenient to add "intelligence" to the system and create communities of "agents" which can then user richer communication languages (ACLs) to communicate information at a semantic level – information about abilities, requirements, commitments, intentions, beliefs, goals, etc. State of the art systems such as Jini provide for networked entities to advertise their capabilities. They are sufficient, for example, for a bluetooth enabled PDA to discover that there is a bluetooth enabled printer in its vicinity. They are probably not sufficient for the printer and the PDA to "negotiate" an agreement about whether and when the printer would print the material given by the PDA. They are certainly not enough for one PDA in the vicinity of another to realize that both their users are trying to find the same conference room in an unfamiliar building, and agree to cooperate in the process. Clearly, the enabling of such ad-hoc collaboration leading to "teams" of devices cooperating to solve common and individual problems will allow for a new class of mobile applications to evolve.

Mechanisms to form ad hoc teams: A key to the realization of our vision is the ability for processes and systems at the application level to be able to easily cooperate and form ad hoc teams. We are building on our earlier work on the theory and practice of agent based systems and focus new work on supporting agent teams with rich negotiation languages and strategies. The DARPA KSE has given rise to a model for developing intelligent cooperating agents based on ontologies, content languages and agent communication languages. Although this approach and its associated components are being adopted as a standard for agent communication by both the research community (KQML, KIF) and by industry (FIPA), it requires significant extensions to support the requirements of the pervasive computing scenario outlined above. Some of the issues to be addressed include:

- Developing a language for defining higher-level protocols for peer-to-peer communication. This new language will build on existing network protocol formalisms to define protocols and associated agent behavior, providing a powerful yet intuitive and pragmatic language with sound theoretical foundation for formal analysis. The unit of communication will be the conversation, which involves not only the ordering and interdependency among exchanged messages, but also proper agent behavior such as actions, constraints and conditions associated with actions, as well as alternative actions and tradeoffs.

- Developing a rich set of negotiation primitives able to support various types of negotiations involving different agents (cooperative, competitive, or a mix of both) and in different fashion (peer-to-peer or multi- party). These primitives will be based on the syntax and semantics of the FIPA Agent Communication Language, ensuring their consistency with emerging software standards.
- Developing prototype communication facilitators and brokers that learn the capabilities and interests of other agents through several active strategies. These include monitoring the conversations, both successful and unsuccessful between agents, performing generalization and specialization reasoning to predict future behavior, capturing uncertainty of agent capabilities by neural networks, fuzzy logic, and probabilistic learning, and initiating experiments to refine its meta data about the interests and capabilities of key agents. In our earlier work, we have shown how a network of such broker type agents can form a recommender system to select appropriate software for scientific computing type tasks[**JRH98**]. In other words, the system accepts a high level definition of the task from the user, and chooses appropriate software components needed to solve the task.

Component oriented support for computational tasks: As mentioned earlier, the individual components (software as well as hardware) that make up the networked computational system can be viewed as agents. We are developing a runtime environment based on a MultiAgent Systems approach that permits mobile computers to access not just information, but also computational resources, across the wireless network.

Unlike some recently proposed software component based systems, we will incorporate both software and hardware components. Software components will be mobile (in the sense of code migration) as well as static, and the hardware components will be adaptive (in the sense of the DARPA ACS program), mobile (in the sense of physical movement) as well as fixed. The runtime environment will accept from the end user a high level description of the task as well as some performance constraints and use a recommender system to automatically and "on-the-fly" map this to the appropriate set of available software components needed for the task and hardware components best suited to run them. The synthesis process will be a result of a sequence of active negotiations between the components and the runtime environment to compose the needed software. Computational tasks will be partitioned across a heterogeneous set of computational resources. This partitioning explicitly incorporates the asymmetry of the system which will have low resource systems such as handheld devices as well as high performance wired machines.

In order to achieve this, we will need to build a fundamental descriptive framework which can capture (i) the semantics of the services that each component agent in our system offers, and (ii) its own requirements in order to perform its declared tasks. We are developing an XML based scheme to encode these features in order to leverage the work on this developing standard, as well as an ontology suitable for our proposed application domains.

Initial prototypes: In our preliminary work, we have modified the Aglets mobile agent system so that each (mobile) agent posses an XML based description of the resources it needs (CPU type, software etc) to execute. Similarly, each aglets server (Tahiti) has an XML based description of the resources it has. When an agent is asked to migrate to a remote site (a dispatch, in Aglet terms), it instead sends a lightweight agent with its requirements to the remote site. This agent compares the resources existing at the remote platform with what the agent to be migrated would need, and then returns back. It then either tells the agent to migrate (if there are adequate resources at the remote end), or stops the migration.

We have also shown how a proxy agent based approach can be used by a mobile client to access remote computational resources[**Tri99**]. Our preliminary system deals with the task of landmine detection and remediation. In particular, we envisage that soldiers seeking to detect landmines will be equipped with Sensors (chemical, Ground Penetrating Radar etc.) connected to laptop type wearable computers which will be wirelessly hooked to one another and to nearby truck mounted high end machines. The novel algorithms being developed to detect landmines typically need more computational power than can be expected from the computers that the soldiers carry. Our system allows the mobile client to specify a task (such as detect mines) to the proxy agent and send it associated data. The proxy then locates a remote server providing the appropriate functionality and sends it the data. The server returns the results to the proxy, which then sends it back to the mobile client. Note that the proxy agent functioned as a broker here – the mobile client did not need to know where to find the service. Of course, the proxy also provided services such as compressing the data stream on the wireless side etc.

4. Discussion

As we approach the turn of the century, mobile devices and wireless data networks are becoming more powerful and affordable. This has lead to the growing importance of mobile access and data management. Mobile technology also seeks to expand the scope of networked computing infrastructure and allow mobile clients to participate in networked applications.

However, existing information systems and distributed applications often make inherent assumptions about the connectivity and topology of the underlying networks, as well as the capability of client devices, that are not valid in mobile environments. These assumptions, which are well understood by the research community, present significant difficulty in meeting nomadic users' expectations of having applications operate seamlessly across wireless and wired networks. While there has been work done to alleviate these problems, it remains nascent. Moreover, developments in ad-hoc networks tend to undermine the proxy oriented model which underlies much of this work.

With rapid advances in wireless communication and networking, and the anticipated deployment of 3G/4G wireless systems, it is our opinion that mobile access and data management will become an increasingly important research area. In this paper, we have discussed how an agent oriented approach can help design solutions. We have illustrated this using existing proxy based approaches to mobile web access, and shown how agents also provide the framework for end–end approaches. We have also outlined our ongoing work in creating a framework and runtime environment, using multiagent approaches, to support mobile information and computation access in next generation/ad-hoc wireless environments.

References

[BJA98] Harini Bharadvaj, A. Joshi, and Sansanee Auephanwiriyakyl, *An active transcoding proxy to support mobile web access*, Proc. IEEE Sumposium on Reliable Distributed Systems, October 1998.

[BKC+98] E.A. Brewer, R.H. Katz, Y. Chawathe, A. Fox, S.D. Gribble, T. Hodes, G. Nguyen, T. Henderson, E. Amir, H. Balakrishnan, A. Fox, V. Padmanabhan, and S. Seshan, *A network architecture for heterogeneous mobile computing*, IEEE Personal Communications Magazine **5** (1998), no. 5, 8–24.

[BMMM96] C. Brooks, M. S. Mazer, S. Meeks, and J. Miller, *Application-specific proxy servers as http stream transducers*, Proc. WWW-4, Boston, http://www.w3.org/pub/Conferences/WWW4/Papers/56Application-Specific, May 1996.

[FB96] A. Fox and E. A. Brewer, *Reducing www latency and bandwidth requirements by real-time distillations*, Proc. Fifth International World Wide Web Conference, May 1996.

[FGG+98] A. Fox, I. Goldberg, S.D. Gribble, D.C. Lee, A. Polito, and E.A. Brewer, *Experience with top gun wingman: A proxy-based graphical web browser for the usr palmpilot*, Proc. IFIP International Conference on Distributed Systems Platforms and Open Distributed Processing (Middleware '98), 1998.

[FZ94] G. Forman and J. Zahorjan, *The challenges of mobile computing*, IEEE Computer **27** (1994), 38–47.

[JdT+95] A. D. Joseph, A. F. deLespinasse, J. A. Tauber, D. K. Gifford, and M. F. Kaashoek, *Rover: A Toolkit for Mobile Information Access*, Proc. 15th Symposium on Operating Systems Principles, ACM, December 1995.

[JK98] Anupam Joshi and R. Krishnapuram, *Robust fuzzy clustering methods to support web mining*, Proc. ACM SIGMOD Workshop on Research Issues in Data Mining and Knowledge Discovery (S. Chaudhuri and U. Dayal, eds.), June 1998.

[Jos99] Anupam Joshi, *On proxy agents, mobility, and web access*, Tech. report, CSEE Department, UMBC, 1999.

[JRH98] A Joshi, N. Ramakrishnan, and E. N. Houstis, *MultiAgent Support of Networked Scientific Computing*, IEEE Internet Computing **2** (1998), no. 3, 69–83.

[JWH97] A. Joshi, S. Weerawarana, and E. N. Houstis, *Disconnected Browsing of Distributed Information*, Proc. Seventh IEEE Intl. Workshop on Research Issues in Data Engineering, IEEE, April 1997, pp. 101–108.

[JWM+96] A. Joshi, R. Weerasinghe, S. P. McDermott, B. K. Tan, G. Benhardt, and S.Weerawarna, *Mowser: Mobile platforms and web browsers*, Bulletin of the IEEE Technical Committee on Operating Systems and Application Environments **8** (1996), no. 1.

[Kat94] R. Katz, *Adaptation and Mobility in Wireless Information Systems*, IEEE Personal Communications **1** (1994), no. 1, 6–17.

[KKW+96] R. Kavasseri, T. Keating, M. Wittman, A. Joshi, and S. Weerawarana, *Web Intelligent Query - Disconnected Web Browsing using Cooperative Techniques*, Proc. 1st. IFCIS Intl. Conf. on Cooperative Information Systems, IEEE, IEEE Press, 1996, pp. 167–174.

[NKJ99] O. Nasraoui, R. Krishnapuram, and A. Joshi, *Mining web access logs using a fuzzy relational clustering algorithm based on a robust estimator*, to appear in Proc. WWW8, August 1999.

[NSN+] B. D. Noble, M. Satyanarayanan, D. Narayanan, J. E. Tilton, J. Flinn, and K. R.Walker, *Agile application-aware adaptation for mobility*, Proceedings of the 16th ACM Symposium on Operating System Principles.

[PE98] M. Perkowitz and O. Etzioni, *Adaptive web sites: Automatically synthesizing web pages*, Proc. National Conference of the American Association of Artificial Intelligence (AAAI), 1998.

[SSK97] S. Seshan, M. Stemm, and R. Katz, *Spand: Shared passive network performance discovery*, Proc. 1st Usenix Symposium on Internet Technologies and Systems (USITS '97, 1997.

[Tri99] Sunil Tripathy, *Distributed mobile system for landmine detection*, Masters Project Report, University of Missouri-Columbia, 1999.

[VYZ+99] J. Vass, J. Yao, S. Zhuang, A. Joshi, and X. Zhuang, *Multimedia information access in wireless environments*, Tech. Report TR-9901, CECS Department, University of Missouri, 1999.

[ZXH98] O.R. Zaine, M. Xin, and J. Han, *Discovering web access patterns and trends by applying olap and data mining technology on web logs*, Proc. Advances in Digital Libraries Conf. (ADL'98), April 1998, pp. 19–29.

Department of Computer Science and Electrical Engineering, University of Maryland Baltimore County, Baltimore, MD 21250

E-mail address: joshi@cs.umbc.edu

DIMACS Series in Discrete Mathematics
and Theoretical Computer Science
Volume **52**, 2000

Multiplexed Serial Wireless Connectivity for Palmtop Computers

*Ibrahim Korpeoglu**, *Pravin Bhagwat,*
Chatschik Bisdikian, Mahmoud Naghshineh

*Computer Science Department
University of Maryland
College Park, MD 20742

IBM T. J. Watson Research Center
P.0.Box 704
Yorktown Heights, NY 10598

ABSTRACT

Palmtop computers which consume very little power, have very small size, and are much cheaper compared to their laptop counterparts are proliferating in the market. However, there exist no satisfactory wireless connectivity solution in terms of cost and power consumption to connect these devices to the network. This paper describes a cost-effective, low-power wireless connectivity solution for these devices and presents the design and implementation of how multiple such devices are connected to an access point over a shared wireless link.

1 Introduction

Palmtop computers are mostly used in disconnected mode for storing address book information, recording appointments, taking notes, etc. Network-based interactive applications have not yet proliferated. The most commonly used networked application is (network) hotsync, which synchronizes the data in a PDA with a desktop computer through an RS-232 serial cable or through modem connected telephone lines (figure 1-a,b). The main reason for unpopularity of networked applications is the lack of good wireless connectivity solutions. Existing wireless solutions are not suitable for networking palmtop devices.

The networking requirements of PDAs differ from those of laptop and desktop computers. Unlike laptop computers, PDAs are designed to operate within strict cost, space, and power budget. Therefore, any communication solution for PDAs need to have similar properties in terms of cost, battery efficiency and form factor. Secondly, since most PDAs only support RS-232 serial interface to communicate, traditional PCMCIA based wireless LAN cards cannot be used.

Although various indoor and outdoor existing wireless network technologies can be adapted for enabling wireless access for PDAs, none of these wireless technologies provide a cost effective solution. Wide area solutions such as CDPD, ARDIS, RAM are not suitable for indoor use. Metricom's Richochet wireless modem design is an attractive alternative. However, Ricochet is intended to be a wide area wireless service, not a plug-n-play, build-your-own-kind of a wireless system. It is fee-based, therefore is not economically attractive, and does not provide a low-cost method of connecting a PDA directly to a LAN. A pair of Ricochet modems can be used to form a wireless RS-232 link, which can be used to connect a PDA directly to a PPP server (figure 1-c). This solution, however, has three limitations: multiple users cannot access the network simultaneously; roaming is not supported; and the configuration is expensive. A Richochet modem also consumes more power than a PDA.

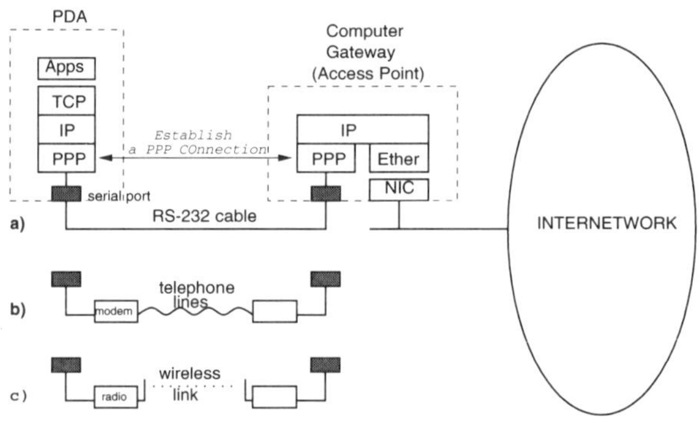

Figure 1: Several current ways to connect a PDA to an internetwork over point-to-point links.

To address the above issues, we designed a new wireless access network (called BlueSky) for PDAs considering low cost, low power consumption, and light-weight protocols as the primary objectives. In this paper, we will first briefly describe the BlueSky architecture, the entities and the protocols. Then we will present our design and implementation to enable multi-access capability for PDAs to share a single wireless link that connects the devices to an access point.

2 BlueSky Approach to Wireless Connectivity

The BlueSky system consists of wireless *BlueSky attachments* that are connected to PDAs and provide short range wireless access to the wired network backbone through the *BlueSky access points* [1]. Multiple PDAs can share a wireless link to an access point using the BlueSky wireless MAC protocol (see figure 2).

A power-conscious, polling based, asymmetric MAC protocol is designed for the wireless link [2]. The portion of the MAC protocol that is running on an access point, called AP-MAC, is more intelligent and coordinates the access to the medium by sending commands (polls) to the PDAs in the range. It includes a scheduler component which schedules MAC-level events such as transport-payload, poll-and-receive-payload, invite-stations, etc. The corresponding MAC protocol that is running on a PDA attachment, called RS-MAC (remote station MAC), just responds to the commands sent from the AP-MAC and is much simpler, requires much less memory and processing power, and has smaller code size than the AP-MAC. The protocol supports sleep modes to conserve energy.

The wireless attachment has been designed as an external attachment so that it can be plugged into any PDA, laptop, or serial port enabled device. It connects to the serial port of a device. It has a radio transceiver component and a microcontroller component. These two are also connected internally through a serial line. Since we did not want to modify the PDA hardware and software to implement the RS-MAC protocol, the microcontroller part is needed to implement it. The RS-MAC layer packetizes the bytes received from a PDA through the serial port and sends them to an access point inside MAC frames.

Since the cost of the radio transceiver is an important factor in the overall cost of the BlueSky attachment and access point, we chose a cost effective radio transceiver which runs in ISM 915 MHz band and uses simple FSK modulation similar to what is used in cordless phones [3]. The attachment and access point can select one of the ten available frequency channels. The link bandwidth is 150 Kbps and the effective range is between 20-30 m depending on the interference and multi-path effects. The bandwidth can be shared among approximately 20 mobile devices.

An access point consists of a PC-104 board (with a 386 CPU on it), a radio transceiver, and an Ethernet card to connect to the internetwork. The AP-MAC protocol is implemented as part of the other access point software, hence a separate microcontroller is not needed. The radio transceiver is connected to the access point board through a serial port. (see figure 3). The byte streams received by the radio unit is given to the AP-MAC protocol through the serial port.

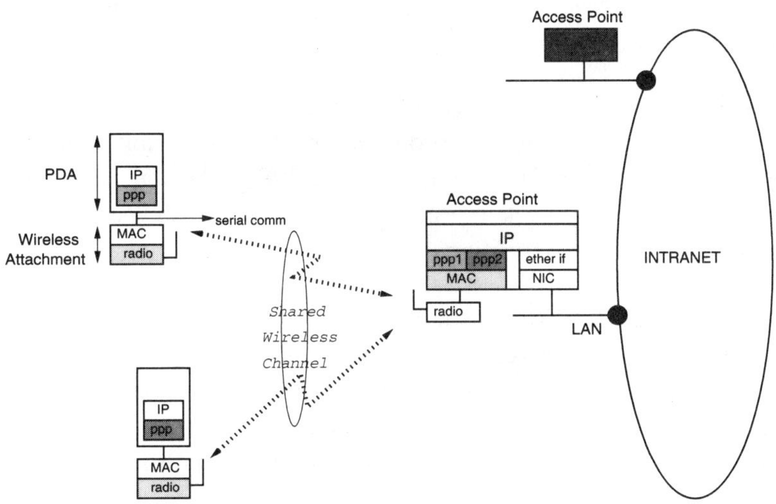

Figure 2: BlueSky architecture

A Unix operating system BSDI/OS 3.0 runs on an access point to implement the AP-MAC protocol, the PPP protocol [4], and the IP stack [5]. The PPP protocol connects a PDA to the access point, and the IP stack connects the access point to the rest of the internetwork. Multiple PDAs can establish PPP connections to the same access point at the same time. PPP protocol exports a packet-oriented network interface to the IP layer so that IP can also run over serial communication lines. There needs to be a separate PPP network interface/server running for each PDA that established a PPP connection to the access point. The challenge is putting an AP-MAC layer under these multiple PPP interfaces (see figure 2) and emulating a serial line communication over this packet-oriented MAC layer. The traffic going to multiple PDAs from multiple PPP interfaces is multiplexed at the AP-MAC layer and then transported over the shared wireless link. Similarly, the traffic coming to the access point from multiple PDAs through the coordination of the AP-MAC layer is demultiplexed to multiple PPP interfaces.

3 Multiplexing PPP Traffic at the MAC Layer

The PPP implementation in BSD Unix consists of two parts (see figure 3):

- *An asynchronous PPP driver* inside the kernel which implements the PPP framing protocol and byte stuffing algorithm. PPP driver is located under the IP layer and exports a network interface to the IP layer. A PPP packet received from the serial line may be a PPP data packet or a PPP control packet (they are distinguished by the protocol identifier field in the PPP header). A PPP data packet contains the payload for the higher network layer (IP in this case). The PPP driver gives the payload to the IP layer after striping off the PPP header and CRC. A PPP control packet carries PPP specific information parameters and is given to the corresponding PPP server daemon after again striping off the PPP header and CRC.

- *A PPP (server) daemon* that runs in user space and implements the PPP control protocols. These control protocols include LCP (link control protocol), NCP (network control protocol), etc. All PPP packets generated by the PPP daemon are control packets that are used to establish, negotiate, maintain, and terminate PPP connections.

An asynchronous PPP driver usually runs on top of a serial communication port driver. In Unix, each port driver (also called terminal device driver) has a line discipline associated with it [6, 7]. A line discipline is located on top of the terminal device driver and has functions such as echoing characters back, assembling characters into lines (a line is a string of characters ending with a newline character), editing the lines, processing the flow control characters, etc. For PPP, these functionalities are not needed, and therefore a line discipline in this sense is not needed. The asynchronous PPP driver acts as a line discipline for PPP.

There is only one actual terminal device driver for a serial port, which can be used by a single PPP server. But, what we need is to be able to run multiple PPP servers and multiplex their traffic at the underlying MAC layer which then transmits the packets one-at-a-time to the corresponding PDAs through the serial port.

In order to achieve this, multiple instances of PPP servers are run on top of virtual communication ports (pseudo-terminal devices) instead of being run on top of an actual serial port hardware. In Unix, a pseudo-terminal device is a virtual

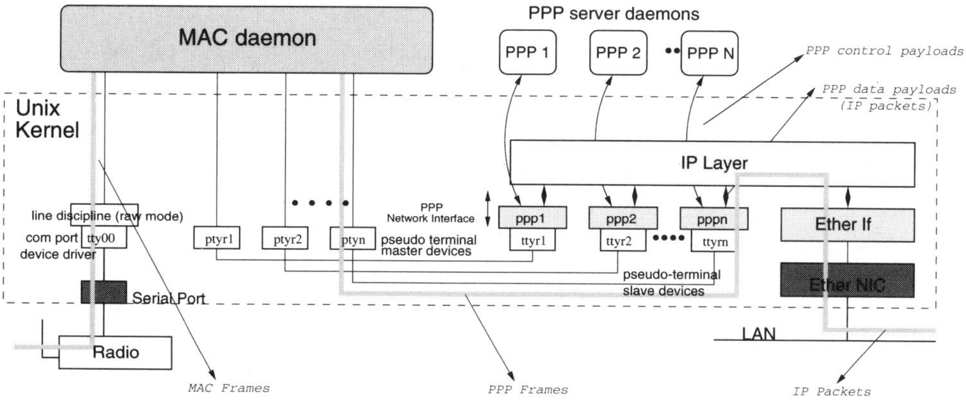

Figure 3: BlueSky access point software architecture

device that is not associated with any physical serial port hardware, but emulates serial line communication. It consists of a pair of devices: one half is called the master and the other half is called the slave.

A process can open a pair of pseudo-terminal devices and can get two file descriptors. Anything written to the master device can be read as input from the slave device and anything written to the slave device can be read as input from the master device. Similarly, two processes can open one half of a pseudo-terminal device pair. In this case, these two processes can communicate with each other using the pseudo-terminal device pair, as if they are communicating through a serial port: they can set the line speed, etc. The slave portion of a pseudo-terminal device presents an interface to the user process that looks like an actual terminal device. It therefore has also a line discipline associated with it.

In our implementation, a PPP server/interface is run on top of a pseudo-terminal slave device. The line discipline associated with the slave device is the PPP framing protocol (see figure 3). Anything that is sent from a PPP interface to the pseudo-terminal slave is given to the respective master device. We implemented the AP-MAC protocol as a user level daemon that can read from (or write to) this master device and thereby communicate with the corresponding PPP side. Up to 64 pseudo-terminal device pairs can be opened at the same time (this number depends on the operating system and its kernel configuration). Therefore up to 64 PPP servers/interfaces can be active at the same time, each one using a different pseudo-terminal slave device. The MAC daemon can open the respective master devices and can communicate with these PPP interfaces. A PPP frame received from a PPP interface by the MAC daemon has the same format as it would have been received at an actual serial port driver. In this way, the traffic from multiple PPP interfaces are multiplexed at the MAC layer.

The MAC daemon communicates with the radio transceiver through a serial port. The bytes received from the serial port are assembled to MAC frames. The extracted payloads, which are PPP packets, are given to the respective PPP interface through a pseudo-terminal device. On the other side, the PPP packets received from a PPP interface are first queued at the MAC daemon. There is a separate queue for each PDA. The scheduler component of the MAC daemon selects the PPP packet for transmission. The selected packet is then transmitted to the respective PDA inside a MAC frame.

4 Conclusion

Palmtop computers which consume very little power, have very small form factor, and are relatively much cheaper compared to their laptop counterparts are proliferating in the market. We described a cost-effective, low-power wireless connectivity solution for these devices and presented our implementation of a multi-access scheme to connect multiple such devices to a single access point over a shared wireless link. The design and implementation of the wireless system shows how existing hardware and software pieces can be utilized to provide a very low-cost wireless connectivity solution with least modifications to the existing palmtop device hardware and software.

References

[1] P. Bhagwat, C. Bisdikian, I. Korpeoglu, M. Naghshineh, and S. K. Tripathi, "Cordless Dialup Networking for Palmtop Computers," Tech. Rep. RC21404(96651), IBM Research, February 1999.

[2] C. Bisdikian, P. Bhagwat, B. P. Gaucher, F. J. Janniello, I. Korpeoglu, M. Naghshineh, and P. Pandoh, "WiSAP: A Wireless Personal Access Network for Handheld Computing Devices," *IEEE Personal Communications*, December 1998.

[3] P. Bhagwat, C. Bisdikian, I. Korpeoglu, A. Krishna, and M. Naghshineh, "System Design Issues for Low-Power, Low-Cost Short Range Wireless Networking," in *IEEE International Conference on Personal Wireless Communications (ICPWC)*, February 1999.

[4] W. Simpson, "The Point-to-Point Protocol (PPP)," July 1994, RFC 1661.

[5] W.R. Stevens, *TCP/IP Illustrated, Volume 1*, Addison-Wesley, 1994.

[6] W. R. Stevens, *Unix Network Programming*, Prentice Hall, 1990.

[7] M.K. McKusick, K. Bostic, M.J. Karels, and J.R. Quarterman, *The Design and Implementation of the 4.4BSD Operating System*, Addison-Wesley, 1996.

DIMACS Series in Discrete Mathematics
and Theoretical Computer Science
Volume **52**, 2000

A Cluster-based Checkpointing Scheme for Mobile Computing on Wide Area Network

Jeng-Ping Lin, Sy-Yen Kuo, and Yennun Huang

Department of Electrical Engineering, Rm 415
National Taiwan University
Taipei, TAIWAN

TEL: (886-2)2363-5251
FAX: (886-2)2363-8247
Email: sykuo@cc.ee.ntu.edu.tw

Abstract

A mobile computing system is a distributed system which consists of mobile and stationary nodes, connected to each other by a communication network. The location of mobile computers may change with time. In such systems, wireless communication channels are less reliable than static communication channels. Transient faults due to power or connectivity problem happen frequently. Conventional fault-tolerant techniques need to be modified so that they could be applied in these systems. If a system is spread over a large network such as WAN, even a remote node fails could impact many nodes or the entire system. Therefore, decomposition of a large network into a set of smaller networks (clusters) is useful since a failed node is isolated in a cluster. Clustering can also be used to facilitate conventional checkpointing techniques in a mobile environment. In this paper, we use cluster-based schemes to decompose the whole network into smaller clusters. In the smaller cluster, we propose a checkpointing scheme, which handles fixed nodes and mobile nodes separately to account for their intrinsic difference in reliability.

Keywords: mobile computing, decomposition, cluster, checkpointing scheme

---

Yennun Huang is with Bell Laboratories, Lucent Technologies, Inc.
This research was supported by the National Science Council under Grant NSC 88-2213-E-002-040.

1. Introduction

Recent advances in technology have provided portable computers with wireless interfaces that allow computers to talk to each other even when it is in motion [19]. As the portable computers gain popularity and the wireless communication technologies advances, mobile computing has become more and more important. A mobile computing system is composed of a mobile part and a static (or fixed) part. The static part contains the computers called fixed hosts (FHs) and the static communication network connecting them, while the mobile part includes the mobile computers called mobile hosts (MHs) and their wireless communication channels consist of the mobile part. Some FHs are augmented with wireless interface to communicate with MHs through wireless links. They are referred to as mobile support stations (MSSs) and act as access points to the static network for the MHs (see Figure 1). Each MSS has a coverage area called a cell, and MHs located in the cell can

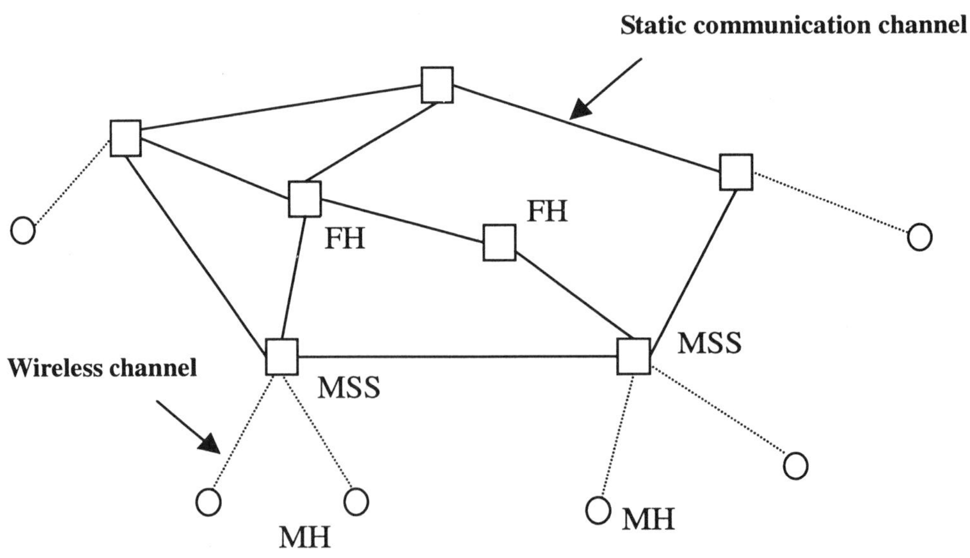

Figure 1. A mobile computing system

directly communicate with the MSS through a wireless channel. A MH can be directly connected to at most one MSS at any given instant of time. However, an MH can migrate from one cell to another. Therefore, the local MSS for a MH may change with time and a wireless channel between a MH and a MSS is referred to a dynamic channel.

Checkpointing and message logging are two fault-tolerant techniques commonly used in real-world applications to reduce service downtime. Roughly speaking, if a node restarts after a failure, each of the processes at the different nodes is rolled back to an earlier state using checkpoint saved in stable storage, so that the recovered global state is consistent [1-3]. In synchronous checkpointing algorithms, there is one and only one consistent snapshot of the system maintained all the time. In asynchronous algorithms, all the nodes take their local snapshots independently. At the time of recovery, a local snapshot must be chosen to build a consistent snapshot of the system.

If message logging is used, each node logs incoming messages to stable storage. Whenever a failure occurs, the message logs are used to compute a consistent global state [13,15]. Message logging algorithms can be divided into two classes: pessimistic logging and optimistic logging. In pessimistic message logging, every message received is logged to stable storage before it is processed [4]. In optimistic message logging, messages received are logged to stable storage asynchronously [5,14]. In other words, message processing can happen before message logging. It imposes less failure-free overhead than pessimistic logging, but some extra work must be done to ensure the recovered states are consistent.

Many snapshot collection algorithms have been developed for static distributed systems [3,6-10]. However, they can not be directly applied to mobile computing systems. In a mobile environment, a MH can disconnect from its local MSS or migrate from one MSS to another. Thus, the topology of the network is dynamic and existing static snapshot collection algorithms can not be used without modification. In addition, checkpointing and message logging require a stable storage for saving state information. Previous papers [16,17] let mobile hosts take their local snapshots in their disk storage. However, the disk storage of a mobile host can not be considered stable and is vulnerable to damages such as dropping [11,12]. Furthermore, some MHs may not be equipped with enough volume of storage disk. This means that the local snapshot of a MH should not be saved on the disks of the MH but should be saved on a FH such as a MSS instead. These issues must be incorporated to design snapshot collection algorithms for mobile computing systems.

Most papers model the system faults as node faults [16,18]. Link faults can be modeled as node faults under some transformation. However, there is a significant difference between the mobile nodes and the fixed nodes in mobile computing systems with respect to reliability. The FHs and the static communication channels are more reliable than the MHs and the wireless channels. The FHs are statonary and do not have an energy consumption problem. There are types of failures that were rare in fixed environments but common with mobile hosts. For example, MHs are vulnerable to a physical damage such as dropping. Besides, transient failures due to power or connectivity problem may happen frequently. In terms of node faults, MHs are different from FHs also. In our paper, we use different protocols to deal with MHs and FHs. Message logging is used on MHs to isolate them from the FHs. This makes checkpointing and recovery on mobile computing systems more efficient.

Another problem occurs when the size of a mobile computing system becomes too large. If synchronous checkpointing is used, any node failure would require the whole large system to rollback. Even if asynchronous checkpointing is used, it may still involve a lot of nodes to rollback. Therefore, it is desirable to partition the whole network into a set of smaller clusters [1]. A natural partition is to decompose the whole network into LANs. There are fixed hosts and mobile hosts in a LAN. The LAN could be decomposed into different clusters to utilize their characteristics.

The rest of the paper is organized as follows. Section 2 presents the system model. In Section 3, we discuss the hierarchical structure to decompose the system to appropriate clusters. Section 4 presents the proposed checkpointing scheme and its restart protocol. Section 5 presents the conclusions.

2. System Model

The mobile system is modeled by a triplet S = (**FH**, **MH**, **CH**). **FH** is a set of static hosts. A Mobile Support Station (MSS) is a special static host which acts as access points for MHs to the static network. **MH** is a set of mobile hosts (MHs), and **CH** is a set of static channels between FHs. Each channel in **CH** is reliable and supports FIFO message communication in both directions. **FH** and **CH** consist of the static network. No assumption is made on the topology of the static network except that it is connected. There is no direct channel between MHs. The MSS forwards messages transmitted from its local MH to the destination hosts and delivers message from other hosts to its local MH. An MH can be directly connected to at most one MSS at any given time. When an MH is connected to an MSS, the communication channel between them also supports FIFO message communication in both directions. An MH is allowed to be connected to no MSS during the movement, but it would connect with one of the MSSs within finite time.

Application in mobile computing systems consists of processes run on different nodes in the system. There are no common clock and shared memory in the system. All the nodes communicate with each other through messages. Each node operates independently of the others, with occasional asynchronous message communication. Message transmission through these links takes an unpredictable, but finite amount of time. For the application to run successfully, all the participating nodes must function properly. Node failures in the system are assumed to be fail-stop.

3. Network Decomposition

Previous discussion restricts the mobile computing system mainly on a small network such as a local area network. If the mobile system grows too large to fit in a LAN, some assumptions and environments may change. For example, message transmission latency becomes variable and thus synchronous checkpointing protocol is not appropriate. The checkpoint initiator may spend too much time to wait for the reply message. This will increase the failure-free overhead and reduce the system service quality.

The concept of hierarchy is widely used in the computer world. Since a large network is beyond our control, it would be natural to decompose it into smaller clusters. Messages for hosts in other LAN must pass the router, while messages for hosts in the same LAN are broadcast (see Figure 2). As a result, messages transmitted to hosts outside the LAN have variable latency. Therefore, a natural selection for a cluster is a LAN. To make these clusters independent from each other, pessimistic logging is used to log messages received from other clusters. In the TCP/IP network, each message has its sender IP address. A mask could be used to judge whether a message belongs to the same cluster or not.

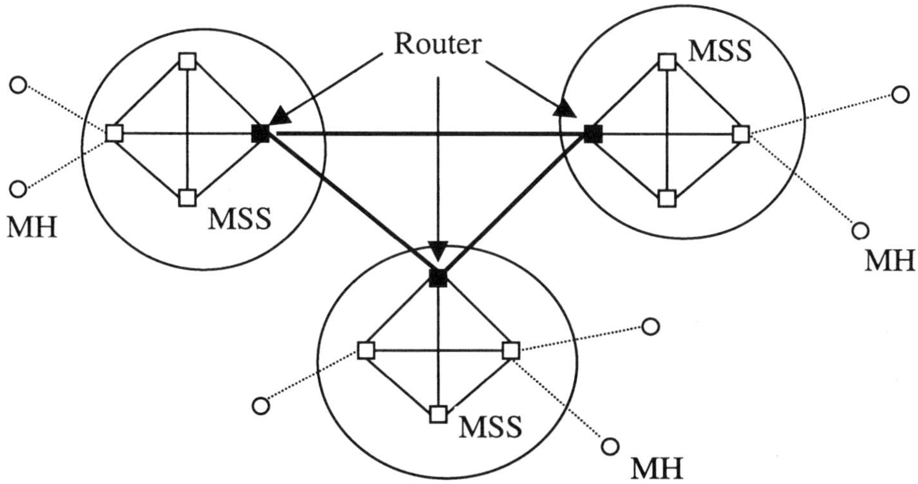

Figure 2. Decomposition of a large mobile computing system on WAN

In the LAN cluster, there are fixed hosts and mobile hosts. However, their characteristics are quite different. The fixed hosts won't change location with time and the static channels connecting them are reliable and have higher bandwidth. The mobile hosts are carried with users while they are moving. Physical damages such as dropping become probable. The mobile host may not have enough power or disconnect suddenly. These transient faults may happen frequently. Therefore, a single checkpointing protocol is not suitable for the two kinds of hosts at the same time. The concept of clusters could be used again. The system is first divided into two clusters: the static cluster and the mobile cluster. The static cluster is composed of the fixed hosts and the static channels. However, there are no direct channels between the mobile hosts. The mobile hosts could not be viewed as a single cluster. The best partition is a single cluster for a single mobile host. Therefore, a mobile cluster is the same a mobile host in the rest of the paper. A LAN is partitioned into the static cluster and the mobile clusters (see Figure 3).

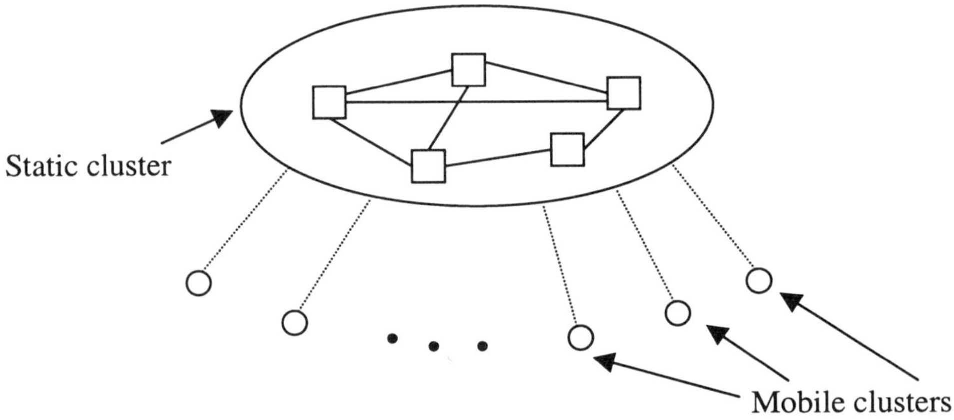

Figure 3. Decomposition of a LAN

4. Checkpointing Schemes

Asynchronous checkpointing protocols save states of processes without exchanging messages. This characteristic is useful for mobile computing systems. The static hosts won't have to wait for the mobile hosts to exchange messages when mobile hosts are disconnected or the wireless links fail temporarily. However, processes need to exchanges messages during recovery to find consistent global states or to obtain state information stored by other processes. On the other hand, most synchronous checkpointing protocols exchange messages when saving the application's checkpoints. Thus, they are not suitable for mobile environments.

Log-based rollback recovery [13,15,20] is widely used in distributed applications to achieve fault tolerance. It is based on *piecewise deterministic (PWD) execution model* [5]: process execution is divided into a sequence of state intervals each of which is started by a nondeterministic event such as message receipt. Message logging algorithms can be divided into two classes: pessimistic logging and optimistic logging. In pessimistic message logging, every message received is logged to stable storage before it is processed [4]. It guarantees that no process failure will revoke any message to force its receiver to roll back. It has the advantage of localized recovery. In optimistic message logging, messages received are first saved in a volatile buffer and later written to the stable storage [5,14]. In other words, message processing can happen before message logging. It imposes less failure-free overhead than pessimistic logging, but some extra work must be done to ensure the recovered states are consistent.

From the above analysis, we find that synchronous checkpointing protocols require the least volume of stable storage and have fast system recovery time. They are suitable for the static cluster but not the mobile clusters. On the contrary, log-based asynchronous checkpointing protocols are suitable for the mobile clusters. Therefore, we wish to utilize the characteristics of different checkpointing protocols to match the static cluster and the mobile clusters separately. And they will be integrated to be a checkpointing scheme suitable for the entire mobile system.

4.1 The Static Cluster

Obviously, synchronous checkpointing protocols are the number one candidate for the static cluster. Since exchanging messages in the static network is fast and the nodes won't change location with time, synchronous checkpointing protocols are chosen due to its fast recovery time and simplicity. The Koo and Toueg's algorithm [3] is chosen in our paper. Other synchronous checkpointing protocols could be used as well, the principle considered here is simplicity.

The fixed hosts take the synchronous checkpoint as follows.

(1) A checkpoint initiator *CI* sends a request message C_{req} to all the other fixed hosts.

(2) Upon receiving C_{req}, each fixed host *FHi* executes the following steps

a). Take a tentative local checkpoint

b). Suspend underlying computation and message transmission

c). Send back a reply message C_{rep} to *CI*.

(3) After receiving C_{rep}s from all the other fixed hosts, *CI* sends a final message C_{fin} to all the other fixed hosts.

(4) Upon receiving C_{fin}, each fixed host *FHi* changes the tentative checkpoint to permanent checkpoint.

4.2 The Mobile Clusters

Log-based checkpointing seems to be the appropriate checkpointing protocols for the mobile clusters. However, several issues must be addressed.

1) The storage disks of mobile hosts could not be considered as stable storage. Therefore, the snapshot and the message logs must be saved on a fixed host's storage. Since the MSS is responsible to receive and deliver messages for its local MH, the MSS's stable storage is the ideal place to be used as an MH's stable storage.

2) User input modifies the state of the mobile hosts and must be treated as messages (external events). Therefore, the mobile host must first forward a copy of the user input to its MSS for logging. The mobile host can process the user input at the same time for the reason of efficiency, but can not send any response. After receiving the acknowledgement for logging from the MSS, the mobile host can send its response.

3) All the messages sent and received by a mobile host have to be recorded in the message log. This ensures that the restart of a mobile host won't make other hosts roll back. In other words, the mobile host is separated from other hosts.

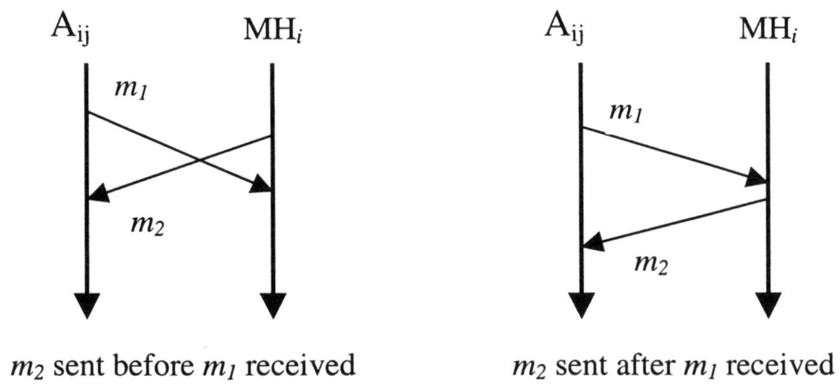

Figure 4. Two possible cases in MH_i for the same sequence in A_{ij}

To achieve the above requirements, there have to be an agent process A_{ij} in the MSS_j to deal with the associated information for MH_i. The agent process records messages sent to and received by a mobile host in the message log in the original sequence happened in MH_i. That is, what comes first is first recorded in the message log. However, the relative sequence sometimes may not be so clear. This happens if A_{ij} sends m_1 to MH_i and MH_i sends m_2 to A_{ij}. A_{ij} sees m_1 before m_2, but there are two possible cases in MH_i (see Figure 4). A_{ij} has difficulty in judging whether m_1 is received before m_2 is sent or not. This could be solved by introducing a variable *counter* to the mobile host and a variable *current_counter* to A_{ij}. Whenever the mobile host receives a message, it increments *counter* by 1. Every message sent by the mobile host carries *counter*. If A_{ij} receives a message m sent by the mobile host, it compares *m.counter* with *current_counter* to determine the message sequence. Then A_{ij} updates *current_counter* with *m.counter*. In the previous example, if

m_2.*counter* > *current_counter* then m_1 is received before m_2 is sent. However, a mobile computing system is a distributed system. In a distributed system, if a message m_1 must be received before another message m_2 is sent, explicit control mechanism must be used to ensure such sequence. If the sequence is not clear to judge as the above case, it means the relative message sequence is not important in terms of correct program execution.

Since the mobile hosts take their local snapshots independently, the required steps are quite easy. On receiving a checkpointing request, the mobile host simply informs the MSS that a snapshot must be taken. Then the mobile host transfers the local snapshot to the stable storage of the MSS. Additional steps must be taken to eliminate previous unnecessary state information. The problem is when and who. Who should issue the checkpoint request? It seems that the mobile host itself is the best candidate. Only the mobile host knows if it is ready to take a local snapshot. A timer or other mechanisms can be used to determine when to take a snapshot.

Another problem to be solved is the mobility problem. A mobile host may migrate from one MSS to another. The mobile host doesn't carry the agent process and the associated state information. If the mobile host has changed its location for several times, the information is scattered and the useless agent processes must be cleared in order not to occupy the system resource. The mobile host could disconnect (using a handoff procedure) or just disappeared and reappeared in a different cell after some time. To accommodate these cases, the mobile host executes the following steps when connecting to a new MSS_{j+1} from MSS_j.

1) MH_i Send T_{req} to A_{ij} in MSS_j.

2) Upon receiving T_{req}, A_{ij} transfers the unprocessed messages and the message log to A_{ij+1} in MSS_{j+1}.

3) After the transfer completed, A_{ij} checks if a snapshot for MH_i exists. A_{ij} is cleared if there exists no snapshot for MH_i.

Through the above steps, at most two nodes (and two agent processes) keep the message log and the snapshot. This ensures no system resource is wasted due to the movement of the mobile host. The snapshot is not transferred because of its size. As long as the snapshot is stayed in the same LAN, the transfer of the snapshot takes almost the same time. Thus, the snapshot is transferred only when the mobile host is required to perform a rollback.

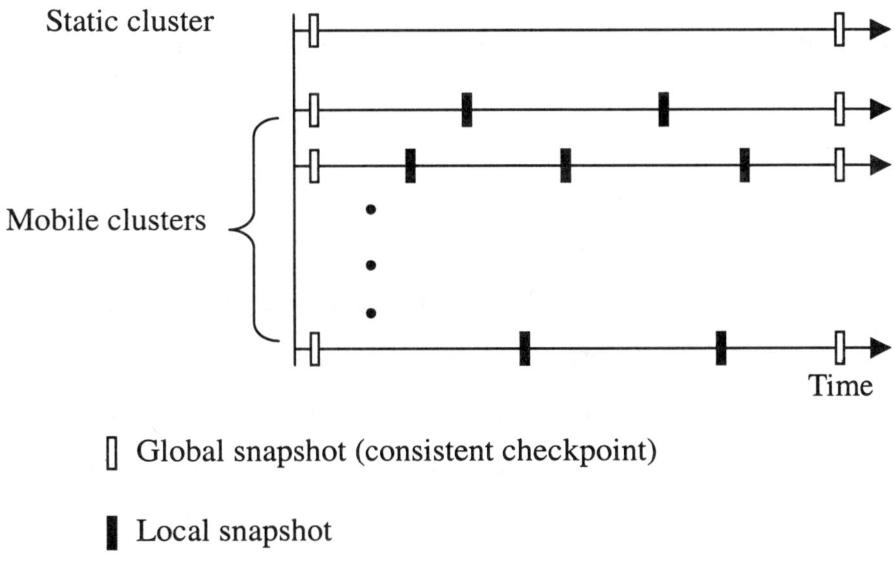

Figure 5. Snapshots on the static cluster and the mobile clusters

4.3 Integration

Now the static cluster uses synchronous checkpointing to take its consistent checkpoint. The mobile clusters take their local snapshots independently (see Figure 5). The advantage is that the restart of a mobile host won't make the static cluster rollback. However, the local snapshots of the mobile clusters may not be consistent with the checkpoint of the static cluster. To integrate the two different parts to get a consistent global checkpoint, a mobile host must keep an extra snapshot which is consistent with the checkpoint of the static cluster. In this way, if one static host fails, the whole system is rolled back to the global consistent checkpoint. This would be rare because the reliability of the static hosts is high. Therefore, the failure-free overhead is reduced. The required modification is listed in the following.

[Modification for the static cluster]

The required modification is simple. The checkpoint initiator *CI* now must send the request message C_{req} to all the other hosts including the mobile hosts. The request message C_{req} is changed to GC_{req} from now on to avoid confusion. The mobile hosts must execute those steps like the static hosts to respond to the *CI*.

[Modification for the mobile clusters]

The mobile clusters may receive two kinds of checkpointing request messages. They are GC_{req} and LC_{req}. LC_{req} is the request message to take the local snapshot. Therefore, two kinds of local snapshots may coexist in the same mobile host. On receiving GC_{req}, the mobile host takes its snapshot and marks it as consistent snapshot. The previous consistent snapshot must be cleared. On receiving LC_{req}, the mobile host takes its snapshot and marks it as local snapshot. The previous local snapshot and its message log must be cleared, but the consistent snapshot should not be cleared. If GC_{req} is received after LC_{req} for not a long time, the local snapshot and its message log could be used to consist of the consistent snapshot. This could save the effort to take an additional snapshot.

4.4 Restart Protocol

If some host is faulty, the fixed hosts and the mobile hosts may need to be restarted to continue the computation. If a fixed host is faulty, the restart protocol in [3] is used because we use the synchronous checkpointing protocol in [3].

1) The rollback initiator *RI* sends a request message R_{req} to all the other fixed hosts and mobile hosts.
2) Upon receiving R_{req}, each fixed host FH_i sends back a reply message R_{rep} to *RI*. However, each mobile host MH_i must transfer the consistent snapshot from some previous MSS. After the transfer is completed, the mobile host MH_i sends back its R_{rep}.
3) If *RI* receives all the R_{rep}s, *RI* sends a final message R_{fin} to all the other hosts.
4) Upon receiving R_{fin}, all the hosts restart from the consistent checkpoint.

Through this way, the system is restarted from a consistent state.

5. Conclusion

In this paper, we consider a mobile computing system on a large network such as WAN. This motivates us to decompose it into clusters which are similar. Each cluster is again decomposed into smaller ones which are different. By using such top-down decomposition, we propose a checkpointing scheme which exploits the characteristics of the fixed and mobile hosts. We are currently investigating on the relationship between the local snapshot and the global snapshot. This helps determine the appropriate checkpoint interval to achieve a reliable and low failure-free overhead system.

References

[1] A. P. Sistla and J. L. Welch, "Efficient Distributed Recovery Using

Message Logging," *Proc. ACM Symp. Principles of Distributed Computing*, pp. 223-238, 1989.

[2] S. Venkatesan, "Optimistic Crash Recovery Without Rolling Back Non-Faulty Processors," *Information Sciences-An Int'l J.*, 1993.

[3] R. Koo and S. Toueg, "Checkpointing and Rollback-Recovery for Distributed Systems," *IEEE Trans. Software Eng,.* vol. 13, no. 1, pp. 23-31, Jan. 1987.

[4] Anita Borg, Jim Baumbach, and Sam Glazer, "A Message System Supporting Fault Tolerance," *Proceedings of the Ninth ACM Symposium on Operating Systems Principles*, October 1983, pp. 90-99.

[5] Robert E. Strom and Shaula Yemini, "Optimistic Recovery in Distributed Systems," *ACM Transactions on Computer Systems*, pp. 204-226, August 1985.

[6] B. Bhargava and S.–R. Lian, "Independent Checkpointing and Concurrent Rollback for Recovery in Distributed Systems – An Optimistic Approach," *Proc. Seventh IEEE Symp. Reliable Distributed Systems*, pp. 3-12, Oct. 1988.

[7] K. M. Chandy and L. Lamport, "Distributed Snapshots: Determining Global States of Distributed Systems," *ACM Trans. Computer Systems*, Vol. 3, no. 1, pp. 63-75, Feb. 1985.

[8] E. N. Elnozahy and W. Zwaenepoel, "Manetho: Transparent Rollback-Recovery with Low Overhead, Limited Rollback, and Fast Output Commit," *IEEE Trans. Computers*, vol. 41, no. 5, pp. 526-531, May 1992.

[9] T. -H. Lai and T. -H. Yang, "On Distributed Snapshots," *Information Processing Letters*, vol. 25, pp. 153-158, 1987.

[10] L. Lamport, "Time, Clocks and the Ordering of Events in a Distributed Systems," *Comm. ACM*, vol. 21, no. 7, pp. 558-565, July 1978.

[11] R. Alonso and H. Korth, "Database System Issues in Nomadic Computing," *SIGMOD*, June, 1993.

[12] A. Acharya and B. R. Badrinath, "Checkpointing Distributed Applications on Mobile Computers," *IEEE Conf. On PDIS*, Sep. 1994.

[13] E. N. Elnozahy and W. Zwaenepoel, "On the Use and Implementation of message logging," *Proc. IEEE Fault-Tolerant Computing Symp.*, pp. 298-307, 1994.

[14] D. B. Johnson, "Efficient Transparent Optimistic Rollback Recovery for Distributed application programs," *Proc. IEEE Symp. Reliable Distributed Syst.*, pp. 86-95, October 1993.

[15] Y. –M. Wang and W. K. Fuchs, "Optimistic Message Logging for Independent Checkpointing in Message-Passing Systems," *Proc. IEEE Symp. Reliable Distributed Systems*, pp. 147-154, October 1992.

[16] R. Prakash and M. Singhal, "Low-Cost Checkpointing and Failure Recovery in Mobil Computing Systems," *IEEE Trans. On Parallel and Distributed Systems*, vol. 7, no. 10, pp. 1035-1048, October 1996.

[17] N. Neves and W. K. Fuchs, "Adaptive Recovery for Mobile Environments," *Communications of the ACM*, vol. 40, no. 1, pp. 69-74, January 1997.

[18] D. K. Pradhan, P. Krishna and N. H. Vaidya, "Recoverable Mobile Environment: Design and Trade-off Analysis," *Proc. IEEE Fault-Tolerant Computing Symp.*, pp. 16-25, 1996.

[19] G. H. Forman and J. Zahorjan, "The Challenges of Mobile Computing," *IEEE Computer*, pp. 38-47, April 1994.

[20] L. Alvisi, B. Hoppe and K. Marzullo, "Nonblocking and Orphan-Free Message Logging Protocols," *Proc. IEEE Fault-Tolerant Computing Symp.*,pp. 145-154, 1993.

DIMACS Series in Discrete Mathematics
and Theoretical Computer Science
Volume **52**, 2000

A GRASP For Frequency Assignment in Mobile Radio Networks

X. Liu, P. M. Pardalos, S. Rajasekaran, and M. G. C. Resende

ABSTRACT. Frequency assignment problem in mobile radio networks refers to the assignment of a limited number of frequencies to cells in the network such that electromagnetic interference is minimized. Interference arises from frequency reuse. This problem, though NP-hard, has been approached with some success by heuristic methods. In this paper, a greedy randomized adaptive search procedure (GRASP) for frequency assignment in mobile radio networks is presented. The new algorithm is a randomized version of saturation degree with a simulated annealing [**17**] local search procedure. Preliminary empirical results indicate that the new heuristic compares favorably to the hybrid of pure saturation degree method with a simulated annealing local search.

1. Introduction

The frequency assignment problem (FAP) is important for resource management in cellular networks [**18, 12, 20**]. Generally speaking, cellular phone communication is accomplished by radio signals, using some frequency of the electromagnetic spectrum. The radio signal from the cellular phone is received by a nearby base station and propagates into a wire-based network. The available radio spectrum is segmented into uniformly sized frequency channels. When it accepts a call request, the cellular system assigns the call to a base station and a channel. Depending on the system, there are different ways that the base stations are associated with channels. In cellular systems, the most used approach is *fixed channel allocation*, managed by a channel plan. Another approach is *dynamic channel allocation*. With this method, all base stations have access to all channels. After selecting a base station to handle a new call, the system searches the entire set of channels for a channel that permits high transmission quality. The channel selection depends on which channels are already used at the base station allocated for the new call and the channels in use at surrounding base stations.

Frequency reuse is a major factor on spectrum efficiency and has strong influence on the cost of providing service in a cellular network. The term *reuse* refers to the fact that within a limited geographical area, several different signals occupy the same frequency band. Since signals travel through air, they are vulnerable to

1991 *Mathematics Subject Classification.* Primary 54C40, 14E20; Secondary 46E25, 20C20.
This work was supported in part by an NSF Grant EIA-9872507.

interference from other signals. Two kinds of interference are possible: *co-channel interference* and *adjacent channel interference.* Either of these two forms of interference can occur if the cellular frequency reuse scheme is not adequate. Co-channel interference results when two base stations within the range of a common subscriber use the same channel simultaneously. Adjacent channel interference happens when two base stations within the range of a common subscriber use adjacent channels simultaneously.

Although different versions of the problem can be found [**8, 13, 16**], the main goal of the problem is to assign a limited number of available frequencies to transmitters minimizing interference due to the frequency reuse. For a given problem with N transmitters, the interference constraints can be represented as either a $N \times N$ constraint matrix C or a constraint graph with N vertices. Each pair of transmitters (i, j) with $1 \leq i < j \leq N$ has a value $c_{i,j}$ which measures the minimal frequency separation required. Two transmitters are connected in the constraint graph if the frequency separation $c_{i,j}$ is greater than zero. Therefore, all the constraints are of the following form:

$$|f_i - f_j| > c_{i,j},$$

where f_i is the frequency assigned to the transmitter i.

There are two types of the frequency assignment problem. The *fixed spectrum version* is to minimize the total interference with a fixed set of frequencies. In the *minimum span version*, the set of frequencies is not fixed and generally is represented by the set of consecutive integers $\{1, 2, \ldots, K\}$. The problem is to determine the minimum K. Both types of the problem can be represented by a constraint matrix, the cost function minimized being:

$$\text{Minimize } f = \sum_{i,j} \delta(|f_i - f_j| > c_{i,j}), \tag{1.1}$$

where $\delta(|f_i - f_j| > c_{i,j})$ is equal to $c_{i,j} - |f_i - f_j|$ if $|f_i - f_j| \leq c_{i,j}$ and 0 otherwise.

The above problem, also known as the T-coloring problem, has long been known to be NP-hard [**3, 4**]. It is an extension of the standard graph coloring problem. Many heuristics have been devised for finding approximate (good but not necessarily optimal) solutions to the problem in a reasonable amount of computation time [**1, 3, 14, 19, 21**].

Some of the heuristics proposed for solving the frequency assignment problem are based on *deterministic sequential coloring* for the graph coloring problem. First, the vertices are sorted by their *saturation degrees* [**2**], i.e. the number of forbidden frequencies for the vertex. The vertex with the largest saturation degree is assigned the available frequency that contributes the least incremental objective function value. The process is repeated until all vertices are assigned a frequency. The assignment generated by the sequential coloring mechanism can be improved by executing interchanges. At each step, a previously assigned vertex exchanges its frequency with another frequency if the exchange decreases the current objective function value. Many deterministic sequential coloring heuristics have been developed for the graph coloring problem, and they can easily be adapted for the frequency assignment problem.

In this paper, we present a *greedy randomized adaptive search procedure*(GRASP) for the frequency assignment problem in mobile radio networks.

```
procedure GRASP()
    InputInstance();
    do while stop criteria are not satisfied→
        ConstructInitialSolution();
        Post-process for LocalSearch(BestSolutionFound);
        UpdateSolution(BestSolutionFound);
    od;
    return BestSolutionFound
end ;
```

FIGURE 1. Pseudo-code for a generic description of GRASP

The remainder of the paper is organized as follows. In Section 2 we sketch how GRASP can be adapted to the frequency assignment problem. Experiments with numerical results are reported in Section 3. We end with some conclusions in Section 4.

2. The GRASP for FAP

A greedy randomized adaptive search procedure (GRASP) [**7, 6**] is a metaheuristic for combinatorial optimization that can be used for frequency assignment problems. The algorithm is inspired by recently developed *randomized heuristics* for solving the graph coloring problem [**23**]. GRASP is a two phase approach. In the first phase, a randomized version of a sequential coloring heuristic based on saturation degrees, generates good, diverse, initial solutions. Then, a simulated annealing procedure is used in the local search phase to attempt to improve the solution generated by the construction procedure. The general procedure of GRASP is described in Fig 1.

The construction phase uses two greedy functions. The first chooses a vertex from a set of unassigned vertices with high saturation degrees. The second greedy function is used to assign a frequency to the chosen vertex. A frequency is selected from a set of permissible frequencies that contribute little additional cost to the objective function. The procedure takes as input the number of GRASP iterations *niters*, the restricted candidate list parameters, *ndex* and α, for the vertex selection list and frequency selection list, respectively, and k, the maximum number of available frequencies to be used. The output is the frequency class partition $\{V_1, \dots, V_{nc}\}$.

The GRASP iterations are repeated *niter* times. Each iteration has a construction phase and a local search phase. The construction phase builds an assignment one vertex-frequency pair at a time. This is repeated until all vertices have been assigned frequencies. First, among the yet unassigned vertices, a restricted candidate list (RCL) is set up with the *ndex* vertices having the largest numbers of forbidden frequencies. A vertex v is selected at random from this list. From the list of unforbidden frequencies for this vertex, another restricted candidate list is set up with those frequencies that contribute at most $(1+\alpha)\bar{c}$ to the objective function, where α is a parameter such that $0 \leq \alpha \leq 1$, and $\bar{c}$ is smallest contribution to the objective function among all unforbidden frequencies. A frequency is selected at random from this list and is assigned to vertex v. The pseudo code of GRASP for the frequency assingment problem is depicted in Fig 2.

```
procedure FAP(V, E, niters, k)
nc := k; F := 0;
for k = 1, ... , niters→
  i := 0; V* = V; V_1 = ∅, ... , V_k = ∅;
  while V* ≠ ∅;
  i := i + 1;
  CL := {ndex unassigned vertices with saturation degrees };
   Select v ∈ CL at random;
  FL = {nf obj function values at v within certain value};
   Select F_i ∈ FL at random;
   F := F + F_i, V_i := V_i + {v}, V* := V* − V_i;
   endwhile
postpro({V_1, ... , V_nc}, V, E, k, F);
rof
return BestSolutionFound
end ;
```

FIGURE 2. GRASP for Frequency Assignment Problem

We now describe the simulated annealing procedure used in the local search phase of GRASP. The objective function is that of original problem. The simulated annealing has new loops. The outer loop runs until the temperature falls below a specified value. The temperature t is reduced by a multiplicative factor at the end of the outer loop, i.e. $t = t{*}a$, where $0 \leq a < 1$. The inner loop runs for a fixed number of iterations using a fixed temperature. Let $s = \{V_1, \dots, V_k\}$ be the partition of the vertex set generated by the construction phase. The neighborhood $N(s)$ of s is the set of all partitions that result from changing the frequency of a vertex. Let the current partition be s. A move in the simulated annealing is generated by randomly choosing a vertex from current assignment and changing it to another unforbidden frequency resulting in partition $s_1 \in N(s)$. Let $\Delta = f(s_1) - f(s)$. If the new partition does not increase the objective function, i.e. $\Delta \leq 0$, then the new partition is accepted and another iteration of the inner loop begins. Otherwise, with probability $\exp(-\Delta/t)$, the new partition is accepted, where t is the current temperature.

The most computationally demanding operation in procedure FAP is ranking the set of unassigned vertices with respect to their saturation degrees. To efficiently implement this step, $G = (V, E)$ is stored using adjacent list structure. The adjacent vertices for each vertex are stored consecutively in an array. Sorting of the vertices in V by their saturation degrees is accomplished using a heap data structure [**15**]. Two integer pointers are implemented to identify the position of each vertex in the heap, and to denote which vertex is in each position of the heap.

Let $n = |V|$, $m = |E|$, and k the maximum number of available frequencies to be used. Consider the time needed to construct one assignment. The initial heap for n vertices can be built in $O(n)$ time[**15**]. sorting takes $O(L \log n)$ time, where L is the length of the candidate list of the saturation degree set. Each heap update takes at most $O(\log n)$ time. In order to count the number of updates, we should also notice that each edge is only estimated when either of its incident vertices are unassigned. This occurs at most $O(m)$ times. So the total time spent in each

n	E	GRASP Best	GRASP average Time	PSD Best	PSD Time
50	373	3751	1.4	4226	1
100	1506	16455	5.95	16725	8
200	6019	45266	50.4	47340	67
300	13501	149775	346.06	151643	118
400	23889	185433	388.95	190530	532
500	37391	293550	765	300816	474
600	53811	425884	1304	435396	1335
700	73410	436251	2051	452212	2244
800	95777	572526	10764	594423	3041
900	121227	731092	12312	753526	4401
1000	149700	904412	14328	931298	5547

TABLE 1. The Final Results from GRASP and PSD

assignment is $O(n + mL \log n)$. This analysis ignores the time spent in the local search. Though in the worst case simulated annealing can take exponential time, stable numerical performance can be obtained by restricting the parameters to be used.

3. Computational Results

Starting with an arbitrary initial solution, simulated annealing may require a large amount of computational time to converge. However, the goal of this test is to quickly search for an improvement to the initial solutions generated by the GRASP pure saturation degree construction procedure. Therefore, the local search procedure should not be allowed to venture too far from the initial solution. In order to achieve this, the parameters are assigned to the simulated annealing procedure in such a way that the procedure should search only a small neighborhood space around the initial starting solution. A fast cooling rate a was assigned to 0.85 and an initial temperature was set to 10. The number of outer iterations is fixed to 15. In each inner iteration, if either the total accepted moves exceeds $0.4 * n$ or the total number of iterations is greater than $2 * n$, then the inner iteration terminates, and the temperature is reduced.

There are two types of test problems. One is randomly generated, and the other is CELAR data set, which are the instances of the European EUCLID project CALMA (Combinatorial Algorithms for Military Applications).

The randomly generated graphs in this paper are the graphs $G_{n,p}$, with n vertices and edge probability p. We use $p = 0.3$ and $n = 50, 100, 200, \ldots, 1000$. The interference constraints were randomly generated as well. The number of available frequencies is set to 20 for the graphs with less than 300 vertices and to 30 for the larger graphs. The forbidden separations are generated at random between 1 and the number of available frequencies. The heuristic is implemented in C++, and the test was performed at a Sun SPARC/10 station running SunOS 5.6 using CC compiler and using `time` routine for collecting the run times.

Table 1 presents the results from GRASP and the method of pure saturation degree combining with the simulated annealing local search (PSD). It turns out that

Name	*Var.*	*Con.*	GRASP Best	Best Reported	GRASP runs	GRASP Time
scen06	200	1322	4273	6787	20	183.3
scen08	916	5744	357	1772	6	2693.7

TABLE 2. The Solutions for CELAR data set by GRASP

the GRASP generated better solutions than the greedy PSD. However, GRASP took more time than PSD for each instance.

Our next numerical experiments are scen06 and scen08 of CELAR data set, because we are interested in the most practical instances where there are not enough frequencies. There is no solution satisfying all the constraints of both problems. There are 6 frequency domains for both problems. Constraints of both problems are either inequalities or equalities. Equal constraints should be mandatorily met, and unequal constraints could be violated with different penalty weights. In the numerical experiment for these two instances, the total number of cycles in the simulated annealing is 20. Table 2 is a summary of the results for the two problems. In this table $Name$ denotes the problem (as identified in the EUCLID project). $Var.$ is the number of vertices, $Con.$ is the number of constraints, GRASP Best denotes the best solution obtained by GRASP, Best Reported stands for the best solution reported by the EUCLID project, GRASP runs is the number of iterations taken by GRASP, and GRASP time is the time taken by GRASP. All the times reported in Table 1 and 2 are in seconds.

4. Concluding Remarks

In this paper we presented a greedy randomized adaptive search procedure for the frequency assignment problem in a mobile radio network. The algorithm is a randomized version of saturation degree method combined with a simulated annealing local search procedure. The preliminary numerical experiment indicated that the algorithm generates better solutions than those obtained by the hybrid of pure saturation degree method with simulated annealing procedure. To increase the efficiency of the algorithm, further research is needed for designing better local search procedures. Current experiment has suggested that the GRASP version of saturation degree method combining efficient local search procedures has strong potential to produce good solutions. Further experiments on more real world data sets will be conducted in future.

References

[1] Borndörfer, R., A. Eisenblätter, M. Grötschel and A. Martin, *Frequency assignment in cellular phone network,* Annals of Operation Research, 76(1998), pp. 73-93.

[2] Brelze, D., *New methods to color vertices of a graph,* Comm. ACM, 22(1979), pp. 251-256.

[3] Costa, D., *On the use of some known methods for T-colorings of graphs,* Annals of Operations Research, 41(1993), pp. 343-358.

[4] Cozzens, M. and F. S. Roberts, *T-colorings of graphs and the channel assignment problem,* Congr. Numer., 35(1982), pp 191-208.

[5] Cvijović, D. and J. Klinowski, *Taboo search: an approach to the multiple minimal problem,* Science, 267(1995), pp. 664-666.

[6] Feo, T. A., and M. G. C. Resende, *Greedy randomized adaptive search procedures,* J. Global. Opt. 6(1995), pp. 106-133.

[7] Feo, T. A., and M. G. C. Resende, *A probabilistic heuristic for a computationally difficult set covering problem,* Operations Research Letters, 8(1989), pp 67-71.

[8] Gamst, A. and W. Rave, *On frequency assignment in mobile automatic telephone systems,* in *Proc. GLOBECOM'82* 1982, pp. 309-315.

[9] Goldberg, D. E., *Genetic Algorithms in Search, Optimization and Machine Learning,* Addison-Wesley, 1989.

[10] Glover, F., *Tabu search,Part 1,* ORSA.J.COMPUT. 1(1989), pp. 190-206.

[11] Glover, F., *Tabu search,Part 2,* ORSA.J.COMPUT. 2(1990) pp. 4-32.

[12] Goodman, David J., *Wireless Personal Communications Systems, Addison Wesley Longman, Inc,* 1997.

[13] Haas, J. Zygmunt, Jack H. Winters and David S. Johnson, *Simulation results of the capacity of cellular systems,* IEEE Trans. Veh. Technol., 46(1997), pp. 805-817.

[14] Hao, Jin-Kao, R. Dorne and P. Galinier, *Tabu search for frequency assignment in mobile radio networks,* Journal of Heuristics, 4(1998), pp. 47-62.

[15] Horowitz, E., S. Sahni and S. Rajasekaran, *Computer Algorithms/C++, Computer Science Press,* New York, 1996.

[16] Hurley, S., D. H. Smith and S. U. Thiel, *FASoft: A system for discrete channel frequency assignment,* Radio Science, 32(1997), pp. 1921-1939.

[17] Kirkpatrick, S. , C. D. Gelatt, Jr., M. P. Vecchi, *Optimization by simulated annealing,* Science, 220(1983), pp. 671-680.

[18] MacDonald, V. H., *Advanced mobile phone system: The cellular concept,* Bell Syst. Tech. J., 58(1979), pp. 15-41.

[19] Murphy, R. A., P. M. Pardalos and M. G. C. Resende, *Frequency assignment problems,* in *Handbook of combinatorial optimization,* D.-Z. Du and P. M. Pardalos, eds., *Kluwer Academic Publishers,* 1999.

[20] Taylor, Mark S., William Waung and Mohsen Banan, *Internetwork Mobility, Prentice-hall, Inc.,* 1996.

[21] Rajasekaran, S., K. Naik and David Wei, *On frequency assignment in cellular networks,* Manuscript, 1998.

[22] Roberts, F. S., *From garbage to rainbows: generalizations of graph coloring and their applications,* RUTCOR Research Report 36-88, Rutgers University(1988).

[23] Smith, Stuart H. and Thomas A. Feo, *A grasp for coloring sparse graphs,* Manuscript, 1990.

Center for Applied Optimization and Department of Industrial and Systems Engineering, University of Florida

E-mail address: `liu@deming.ise.ufl.edu`

Center for Applied Optimization and Department of Industrial and Systems Engineering, University of Florida

E-mail address: `pardalos@ophelia.ise.ufl.edu`

Department of Computer and Information Science and Engineering, University of Florida

E-mail address: `raj@cise.ufl.edu`

Information Science Research Center,AT&T Labs Research

E-mail address: `mgcr@research.att.com`

DIMACS Series in Discrete Mathematics
and Theoretical Computer Science
Volume **52**, 2000

Multicriteria Optimization for Frequency Assignment

Robert A. Murphey, Panos M. Pardalos, and Eduardo Pasiliao

ABSTRACT. The frequency assignment problem (FAP) concerns the allocation of a limited frequency spectrum to a group of transmitters. Usually the frequency spectrum is partitioned into even intervals, termed channels, and one of three objectives is considered: (1) minimize the number of unique channels used, called a minimum *order* assignment; (2) minimize the difference between the maximum and minimum channels used, called a minimum *span* assignment; (3) minimize the level of interference in the system. *Restricted order* and *restricted span* assignments are defined as minimum span (order) assignments such that the order (span) is fixed at it minimum value. The relationship between restricted order and restricted span assignments is explored and is placed in a multicriteria optimization framework that utilizes lexicographic preference ordering. Two techniques for obtaining restricted assignments using approximate methods such as GRASP and DSATUR are described. A three dimensional multicriteria optimization problem is also posed and approximate search methods applied to obtain assignments that are efficient in order, span and interference.

1. Introduction

Frequency assignment problems model the allocation of a limited frequency spectrum to a group of transmitters. Usually the frequency spectrum is partitioned into even intervals, termed channels, and one of three objectives is considered: (1) minimize the number of unique channels used, called a minimum *order* assignment; (2) minimize the difference between the maximum and minimum channels used, called a minimum *span* assignment; (3) minimize the level of interference in the system. Since, in general, frequency assignment is a hard problem, much research has been done on designing efficient heuristics that obtain nearly optimal solutions. However, the authors have observed that in all of this rich literature, the proper choice of an objective function is never addressed. That is, should we minimize order, span or interference? Indeed, many examples of each approach may be found in the literature yet in each instance, the reasoning for their choice of a particular objective is rarely alluded to. In this paper, we attempt to address that issue head-on by considering the FAP as a multicriteria decision model. Only then can the trade-off between objectives be properly illuminated.

The remainder of the paper is organized as follows. In Section 2, a general model of the frequency assignment problem is presented that utilizes the concept of unit-disk graphs to represent minimum separation of channels assigned between pairs of transmitters. In Section 3 a multicriteria model of frequency assignment is presented that builds on the model of Section 2. Section 4 briefly covers the

2000 *Mathematics Subject Classification.* Primary 90C35.

state-of-the-art in single objective frequency assignment methods. These methods will be used in the design of the multi objective frequency assignment methods to be presented in Sections 5, 6 and 7. Experimental results are provided in Section 8 with discussion of those results in Section 9.

2. Frequency Assignment Models

Consider the graph G with vertex set $V(G)$ and a function, ϕ that maps the vertices of the graph to points in Euclidean j-space: $\phi(v) \in \mathbb{R}^j, \forall v \in V(G)$. Define the edge set $E(G)$ such that

$$(v_i, v_j) \in E(G) \Longleftrightarrow ||\phi(v_i) - \phi(v_j)|| < D(i,j)$$
$$\text{for all } v_i, v_j \in V$$

where $D(i,j) \in \mathbb{R}_+$ and $|| \cdot ||$ denotes the Euclidean distance operation. The graph G is called a *k-unit sphere* graph. If the vertices of the graph represent transmitters, $\phi(v)$ is the physical location of transmitter v, $D(v_i, v_j)$ is the minimum distance between transmitters v_i and v_j for operation without interference, and $k = 1, 2$ or 3, then graph G models many frequency assignment problems. In particular, when $k = 2$, G is referred to a *unit disk* graph and models transmitters distributed on a plane (approximately the Earth's surface).

Now suppose that the spectrum is partitioned into integer valued *channels.* Consider a frequency assignment to be the rule $f: \ V \to \mathbb{Z}_+ \setminus \{0\}$ with $\mathbb{Z}_+ \setminus \{0\}$ denoting the set of all positive integers. Define the *channel distance* d_{ij} as the minimum number of channels that must separate v_i and v_j for no interference, where $v_i, v_j \in V$.
A pair of vertices is adjacent in G if interference results when the absolute number of channels which separates them is less the channel distance. That is, $(v_i, v_j) \in E(G)$ if $|f(v_i) - f(v_j)| < d_{ij}$ would cause interference. When using this definition, the graph G is often referred to as an *interference* graph. For a unit disk graph, if $(v_i, v_j) \in E(G)$ then $||\phi(v_i) - \phi(v_j)|| < D(i,j)$ implies that $|f(v_i) - f(v_j)| > d_{ij}$ and vise versa. Since it is typically easier to work directly with the interference graph instead of Euclidean distance, the FAP feasibility problem is defined as:

$$\begin{array}{ll} \texttt{Instance:} & G(V,E),\ d \\ & \texttt{find}\ f:\ V \to \mathbb{Z}_+ \setminus \{0\} \\ & s.t.\ \ (v_i, v_j) \in E \longleftrightarrow |f(v_i) - f(v_j)| \geq d_{ij},\ d_{ij} \in \mathbb{Z}_+. \end{array}$$

Hale [**10**]defines two figures of merit for any feasible frequency assignment; the number of unique channels used in the assignment, which he calls *order*, and the largest channel assigned minus the smallest channel assigned, which he calls *span.* Cozzens and Roberts [**4**] define the notation for minimum order and minimum span to be $\chi_T(G)$ and $sp_T(G)$ respectively. Hale points out that when $d_{ij} = 1$, the FAP is equivalent to ordinary graph coloring and termed this the *co-channel* FAP. He proved that the minimum span and minimum order co-channel FAP are equivalent to graph K-colorability, which is known to be $\mathcal{NP}$-complete [**9**], so we expect that the unit disk FAP is at least as hard. Hale further showed that for the co-channel problem, $sp_T(G) + 1 = \chi_T(G) = \chi(G)$, where $\chi(G)$ denotes the chromatic number of graph G.

Raychaudhuri [**15**] defines more complete measures of optimality as follows: A *restricted order* assignment is the assignment of minimum span such that order is restricted to $\chi_T(G)$. Conversely, a *restricted span* assignment is the assignment of minimum order such that span is restricted to $sp_T(G)$. In this treatment the span of a restricted order assignment will be denoted by $sp_{ro}(G)$ and the order of a restricted span assignment by $\chi_{rs}(G)$. Clearly, $sp_{ro}(G) \geq sp_T(G)$ and $\chi_{rs}(G) \geq \chi_T(G)$.

3. A Multicriteria Optimization Model For The FAP

The relationships between minimum order, minimum span and restricted assignments are perhaps best illustrated by considering the FAP a multicriteria optimization problem where two objectives exist: order and span.

Let the set of all feasible assignments be represented by the set $\mathcal{F}$. Define the vector $f(s) \in \mathbb{Z}_+^2$ to be the objective values for a solution $s \in \mathcal{F}$ with $f_1(s)$ being the objective most desired. For example, $f_1(s)$ and $f_2(s)$ are the order and span of solution s respectively if a small order is more important than a small span. In this case, there is said to be a *preference* for order over span.

A solution $s \in \mathcal{F}$ is termed *lexicographically* less than a solution $t \in \mathcal{F}$ if $f_1(s) = f_1(t)$ and $f_2(s) < f_2(t)$ and is denoted by $f(s) \prec f(t)$. Hence, if two solutions s and t have the same value for the preferred objective, the solution which is lexicographically less than the other will be most desirable. This concept is easily extended to $f \in \mathbb{Z}_+^n$ for arbitrary $n = |V(G)|$ [**11**].

Define a *lexicographic optimal* solution s to be the solution to the following problem:

(LOP) $s \in \mathcal{F}$ such that $f(s) \prec f(t)$ for all $t \in \mathcal{F}$ with $s \neq t$.

If order is preferred to span, then a lexicographic optimal solution s has $f_1(s) = \chi_T(G)$ and $f_2(s) = sp_{ro}(G)$ hence s is a restricted order solution. If span is preferred to order, the lexicographic optima has $f_1(s) = sp_T(G)$ and $f_2(s) = \chi_{rs}(G)$ and s is a restricted span solution.

The important observation here is that *restricted solutions for the FAP are lexicographic optimal solutions of a multicriteria optimization problem.*

Typically, minimum order and minimum span solutions are not unique. In Figure 1 the solutions that minimize order and span lie on the lines $order = \chi_T(G)$ and $span = sp_T(G)$ respectively.

The lexicographic optima are identified in Figure 1 as the points closest to the origin on the lines $order = \chi_T(G)$ and $span = sp_T(G)$. Clearly, the lexicographic optima are more desirable than the other minimum order and minimum span solutions.

This lexicographic ordering can be extended to 3 criteria by adding "minimize network interference" as a third objective.

In the following sections, techniques for finding lexicographic optimal solutions will be explored. During this process, many efficient points will also be found. A direct technique for solving **LOP** is described by Burkard and Rendl in [**3**] and should be explored further for the FAP. In this paper, two approximate techniques are developed. For obtaining a restricted order solution, a *two-phased* algorithm is presented that first finds a minimum order (span) solution and then improves

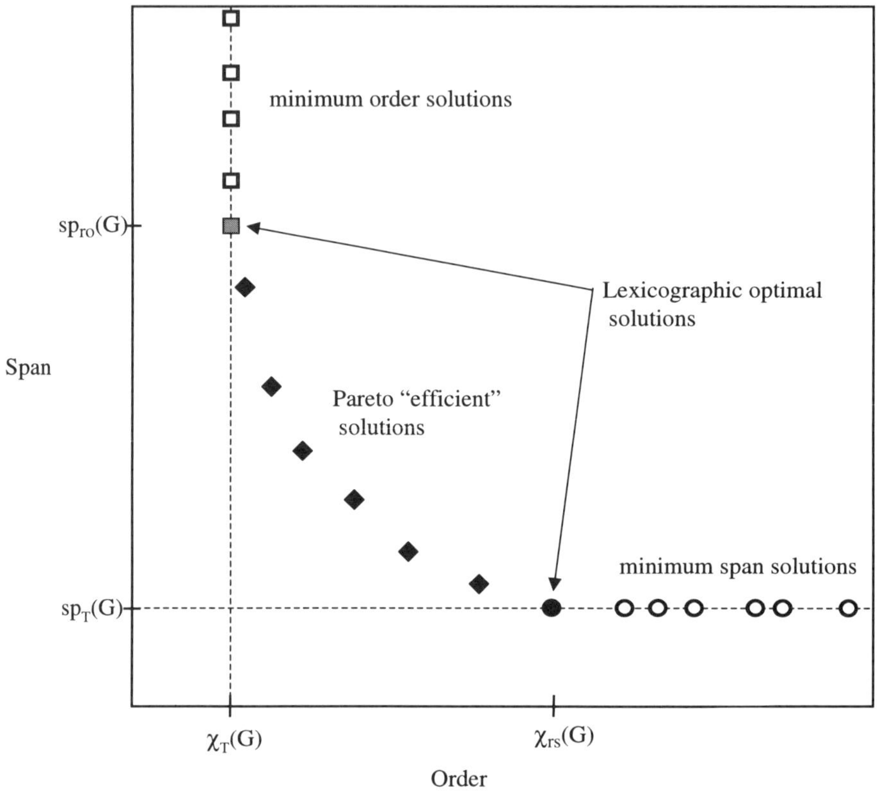

FIGURE 1. Multiobjective Optimization Representation of FAP

the solution via local search to obtain the restricted order solution. Restricted span solutions are obtained using an *iterated domain restriction* approach which finds low order solutions on domains of successively smaller and smaller span. This last technique may be expanded to consider a three objective problem, the third objective being to minimize interference.

Before proceeding to the multi-objective discussion, methods for obtaining single objective minimum order and minimum span assignments will be presented.

4. Minimum Order and Minimum Span Assignments

The easiest FAP, the co-channel FAP, is equivalent to K-colorability. It is no small wonder that graph coloring algorithms are often used when designing algorithms for the FAP. Indeed, re placing the coloring constraint of the unit disk FAP $(v_i, v_j) \in E \longleftrightarrow |f(v_i) - f(v_j)| \geq 1$ with $(v_i, v_j) \in E \longleftrightarrow |f(v_i) - f(v_j)| \geq d_{ij}$ should motivate this practice. However, graph coloring algorithms are designed to obtain colorings with χ colors which means that they are much better at finding minimum order assignments and aren't as successful for minimum span objectives.

4.1. Exact Methods. Several exact methods exist for coloring graphs. Brelaz [**1**] developed a widely used method which uses a depth-first backtracking scheme to improve a solution obtained by some primal heuristic. Brelaz's algorithm was later improved by Peemöller [**14**]. Lanfear [**12**] presented an adaptation of Peemöller's algorithm specifically for the FAP.

None of the exact techniques work efficiently on large FAPs since in the worst case, they may enumerate an exponential number (in $n = |V|$) of solutions.

4.2. Sequential Heuristics. Since exact methods are, in general, not efficient for most frequency assignment problems, approximate techniques have usually been the focus of the research. The most popular approximation approach is based upon what are known as *sequential heuristics* for graph coloring. All sequential methods have 2 steps; selection of a vertex to color and selection of a color to assign to the selected vertex. Usually the color selection is a greedy one; the smallest color is assigned. For this reason, all sequential methods that use a greedy color selection are jointly referred to as *greedy*. The selection of a vertex via an ordering of the vertices is what typically distinguishes different sequential methods. Given a vertex order, it can easily be shown that a greedy sequential technique runs in $O(n^2t)$ time, where $n = |V(G)|$ and t is the number of unique values in d [**4**].

Many vertex ordering schemes have been studied for the greedy sequential methods, including largest first, smallest last, uniform and random, [**17**] with mixed results. We will describe a method called DSATUR which is of particular interest due to it's successful application on many FAPs. Then we will introduce a new method based on a Greedy Randomized Adaptive Search Procedure (GRASP).

4.3. DSATUR. Consider a graph $G(V, E)$ with a coloring f. That is, f is a co-channel FAP on G. The *saturation degree* of a vertex $v \in V(G)$ is defined to be the number of unique colors (equivalently channels) that exist on the vertices that are adjacent to v. During a sequential assignment, of all the unassigned vertices, the vertex with the highest saturation degree is considered to be the "most denied" since there are fewer colors to choose from.

Based upon this notion, Brelaz [**1**] developed the DSATUR algorithm. DSATUR is a greedy sequential heuristic that uses the saturation degree of the vertices to specify an ordering. The algorithm begins (step 1) by randomly selecting a vertex from $V(G)$ and assigning it the smallest color. At step i, the saturation degree of the $n - i$ uncolored vertices is computed and the vertex with the largest saturation degree is selected to be colored the smallest feasible color. DSATUR terminates when all n vertices have been colored. The algorithm may construct K such solutions using K random starts and select the best one as the global solution.

For the minimum order or minimum span problem, DSATUR may be implemented for general unit disk graphs in $O(n^2t)$ time by using a bookkeeping method to update the saturation degree of the assigned vertices.

4.4. GRASP Heuristics. A Greedy Randomized Adaptive Search Procedure (GRASP) is an iterative sequential heuristic with each iteration consisting of two stages: a construction phase and a local search phase. At each iteration a solution is found and the best solution of all iterations is stored as the final solution. It is a heuristic algorithm which has been shown to produce good quality solutions with relatively short computation times for many nonlinear assignment problems [**6, 7, 8, 16**]. The generic GRASP procedure is presented in pseudo-code in Figure 2.

```
PROCEDURE GRASP(ListSize,MaxIter,RandomSeed)
InputInstance( );
DO k = 1,... , MaxIter →
  Construction(ListSize,RandomSeed);
  LocalSearch(BestSolutionFound);
  UpdateSolution(BestSolutionFound);
OD;
RETURN BestSolutionFound
END GRASP;
```

FIGURE 2. GRASP pseudo-code

The construction phase builds a feasible solution through the use of a greedy randomized function. The "greedy" part of the function makes sure the feasible solution generated is of low cost. The random component is included to prevent entrapment in any local minima and is therefore one of the keys to the performance of GRASP. The construction of a feasible solution begins with a selected number of the best of all costs gathered in a restricted candidate list (RCL). From the RCL, a candidate is randomly selected. Then the RCL is updated, i.e. all costs that are no longer feasible are removed from the RCL and new ones are entered. This procedure is repeated until the last assignment is made and a feasible solution is obtained.

The local search phase attempts to improve upon the greedy feasible solution by searching some predefined neighborhood of the current solution to see if a better solution can found. If an improvement is found, this solution is then kept as the current solution and a new neighborhood is searched. When no better solution can be found, the search is terminated and a local minimum is returned.

GRASP runs numerous main steps, each composed of a construction and local search. Since each step depends to some degree upon a randomization effect, each step may result in a different locally optimal solution. Consequently, after K main steps have run, The best local solution is selected as the global optimum to the problem and GRASP terminates.

In the following, GRASP is applied to a minimum order FAP. We will then show how GRASP can also be applied to minimum span FAPs. In Sections 6 and 7 we will examine the relevance of these GRASP methods to multicriteria optimization. The remainder of this section describes in detail the GRASP implementation for the single criteria problems. In section 4.4.1, the first of the two GRASP stages, the construction phase, is presented. The second GRASP stage, the local search is described in section 4.4.2.

4.4.1. *GRASP Construction Phase.* The GRASP construction phase for the RLFAP in Figure 3 begins by initializing all the transceiver assignments to the empty set or simply zero. This process, in addition to a randomized starting vertex, ensures that each GRASP iteration always starts from scratch and provides a high probability that the resulting solution from the current iteration will be different from any prior solution obtained thus far. A candidate list (RCL) is generated by selecting the α unassigned vertices of highest "prior-degree". The prior degree is simply the degree of the vertex with respect to only those vertices that have

```
PROCEDURE Construction( )
{  Initialize all the vertices with frequency value of zero
   While there are unlabeled vertices and iter < max
   {  Create RCL according to the following priorities:
        [ α unlabeled vertices linked to the largest number
          of labeled vertices first.
          If ties exist → list unlabeled vertices with a duplex link
          to a labeled vertex first.  ]
      s=randomly select a vertex from the RCL
      Assign s a random but non-interfering channel
        [ If all choices of channels lead to interference →
          randomly assign s a channel
          and reinitialize every vertex that interferes with s.  ]
} }
```

FIGURE 3. GRASP Construction Phase for FAP

assignments. A vertex is selected from this list at random and a non-interfering channel assigned at random. Whenever a channel assignment to a transceiver interferes with a neighboring transceiver, the algorithm backtracks to the interfering neighbor(s), reinitializes it to zero, and removes it from the current solution set. When all communication and interference constraints are satisfied, the construction step returns a feasible assignment solution with reliable communication links. The iteration is repeated until all the transceivers or vertices have been assigned a frequency channel and there are no violated constraints. This solution is then processed by the local search phase for further improvements.

An example of the prior-degree and how the candidate lists are obtained is shown in Figure 4. The hollow circles are vertices that are still awaiting a channel assignment and the dark edges indicate a duplex (bidirectional) link. In the leftmost figure, vertex 5 is already assigned and vertex 8 is ranked highest in the candidate list. However, vertex 8 need not be selected as the next iterate since the selection is randomized within the top few candidates, $\{8, 2, 4\}$. Suppose vertex 8 is chosen and receives a channel assignment. The rightmost figure illustrates the subsequent RCL.

4.4.2. *GRASP Local Search Phase.* The GRASP local search phase in Figure 5 is is now presented. A solution generated by the construction phase is always feasible but due to the randomness in the vertex and channel selection rules it is usually not optimal. Since the objective is to minimize order (the total number of frequency channels in use), at each iteration the local search attempts to eliminate a channel from the construction assignment while not allowing the introduction of any new channels into the assignment. This process begins by choosing a channel currently in use and re-initializing all transceivers that are allocated that channel. For minimum order problems the channel that is feasible to the least number of transceivers is removed. If alternate channels are not available for each such vertex, the removed channel is reinstated into the solution set and another channel is chosen to be removed from the set. If no feasible solution can be found with the diminished set of available frequency channels, then the vertices that cause the interference are backtracked to and reassigned new channels that maintain feasibility. This process

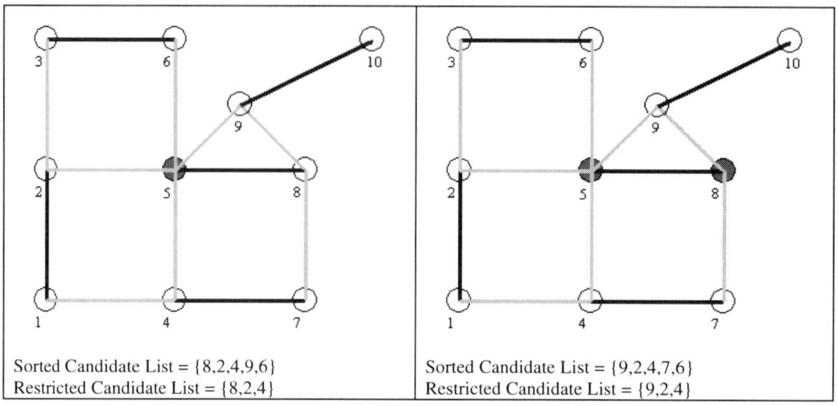

FIGURE 4. Candidate List Examples

```
PROCEDURE LocalSearch( )
{ For every channel f ∈ F
   { Select channel f_i using designated selection rule
    Re-initialize all vertices that are assigned channel f_i
      While there are unlabeled vertices and iter < max
      { Create RCL according to the phase 1 process
         s=randomly selected vertex from the RCL
         Assign s a non-interfering random channel
           [ If all choices of channels lead to interference →
             randomly assign s any feasible channel
             and re-initialize every vertex that interferes with s.  ]
      }
      If all vertices are labeled → update current solution
} }
```

FIGURE 5. Two Criteria Local Search Phase for FAP

is repeated until an attempt to eliminate every single channel in the solution set has been made.

The GRASP may be adapted to obtain a minimum span assignment by changing the channel selection rule in the local search. For the minimum span problem, the largest channel is removed. The minimum span local search phase requires less computation time since the elimination process terminates as soon as it encounters a channel that cannot be eliminated from the solution set whereas the minimum order search has to check every channel in the entire set.

5. Restricted Order Assignments

Local search is an effective way to improve solutions obtained by primal (sequential) heuristics. For minimum order FAPs, sequential methods like DSATUR often yield solutions of low or minimum order but large span. A local search method can be used to decrease the span of a feasible assignment while restricting the order. Two local search neighborhoods are studied in this paper. Each is motivated by the observation that the only way to decrease the span of an assignment is to reassign those vertices with the largest (highest) channel.

Neighborhood $\mathcal{N}1$. Begin with an assignment $f \in \mathbb{Z}_+^n$ obtained for $G, \mathbf{d}$ with highest channel $F = \max\{f\}$ and order X. Define the set $\mathcal{V}$ to be all vertices with channel F assigned. Consider all of the solutions that result when all of the vertices in $\mathcal{V}$ are reassigned to lower channels such that the order of the new solutions are each maintained at X. In this procedure, if $|\mathcal{V}| > 1$, then the vertex of largest degree is chosen to be reassigned first. Ties of degree are broken arbitrarily. This procedure is repeated on the best new solution and iterates in this manner until reassignment of all vertices in $\mathcal{V}$ is not possible. All of the solutions constructed during this process together comprise the neighborhood $\mathcal{N}1$. This neighborhood is not exact for the reasons discussed next.

Neighborhood $\mathcal{N}2$. This neighborhood is identical to $\mathcal{N}1$ with the exception that if reassignment of all vertices in $\mathcal{V}$ is not possible, then those vertices that cause the conflict are backtracked to and an attempt is made to reassign them. If complete backtracking is used, this neighborhood is exact. Hence if the initial solution is a minimum order solution, complete backtracking of $\mathcal{N}2$ will yield a restricted order solution. Realistically, complete backtracking is not time effective. Leung [**13**] found that great improvement in graph coloring problems may be made even with just one level of backtracking and that subsequent levels offer diminishing levels of improvement. Consequently, a *partial backtracking* scheme is used in this neighborhood where only one level of backtracking is permitted. Since searching $\mathcal{N}2$ is efficient and always yields solutions that are better than those obtained using $\mathcal{N}1$, we will not pursue $\mathcal{N}1$ any further.

Restricting order in the search. Searching neighborhoods $\mathcal{N}1$ and $\mathcal{N}2$ requires that the order not be increased beyond X. Ensuring that order does not increase requires a strategy that will force exactly one new channel into the new assignment and exactly one old channel out. Obviously, the largest channel should exit the solution to decrease span. However, there is no obvious rule for which channel should enter the solution or which vertices should use the new channel. To force such a rule, we recognize that a feasible co-channel assignment with order X is equivalent to a collection of X independent sets $y_1, y_2, \ldots, y_X$ which partitions $V(G)$. If the sets $y_1, y_2, \ldots, y_X$ are represented as vertices on a graph G^A, then reassigning the channel on vertex y_j results in the reassignment on all vertices x_i in the set y_j. Now consider the local search applied to graph G^A. Clearly, any solution obtained will always have order restricted to X as we desire. One issue remains; the edges of G^A must be defined. The channel distances $\mathbf{d}^A$ for G^A may be found as:

$$d_{ij}^A = \max_{k,l}\left\{ d_{kl} : \text{for all } k \in y_i, l \in y_j \right\}.$$

Therefore the edges of the alternate graph are:

$$(y_i, y_j) \in E(G^A) \text{ if } d_{ij}^A \neq 0 \text{ for all } y_i, y_j \in V(G^A),\ i \neq j.$$

Suppose that sets y_i and y_j are mutually independent for $i \neq j$, that is, there exists no vertex in y_i that is adjacent to any vertex in y_j. If this is true, then y_i and y_j may be combined as a single, larger independent set. It is easy to see that if

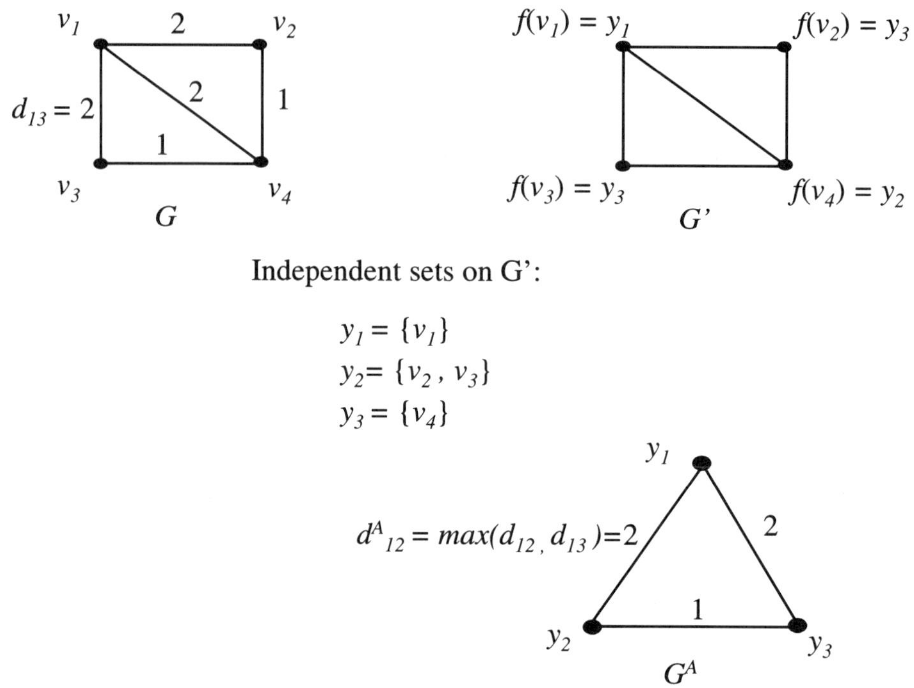

FIGURE 6. Example of Construction of G^A

sets y_i and y_j cannot be combined for all $i \neq j$, then G^A is complete. An example of the construction of G^A from G is given in Figure 6.

6. Restricted Span Assignments

A GRASP heuristic is used to generate approximate restricted span solutions through a process termed *iterated domain restriction*. Indeed, if we begin with a minimum order solution, this technique is able to approximately construct the order–span efficient frontier beginning with minimum order solutions, then the restricted order solution, the efficient solutions and finally ending with the restricted span solution. To create the efficient solutions, we begin with a minimum order solution and restrict the span by removing the largest channel. The minimum order GRASP algorithm is applied to the problem with the assignment made on the restricted domain. If feasible, the domain is restricted by removing the remaining largest channel and GRASP reapplied. This procedure continues until the span can no longer be decreased. This final solution is an approximation to the restricted span assignment. All of the solutions generated through this process approximate the efficient frontier.

Clearly this process need not be constrained to using GRASP for obtaining the minimum order solutions. However, as will be shown in Section 8.1, GRASP is able to obtain very good minimum order solutions in minimal time.

7. Three Criteria Problems

The multicriteria framework need not constrain the objectives to span and order. A three criteria model is now presented which adds "minimize network interference" as a third objective.

Recall that an edge is placed on the interference graph if the frequencies assigned to the end vertices have the potential to interfere with one another. That is $(v_i, v_j) \in E(G)$ if $|f(v_i) - f(v_j)| < d_{ij}$ where d_{ij} is the minimum channel distance required to mitigate interference. The values of d_{ij} are established based upon some notion of what "acceptable" interference is. This level may be influenced by many factors, including spatial separation of transceivers, pattern and polarization of antennas, and filtering, all with associated costs to the system. Indeed, adding filtering and special antenna features to each transceiver will lower the potential for interference and d_{ij} will increase, but at the price of adding complexity and cost to each transceiver.

Consequently, a three criteria model of the frequency assignment problem that includes interference allows the decision maker to examine the trade-offs associated with three fundamental cost drivers: total spectrum used (span), channel diversity (order), and transceiver complexity (interference). The three criteria model is clearly superior since it permits the decision maker to directly trade-off hardware cost and spectrum cost/availability.

There are potentially many approaches to representing interference levels. The approach used in this study defines the interference level as the total number of violated inequality constraints. We will assume that equality constraints must always be satisfied since parallel links are understood to be set at the absolute minimum channel distance for which the transceiver pair can reliably communicate. Unfortunately, this model of interference level does not completely represent the overall interference experienced by a system of transceivers. For example, no distinction is made as to how much actual communication deterioration is experienced by the transceiver pairs. It does not matter if the channel distance between two transceivers almost satisfies the constraint between them or the same channel had been assigned to both transceivers. This lack of discrimination may lead to a change in a transceiver's channel assignment from one which causes a minor and workable interference with it's neighbors to one which could completely shutdown communication to a nearby transceiver pair. However, the simplified definition of the system interference level is sufficient to show how a real life based problem may behave when interference is allowed.

The process of finding the three criteria solutions is essentially the same as the GRASP restricted span method with the relaxation of allowing an arbitrary number of interference constraints to be violated. To draw the three dimensional efficient surface, GRASP is repeatedly applied to find the minimum order given a restricted span and a maximum number of violated constraints. Unfortunately, the backtracking method used so successfully in the GRASP for finding minimum order and restricted span solutions is not applicable to problems where interference constraints are allowed to be violated. This is due to the fact that in the GRASP, backtracking is used to remove a channel assignment to a transceiver if that assignment violates the constraints with the last assigned transceiver. Clearly, this process does not allow any kind of interference and will only return solutions which satisfies all constraints.

The GRASP construction phase is slightly different from that for which no interference is allowed. Instead of backtracking whenever interference is encountered, the construction phase does not reinitialize any previously labeled vertices. Once all the vertices have been assigned an available frequency channel, regardless of whether interference is incurred or not, the construction phase attempts to find a solution which has a total number of violated constraints that is less than or equal to the maximum number allowed. If a solution is found with acceptable interference, GRASP begins the local search phase.

The local search phase of the GRASP algorithm is only called if the construction phase solution has an acceptable interference level. As in the construction phase, backtracking is not possible in the local search. Besides not reinitializing any vertex that causes a constraint violation, the second version of the local search phase is exactly the same as the original. This phase attempts to eliminate any unnecessary frequency channels one at a time from the solution set.

8. Experimental Results

The test problems for all the experiments were obtained from the *European Cooperation on the Long term in Defense* (EUCLID) program *Combinatorial Algorithms for Military Applications* (CALMA) project at web site

ftp://ftp.win.tue.nl/pub/techreports/CALMA/index.html.

8.1. Minimum Order GRASP Results. A GRASP for solving minimum order FAPs was coded in C and run on a SUN SPARCstation. The test instances used in the analysis of the GRASP algorithm for both minimum order and multicriteria problems are three empirical instances provided by the EUCLID CALMA project and five artificially generated instances with the same characteristics as the empirical instances. All eight instances are known to have feasible assignments, that is an assignment with no interference. The stated objective of these instances is minimum order but multicriteria solutions were also obtained for them.

Simulated Annealing and Taboo Search were applied to the same test problems and were used as benchmark algorithms for the minimum order problems. The benchmark results come from the T.U. Eindhoven FAP Group and were tested on SUN SPARCstations. Because the two benchmark algorithms are tested on different versions of the same type of machine, care should be taken in any running time comparisons. The benchmarks, however, do provide a good basis for qualifying the GRASP results.

Compared to the Simulated Annealing and Taboo Search benchmarks, GRASP proved to be a more effective method of solving minimum order FAPs. Table 1 compares the performance of all three algorithms on the 8 test problems. The size of instances ranged from a modest graph of 200 vertices with 1,235 constraints to an extremely complex graph of 916 vertices with 5,548 constraints. The average local optimum values and the average computational time obtained from 100 GRASP main steps are reported.

For the empirical instances, the GRASP results compare favorably to both the Simulated Annealing and Tabu Search algorithms, especially as the size of the FAP instance increases. GRASP running times did not increase as dramatically as the two benchmarks when the communication network size became larger. Even in the artificially generated instances, GRASP running times remained relatively

TABLE 1. Minimum Order GRASP Solutions For Test Instances

Empir. Instances			SA		TS		GRASP			
								order		time[s]
Inst	var.	ctr.	order	time[s]	order	time[s]	lo	ave.	hi	ave.
1	916	5548	*16	396	*16	705	*16	16.16	18	97.48
2	200	1235	*14	2	*14	7	*14	14.10	16	27.30
3	400	2760	16	6	16	32	*14	15.88	18	83.22
Gener. Instances										
Inst	var.	ctr.	order	time[s]	order	time[s]	lo	ave.	hi	ave.
1	200	1134	*18	—	*18	—	*18	18.00	18	11.19
2	400	2245	*14	—	*14	—	*14	14.38	16	15.03
3	680	3757	*18	—	*18	—	*18	19.88	24	94.24
4	916	5245	*18	—	22	—	*18	19.66	26	190.19
5	916	4638	*10	—	*10	—	*10	11.60	12	31.71

TABLE 2. Restricted Order Solutions

problem	algorithm	order	span	run-time (sec)
TUD200.3	DSATUR	12	1637	41
	with $\mathcal{N}2$	12	1566	117
TUD916.1	DSATUR	18	2665	396
	with $\mathcal{N}2$	18	2632	476
TUD916.3	DSATUR	18	2765	390
	with $\mathcal{N}2$	18	2584	471

short and predictable. The GRASP average optimal values also stayed close to the true optimal solutions, denoted with asterisks, while the maximum optimal value returned from all the iterations still provided a reliable approximation.

8.2. Restricted Order Results. For the restricted order solutions, the test instances are all from the EUCLID CALMA ftp site *DUTtest1.tar.gz* file with the domains removed and all equality constraints replaced with $\geq$ constraints. This should result in problems with much larger feasible search spaces than those of the domain constrained instances.

For each test problem, a minimum order solution was obtained using DSATUR with $\lceil 0.01 \times n \rceil$ random starts as a starting heuristic and then improved (order reduced) using an iterated 1-opt (vertex 2-exchange). The level of improvement that the iterated 1-opt yielded was usually slight if at all due to the high quality of the DSATUR solution. These solutions were used as the starting point for the restricted order local search algorithm for $\mathcal{N}2$. The local search algorithm was coded in Matlab 5.1 and run on a Pentium 166 with 24Mbytes of memory. Ten replications were run for each instance/algorithm combination and the sample mean values recorded in Tables 2. As can be seen, the local search approach is able to achieve modest decreases in span without any increase in order in each of the 3 test problems.

TABLE 3. Restricted Span Solutions

Empir. Instances	span	order
1	680	16
2	394	14
3	680	14
	666	16
	652	18
Numer. Instances	**span**	**order**
1	652	18
2	652	14
3	750	18
	736	20
	722	24
4	764	18
	750	20
5	390	10
	380	12

8.3. Restricted Span Results. Using the iterated domain restriction with GRASP to obtain approximate restricted span solutions and, in so doing, an approximation to the efficient frontier yielded mixed results. The solutions are given in Table 3. The table reports data from the smallest span at which the optimal order was still obtained (approximately the restricted order solution) through the smallest span at which a feasible solution still existed (approximately the restricted span solution). Unlike the test problems for restricted order, we maintained the channel domains and equality constraints for the restricted span experiments. This was done to reduce the number of iterations of span restriction. However an unexpected result of using limited domains was the inability to significantly reduce the span simply due to the lack of alternate channels that can be assigned to any given transceiver. The best result was for the third generated instance which is plotted in Figure 7. The figure shows the efficient solutions with span as the ordinate and order as the abscissa. From this plot, a decision may be made as to which assignment solution would be most appropriate and beneficial.

8.4. Three Criteria Results. Because of the significant amount of computation time involved in finding efficient curves, three-dimensional optimization was only applied to two empirical test problems. For these problems, an interference cost of zero, ten, and twenty violated constraints were used. The three criteria FAP results are shown in Table 4. For each interference level, the efficient frontier is constructed by iteratively restricting the domain, ending with the restricted span solution. At any iteration, if the spans of the current solution and restricted order solutions are not the same, then the span is incrementally increased and a minimum order problem is solved. For the zero interference solutions of test problem 3, the minimum span solution has a span of 722, the minimum order solution has an order of 14, and the minimum order solution given that the span is restricted to 666 has order 16. Plotting these three points will give a very brief two dimensional efficient curve. If the same procedure is done for increasing interference levels, a

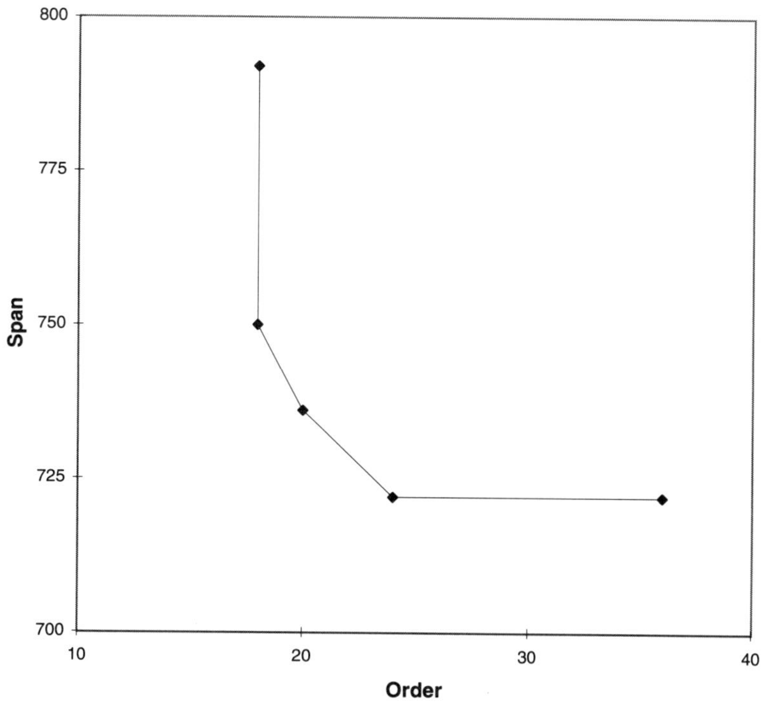

FIGURE 7. Two Criteria Solutions For Generated Data #3

TABLE 4. Three Criteria Results

	Cost = 0		Cost = 10		Cost = 20	
Empir. Instances	Span	Order	Span	Order	Span	Order
2	394	14	652	10	352	10
			366	12	338	12
			352	14		
3	722	14	680	12	652	12
	666	16	652	14	380	14
	652	18	394	16		

set of nonintersecting efficient curves may be plotted. The solutions for the third empirical test problem are plotted in Figure 8.

Plotting the three criteria assignment problem should produce non-intersecting two-dimensional curves corresponding to different interference levels. However, the approximate efficient curves constructed in our experiments are not symmetrical and sometimes overlap. Nonetheless, the figure does show how simply allowing 10 out of 2,760 constraints to be violated can dramatically drop the minimum span from 652 to 394. Even the minimum order drops from 14 to 12 channels. Increasing the number of violated constraints from 10 to 20 does not result in appreciable improvement. From this figure, it may be worthwhile to consider allowing a limited amount of interference to save cost.

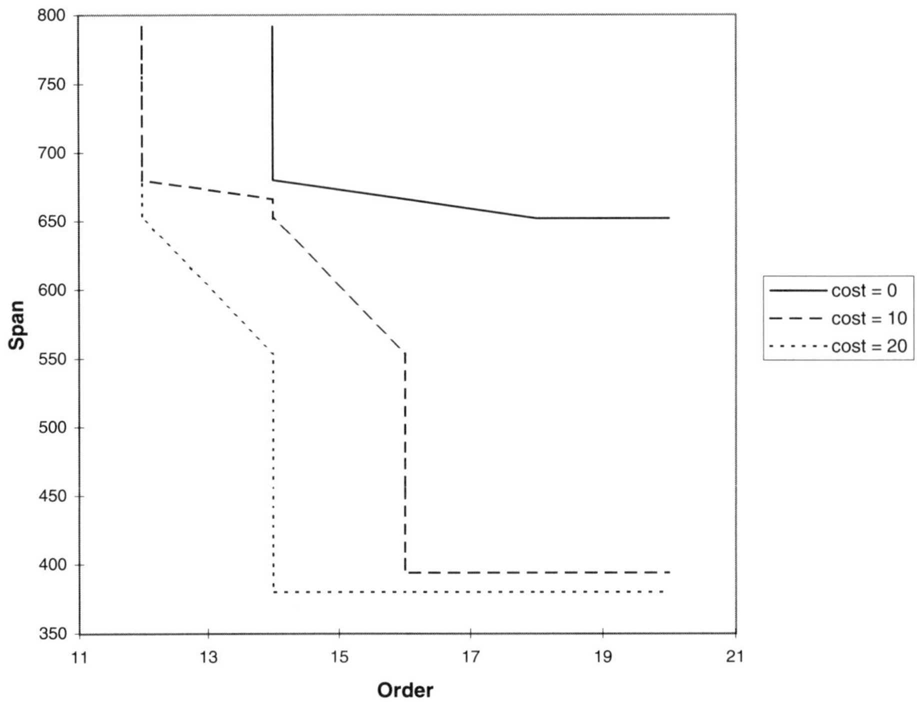

FIGURE 8. Three Criteria Solutions for Empirical Data #3

9. Conclusions

GRASP was shown to be an effective minimum order solver due to the quality of the solutions it obtains and also due to the ability to constrain running time by fixing the number of major iterations. Development and test of this algorithm, perhaps on GSM network data, should continue.

The multicriteria solvers illustrate that trade-offs between spectrum utilization (span) and transceiver complexity (order and interference) can be made effectively within a multicriteria framework. The solutions were not of the best quality due in a major part to the structure of the test problems. The CELAR test instances all assume a given level of transceiver complexity by limiting channels to small domains and by fixing interference levels. New test problems that permit the system designer to easily and accurately adjust domains and interference levels are needed to properly test the benefits of the multicriteria solvers.

The local search method discussed in Section 5 for finding restricted order solutions appears to reduce span fairly effectively. However, since the test problems have unknown restricted order, it is difficult to quantify the results. Clearly test problems are needed that have known efficient solutions.

The efficient and restricted span solutions obtained by the iterated domain restriction approach in Section 6 shows great promise. To allow greater detail in the efficient curve, the channel domains need to be removed. New test problems may help make a better case for this algorithm.

The 3 criteria algorithm also shows great promise but was primarily limited by the inability to backtrack with interference present. Consequently, backtracking techniques in the presence of interference are needed. Finally, as was previously mentioned, our measure of interference is not adequate for accurately representing the trade-off between spectrum and interference. Other measures of interference level should be developed.

References

[1] D. Brelaz , *New methods to color the vertices of a graph*, Communications ACM, Vol. **22**, (1979), 251–256.

[2] R. Borndörfer, A. Eisenblätter, M. Grötschel and A. Martin, *Frequency assignment in cellular phone networks*, Annals of Operations Research **76** (1998) 73–93.

[3] R.E. Burkard and F. Rendl, *Lexicographic Bottleneck problems*, Operations Research Letters, **10**, (1991) 303–308.

[4] M.B. Cozzens and F.S. Roberts, *T-colorings of graphs and the channel assignment problem*, Congressus Numerantium, Vol 35 (1982), 191–208.

[5] J. Culberson, A. Beacham and D. Papp, *Hiding our colors*, CP'95 Workshop, Studying and Solving Really Hard Problems, Cassis, France, September (1995), 31–42.

[6] T.A. Feo and M.G.C. Resende, *A probabilistic heuristic for a computationally difficult set covering problem*, Operations Research Letters **8** , (1989), 67–71.

[7] T.A. Feo, M.G.C. Resende, and S.H. Smith, *A greedy randomized adaptive search procedure for a maximum independent set*, Operations Research **42**, (1994), 860–876.

[8] T.A. Feo and M.G.C. Resende, *Greedy randomized adaptive search procedures*, Journal of Global Optimization **6**, (1995), 109–133.

[9] M. R. Garey and D. S. Johnson, *Computers and Intractability, A Guide To The Theory of $\mathcal{NP}$-Completeness*, W.H. Freeman and Company, New York, NY (1979).

[10] W.K. Hale, *Frequency assignment: theory and applications*, Proc. of The IEEE, Vol 68, No. **12**, (1980), 1497–1514.

[11] R.L. Keeney and H. Raiffa, *Decisions With Multiple Objectives: Preference and Value Trade-offs*, John Wiley and Sons, Inc., (1976).

[12] T.A. Lanfear, *Graph Theory and Radio Frequency Assignment*, Technical Report, Allied Radio Frequency Agency NATO Headquarters, B-1110, Brussels, Belgium, (1989).

[13] D.S.P. Leung, *Application of the partial backtracking technique to the frequency assignment problem*, IEEE International Symposium on Electromagnetic Compatibility, Boulder, CO (1981), 70–74.

[14] J. Peemöller, *A correction to Brelaz's modification of Brown's coloring algorithm*, Communications of the ACM, Vol. 26, No. **8**, (1983), 595–597.

[15] A. Raychaudhuri, *Intersection assignments, T-coloring, and powers of graphs*, Ph.D. Thesis, Department of Mathematics, Rutgers University, New Brunswick, NJ, (1985).

[16] L. Yong, P.M. Pardalos, and M.G.C. Resende. *A greedy randomized adaptive search procedure for the quadratic assignment problem*, Quadratic Assignment and Related Problems, Panos M. Pardalos and Henry Wolkowicz (eds.), DIMACS Series on Discrete Mathematics and Theoretical Computer Science, **16**, American Mathematical Society, (1994), 237–262.

[17] J.A. Zoellner and C.L. Beall, *A breakthrough in spectrum conserving frequency assignment technology*, IEEE Transactions on Electromagnetic Compatibility, Vol EMC-19, (1977), 313–319.

AIR FORCE RESEARCH LABORATORY MUNITIONS DIRECTORATE, EGLIN AFB, FL 32542 USA
E-mail address: murphey@eglin.af.mil

CENTER FOR APPLIED OPTIMIZATION, ISE DEPARTMENT, UNIVERSITY OF FLORIDA, GAINESVILLE, FL 32611 USA
E-mail address: pardalos@ufl.edu

CENTER FOR APPLIED OPTIMIZATION, ISE DEPARTMENT, UNIVERSITY OF FLORIDA, GAINESVILLE, FL 32611 USA
E-mail address: eduardo@ufl.edu

DIMACS Series in Discrete Mathematics
and Theoretical Computer Science
Volume **52**, 2000

Randomized Initialization Protocols for Packet Radio Networks *

Tatsuya Hayashi† Koji Nakano† Stephan Olariu‡

Abstract

The main contribution of this work is to propose efficient randomized initialization protocols for Packet Radio Networks (PRN, for short). As a result of the initialization protocol, the n stations of a PRN are assigned distinct integer IDs from 1 to n.

First, we show that if the number n of stations is known beforehand, the single-channel PRN can be initialized in $O(n)$ broadcast rounds with probability at least $1 - \frac{1}{2^n}$, regardless of whether or not the PRN has the collision detection (CD) capability.

We then go on to show that even if n is not known in advance, the single-channel PRN with CD can be initialized in $O(n)$ broadcast rounds with probability at least $1-\frac{1}{2^n}$. Using this protocol as a stepping stone, we then present an initialization protocol for the k-channel PRN with CD that terminates in $O(\frac{n}{k})$ broadcast rounds with probability at least $1 - \frac{1}{n}$, whenever $k \leq \frac{n}{3\log n}$.

Finally, we look at the case where the CD capability is not present. Our first result in this scenario is to show that the well-known *leader election* problem can be solved on the single-channel PRN in $O((\log n)^2)$ broadcast rounds with probability at least $1 - \frac{1}{n}$. This leader election protocol allows us to design an initialization protocol for the single-channel PRN with no CD that terminates in $O(n)$ broadcast rounds with probability at least $1 - \frac{1}{2\sqrt{n}}$. We then discuss an initialization protocol for the k-channel PRN with no CD that terminates in $O(\frac{n}{k})$ broadcast rounds with probability at least $1-\frac{1}{n}$, whenever $k \leq \frac{n}{4(\log n)^2}$.

AMS Subject Classification: 68W15, 68W20, 68W40.

1 Introduction

A Packet Radio Network (PRN, for short) $\mathcal{S}$ consists of n radio transceivers, henceforth referred to as *stations*. The stations are identical and cannot be

*Work supported in part by the NSF grant CCR-9522093 and by ONR grant N00014-97-1-0526.

†Department of Electrical and Computer Engineering, Nagoya Institute of Technology, Showa-ku, Nagoya 466, JAPAN, `nakano@elcom.nitech.ac.jp`

‡Department of Computer Science, Old Dominion University, Norfolk, Virginia 23529, USA, `olariu@cs.odu.edu`

distinguished by serial or manufacturing number. We refer the reader to Figure 1 depicting a 10-station PRN.

The classical *leader election* problem asks to designate one of the station as *leader*. In other words, after performing the leader election protocol, exactly one station learns that it was elected leader, while the remaining stations learn the identity of the leader elected. The leader election problem is fundamental, for many other problems rely directly or indirectly on the presence of a leader in a network [4, 19, 20].

The *initialization* problem is to assign to each of the n stations in $\mathcal{S}$ an integer ID number in the range $[1, n]$ such that no two stations are assigned the same ID. The initialization problem is fundamental in both network design and in multiprocessor systems [4, 17].

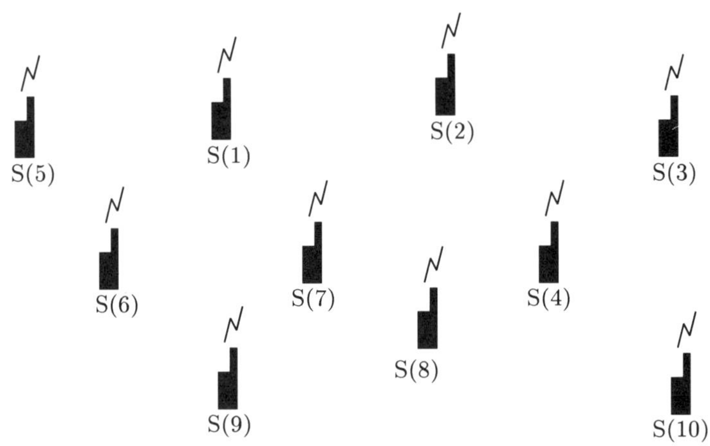

Figure 1: *Illustrating a 10-station PRN.*

As customary, the time is assumed to be slotted and all the stations have a local clock that keeps (synchronous) time. Also, all broadcast operations are performed at time slot boundaries.

The stations are assumed to have the computing power of a usual laptop computer; in particular, they all run the same protocol and can generate random bits that provide "local data" on which the stations may perform computations. The stations communicate using k radio channels denoted by $C(1), C(2), \ldots, C(k)$. We assume that in any time slot, a station can tune in to one such channel and/or broadcast on at most one (possibly the same) channel. A broadcast operation involves a data packet whose length is such that the broadcast operation can be completed within one time slot.

We employ two assumptions in terms of the capability of the system. In the PRN with *collision detection* (CD, for short), the status of a radio channel is:

- *NULL:* if no station broadcasts on the channel in the current time slot,
- *SINGLE:* if exactly one station broadcasts on the channel in the current time slot, and
- *COLLISION:* if two or more stations broadcast on the channel in the current time slot.

In the PRN with no collision detection the status of a radio channel is:

- *NULL:* if either no station broadcasts or two or more stations broadcast on the channel in the current time slot, and
- *SINGLE:* if exactly one station broadcasts on the channel in the current time slot.

Several workers have argued that from a practical standpoint the no CD assumption makes a lot of sense since in many situations, especially in the presence of noisy channels, the stations cannot distinguish between the no broadcast case and the collision of several packets that arises when several stations attempt to broadcast at once [5, 6]. On the other hand, many other radio or cellular networks including AMPS, GSM, ALOHA-net, as well as the well-known Ethernet are systems where collision detection is possible [1, 2, 7, 8, 9, 14].

Recent advances in wireless communications and mobile computing have exacerbated the need for efficient protocols for Packet Radio Networks. Indeed, a large number of such protocols have been reported in the literature [3, 11, 12, 16, 18]. However, many of these protocols function under the assumption that the n stations in the PRN have been initialized in advance. The highly non-trivial task of assigning the stations distinct ID numbers, i.e. initializing the stations, is often ignored in the literature. It is, therefore, of importance to design efficient initialization protocols for PRNs both in the case where the system has a collision detection capability and for the case where this capability is not present.

The main contribution of this work is to propose efficient randomized initialization protocols for Packet Radio Networks. To begin, we show that if the number n of stations is known in advance, the single-channel PRN can be initialized in $O(n)$ broadcast rounds with probability at least $1 - \frac{1}{2^n}$, regardless of whether or not the collision detection capability is present.

We then go on to address the initialization problem in the more realistic case where the number n of stations is not known ahead of time. We show that the single-channel PRN with CD can be initialized in $O(n)$ broadcast rounds with probability at least $1 - \frac{1}{2^n}$. We then generalize this result by showing that the k-channel PRN with CD can be initialized in $O(\frac{n}{k})$ broadcast rounds with probability at least $1 - \frac{1}{n}$, whenever $k \leq \frac{n}{3 \log n}$.

Next, we design a *leader election* protocol for the single-channel PRN with no CD that terminates in $O((\log n)^2)$ broadcast rounds with probability at least $1 - \frac{1}{n}$. This leader election protocol is key in designing an initialization protocol for the single-channel PRN that terminates in $O(n)$ broadcast rounds with probability at least $1 - \frac{1}{2\sqrt{n}}$. Using this protocol, we design an initialization protocol for the k-channel PRN with no CD that runs in $O(\frac{n}{k})$ broadcast rounds with probability at least $1 - \frac{1}{n}$, provided that $k \leq \frac{n}{4(\log n)^2}$.

The remainder of this extended abstract is organized as follows. Section 2 discusses the initialization problem for known n. Section 3 discusses the more realistic case where the number of stations in the PRN is not known beforehand. Finally, in Section 4 we address the problems of leader election, and initialization in the PRN with no collision resolution in case the number n of stations in not known beforehand.

2 Initializing the single-channel PRN for known n

We begin this section by reviewing basic probability theory results that are useful for analyzing the performance of our protocols. For a more detailed discussion of background material we refer the reader to [13, 15]. We then go on to discuss a simple initialization protocol for the single-channel PRN in case the number n of stations is known. Our protocol does not rely on collision detection and, therefore, works regardless of whether or not the PRN has collision detection capabilities.

Throughout, $\Pr[A]$ will denote the probability of event A. For a random variable X, $E[X]$ denotes the expected value of X. Let X be a random variable denoting the number of successes in n independent Bernoulli trials with parameters p and $1-p$. It is well known that X has a *binomial distribution* and that for every r, $(0 \leq r \leq n)$,

$$\Pr[X = r] = \binom{n}{r} p^r (1-p)^{n-r}. \tag{1}$$

Further, the expected value of X is given by

$$E[X] = \sum_{r=0}^{n} r \cdot \Pr[X = r] = np. \tag{2}$$

To analyze the tail of the binomial distribution, we shall make use of the following estimates, commonly referred to as *Chernoff bounds*:

$$\Pr[X \leq (1-\epsilon)E[X]] \leq e^{-\frac{\epsilon^2}{2}E[X]} \qquad (0 \leq \epsilon \leq 1). \tag{3}$$

$$\Pr[X \geq (1+\epsilon)E[X]] \leq e^{-\frac{\epsilon^2}{3}E[X]} \qquad (0 \leq \epsilon \leq 1). \tag{4}$$

Consider a single-channel PRN where the number n of stations is known to all the participants. Let $C(1)$ stand for the unique channel available. The idea of our protocol is to assign ID numbers sequentially starting with 1. The details are spelled out as follows:

Protocol `Initialization-for-known-n`
for $m \leftarrow n$ **down to** 1 **do**
  repeat
    each station broadcasts on channel C(1) with probability $\frac{1}{m}$; ($\star$)
  until the status of channel C(1) is SINGLE;
  the station that has broadcast in the previous round is declared
  the $(n - m + 1)$-th station;
endfor

The correctness of protocol `Initialization-for-known-n` being easily seen, we now turn to the task of evaluating the number of broadcast rounds it takes the protocol to terminate. We say that the current broadcast round in step ($\star$) is *successful* if the status of channel C(1) is SINGLE. Let X be the random variable denoting the number of stations broadcasting in a given round. Then, by virtue of (1), at the end of this round the status of the channel is SINGLE with probability

$$\begin{aligned}\Pr[X = 1] &= \binom{m}{1}\left(\frac{1}{m}\right)^1\left(1 - \frac{1}{m}\right)^{m-1} \\ &= \left(1 - \frac{1}{m}\right)^{m-1} > \frac{1}{e}.\end{aligned}$$

Clearly, the protocol `Initialization-for-known-n` requires n successful rounds to terminate. Let Y be the random variable denoting the number of successful rounds among the first $4en$ broadcast rounds in step ($\star$) of the protocol. It is clear that $E[Y] > 4n$. By virtue of (3) with $\epsilon = \frac{3}{4}$, we have

$$\Pr[Y < n] < \Pr[Y \leq (1 - \frac{3}{4})E[Y]] \leq e^{-\frac{9}{32}E[Y]} < e^{-\frac{9}{8}n} < 2^{-\frac{9}{8}n} < \frac{1}{2^n}.$$

We just proved that with probability at least $1 - \frac{1}{2^n}$ among the first $4en$ broadcast rounds there are at least n successful rounds. Moreover, our protocol does not rely on collision detection. Therefore, we have the following result.

Lemma 2.1 *Let S be an n-station, single-channel PRN with no collision detection. If n is known beforehand, Protocol* `Initialization-for-known-n` *terminates in $O(n)$ broadcast rounds with probability at least $1 - \frac{1}{2^n}$.*

It is interesting to note that Lemma 2.1 asserts that if the number of stations in the PRN is known, for the purpose of initialization it does not matter whether the PRN has collision detection or not. In the remainder of the paper we shall see that if the number of station in the PRN is *not known* in advance the collision detection capability makes a difference.

3 Initializing the PRN with CD and unknown n

This section is devoted to presenting an initialization protocols for the PRN with CD, in the case where the number n of stations is not known beforehand. We begin by presenting an initialization protocol for the single-channel PRN that terminates in $O(n)$ broadcast rounds with probability at least $1 - \frac{1}{2^n}$. Next, we generalize this result to the case of the k-channel PRN with CD. Specifically, we discuss an initialization protocol for this case that terminates in $O(\frac{n}{k})$ broadcast rounds with probability at least $1-\frac{1}{n}$, whenever $k \leq \frac{n}{3\log n}$.

3.1 Initializing the single-channel PRN

In outline, our protocol partitions the stations into non-empty subsets $\mathcal{P}_1$ and $\mathcal{P}_2$. In turn, $\mathcal{P}_2$ is also partitioned into two non-empty subsets $\mathcal{P}_2$ and $\mathcal{P}_3$. The same partition is then applied to $\mathcal{P}_3$. In general, $\mathcal{P}_{i-1}$ is partitioned into non-empty subsets $\mathcal{P}_{i-1}$ and $\mathcal{P}_i$. This procedure is repeated until, at some stage, some $\mathcal{P}_l$ contains a single station. This station receives the ID of 1 and quits the protocol. After that, the same partition procedure is applied to $\mathcal{P}_{l-1}$ and so on. This is then repeated until all stations in the PRN have been assigned IDs. In order to perform the partitioning above, we use the protocol `Partition-with-CD` whose details follow.

Protocol `Partition-with-CD`
repeat
  each station selects 0 or 1 with probability $\frac{1}{2}$;
  all the stations that selected 1 broadcast on channel C(1);
  let $Status(1)$ be the resulting status of C(1);
  all the stations that selected 0 broadcast on channel C(1);
  let $Status(0)$ be the resulting status of C(1)
until neither $Status(1)$ nor $Status(0)$ is NULL.

It is easy to see that protocol `Partition-with-CD` correctly partitions the set of stations into two non-empty subsets. We now present the details of our initialization protocol using `Partition-with-CD`. Each station maintains local variables l, L and N. Let $\mathcal{P}_i$ denote the set of stations whose local variable l has value i. Notice that the collision detection capability allows us to determine whether $|\mathcal{P}_i| = 0$, $|\mathcal{P}_i| = 1$, or $|\mathcal{P}_i| \geq 2$. This is done, simply, by mandating the stations in $\mathcal{P}_i$ to broadcast and by recording the resulting status of the channel.

Protocol `Initialization-with-CD`
Every station sets $l \leftarrow L \leftarrow N \leftarrow 1$;
while $L \geq 1$ **do**
  if $|\mathcal{P}_L| = 1$ **then**
    the unique station in $\mathcal{P}_L$ is declared the N-th station and leaves the protocol;
    $N \leftarrow N + 1$; $L \leftarrow L - 1$;
  else
    use protocol `Partition-with-CD` to partition $\mathcal{P}_L$ into two non-empty sets $\mathcal{P}_L$ and $\mathcal{P}_{L+1}$;
    $L \leftarrow L + 1$;
    each station in $\mathcal{P}_L$ sets $l \leftarrow L$;
  endif
endwhile

The correctness of protocol `Initialization-with-CD` being easily seen, we now turn to the task of evaluating the number of broadcast rounds it takes the protocol to terminate. To begin, the task of checking whether $|\mathcal{P}_L| = 1$ can be done in one broadcast. Let us estimate how many times the protocol checks if $|\mathcal{P}_L| = 1$. Note that in case $|\mathcal{P}_L| = 1$ the unique station in $\mathcal{P}_L$ is assigned an ID and leaves the protocol. Thus, exactly n times the condition $|\mathcal{P}_L| = 1$ must evaluate to "true" in the **if** statement. The protocol `Partition-with-CD` partitions a set of stations into two non-empty subsets. Hence, protocol `Partition-with-CD` must be executed exactly $n - 1$ times. Thus, exactly $n-1$ times the condition $|\mathcal{P}_L| = 1$ is "false" in the **if** statement. Therefore, the task of checking $|\mathcal{P}_L| = 1$ in the **if** statement requires $2n - 1$ broadcasts.

Next, we evaluate the number of broadcast rounds involved in protocol `Partition-with-CD`. Suppose that m, ($m \geq 2$), stations are to be partitioned into two non-empty sets using protocol `Partition-with-CD`. We say that an iteration of the `repeat-until` loop in protocol `Partition-with-CD` is *successful* if it succeeds in partitioning the set into two non-empty subsets. Let X be the random variable denoting the number of stations that selected a 1. Then, since $m \geq 2$, the probability of a successful iteration is

$$\Pr[1 \leq X \leq m - 1] = 1 - \Pr[X = 0] - \Pr[X = m] = 1 - \frac{1}{2^{m-1}} \geq \frac{1}{2}.$$

Since a successful iteration produces two non-empty sets, it is clear that `Initialization-with-CD` must perform, overall, $n-1$ successful iterations. Let Y be the random variable denoting the number of successes among $8n$

Bernoulli trials, with parameter $p = \frac{1}{2}$. It is clear that $E[Y] = 4n$. By virtue of (3) with $\epsilon = \frac{3}{4}$,

$$\Pr[Y < n-1] < \Pr[Y \le (1-\frac{3}{4})E[Y]] \le e^{-\frac{9}{32}E[Y]} = e^{-\frac{9}{8}n} < 2^{-\frac{9}{8}n} < \frac{1}{2^n}.$$

We just proved that with probability at least $1 - \frac{1}{2^n}$, among the first $8n$ iterations of the `repeat-until` loop in protocol `Partition-with-CD` there must exist at least $n-1$ successful iterations. Thus, with probability at least $1 - \frac{1}{2^n}$, `Initialization-with-CD` terminates in $O(n)$ broadcast rounds.

Theorem 3.1 *Even if n is not known beforehand, an n-station PRN with CD can be initialized in in $O(n)$ rounds with probability at least $1 - \frac{1}{2^n}$.*

Now, consider a PRN with m stations, where $m \le n$. For later reference, we now evaluate the probability that protocol `Initialization-with-CD` will take $O(n)$ broadcast rounds to initialize the PRN. Let Y be the random variable defined above. Then, it is clear that

$$\Pr[Y < m-1] < \Pr[Y < n-1] \le e^{-\frac{9}{8}n} < 2^{-\frac{9}{8}n} < \frac{1}{2^n}.$$

Corollary 3.2 *Protocol* `Initialization-with-CD` *initializes an m-station PRN with CD and unknown m, $m \le n$, in $O(n)$ rounds with probability at least $1 - \frac{1}{2^n}$.*

3.2 Initializing the k-channel PRN with CD

In this subsection we present an efficient initialization protocol for the k-channel PRN when the number n of stations is not known beforehand.

Protocol `Initialization-k-channel-PRN`

Stage 1 Each station selects one of the channels at random. Let P_i, $(1 \le i \le k)$, denote the set of stations that selected channel $C(i)$. For every i, $(1 \le i \le k)$, dedicate channel $C(i)$ to P_i and use the protocol `Initialization-with-CD` to initialize P_i. Let $p_{i,j}$, $(1 \le i \le k, 1 \le j \le |P_i|)$, denote the j-th station in P_i. At the end of this stage, each station $p_{i,j}$ knows its local ID number j within P_i.

Stage 2 Compute the prefix-sums of $|P_1|, |P_2|, \ldots, |P_k|$. That is, compute for every i, $(1 \le i \le k)$, the sum $|P_1| + |P_2| + \cdots + |P_i|$. Next, for every i, $(1 \le i \le k-1)$, broadcast the value $|P_1| + |P_2| + \cdots + |P_i|$ on channel $C(i+1)$. Now, each station $p_{i,j}$, $(2 \le i \le k)$, computes its ID whose value is $|P_1| + |P_2| + \cdots + |P_i| + j$.

It is worth noting that Stage 1 uses the randomized protocol developed in Subsection 3.1, while Stage 2 uses a deterministic protocol inspired by the well-known prefix sums algorithm for the PRAM [13]. We begin by evaluating the number of rounds involved in completing Stage 1. For this purpose, we show that no set P_i is likely to contain too many stations.

Fix a channel and let X be the random variable denoting the number of stations that selected that channel. It should be clear that $E[X] = \frac{n}{k}$. By using the Chernoff bound in (4) with $\epsilon = 1$, we can bound the probability $\Pr[X \geq \frac{2n}{k}]$ that the channel was selected by $\frac{2n}{k}$ or more stations as follows:

$$\Pr[X \geq \frac{2n}{k}] \leq e^{-\frac{2n}{3k}}.$$

Thus, the probability that some channel is chosen by at least $\frac{2n}{k}$ stations is less than $k \cdot e^{-\frac{2n}{3k}}$. It follows that, with probability at least $1 - k \cdot e^{-\frac{2n}{3k}}$, all the channels are chosen by fewer than $\frac{2n}{k}$ stations.

Let us assume that each of the k channels was selected by fewer than $\frac{2n}{k}$ stations. By Corollary 3.2, protocol `Initialization-with-CD` initializes an m-station, single-channel PRN, where $m < \frac{2n}{k}$, in $O(\frac{n}{k})$ rounds with probability at least $1 - e^{-\frac{9n}{4k}}$.

Note, however, that before the protocol can advance to Stage 2, it must check whether each of the k instances of `Initialization-with-CD` running in the k channels has finished executing. For this purpose, the execution of `Initialization-with-CD` is suspended every, say, 10 rounds. Using one broadcast in this suspended round, every station that has not yet been assigned a local ID number broadcasts on channel C(1). If the status of C(1) is NULL, we know that Stage 1 must be complete and Stage 2 may begin.

The above discussion implies that with probability at least

$$1 - k \cdot e^{-\frac{2n}{3k}} - ke^{-\frac{9n}{4k}} \quad \geq \quad 1 - 2k \cdot e^{-\frac{2n}{3k}}$$

Stages 2 is ready to begin at the end of $O(\frac{n}{k})$ broadcast rounds.

Recall that at the end of Stage 1, every station $p_{i,1}$, $(1 \leq i \leq k)$, knows the number $|P_i|$ of stations in P_i. Thus, Stage 2 can, in principle, use the stations $p_{i,1}$, $(1 \leq i \leq k)$, to compute the prefix-sums of $|P_1|, |P_2|, \ldots, |P_k|$. If every set P_i, $(1 \leq i \leq k)$, is non-empty, the prefix sums are computed by a trivial implementation of the prefix-sums algorithm for the PRAM [13]. However, in our case, some of the P_is may be empty and, consequently, some of the stations $p_{i,1}$ may not exist. Thus, we need to adapt the prefix-sums algorithm for the PRAM to work even if some of the stations do not exist.

The protocol to compute the prefix-sums is executed in a recursive manner. After the execution of the prefix-sums, for every i, $(1 \leq i \leq k)$, the following conditions are satisfied:

(ps1) Every existing station $p_{i,1}$, knows the prefix-sum $|P_1| + |P_2| + \cdots + |P_i|$,

(ps2) A leader is elected in $P_1 \cup P_2 \cup \cdots \cup P_i$ if at least one of $P_1, P_2, \ldots, P_i$ is non-empty, and (ps3) The elected leader knows the value of the sum $|P_1| + |P_2| + \cdots + |P_i|$.

Our prefix-sums protocol is as follows. If $k = 1$, elect station $p_{i,1}$ as the leader of P_1. Clearly, station $p_{i,1}$ knows the prefix sum $|P_1|$. Thus above conditions are satisfied.

If $k \geq 2$, partition $G = \{P_1, P_2, \ldots, P_k\}$ into groups $G_1 = \{P_1, P_2, \ldots, P_{\frac{k}{2}}\}$ and $G_2 = \{P_{\frac{k}{2}+1}, P_{\frac{k}{2}+2}, \ldots, P_k\}$. Recursively compute the prefix sums in G_1 and G_2 using channels $C(1), C(2), \ldots, C(\frac{k}{2})$ and $C(\frac{k}{2}+1), C(\frac{k}{2}+2), \ldots, C(k)$, respectively.

By the inductive hypothesis, when these prefix sums protocols terminate the conditions (ps1)–(ps3) above are satisfied in both G_1 and G_2. Notice that since every channel was chosen by fewer than $\frac{2n}{k}$ stations, both G_1 and G_2 must be nonempty and by (ps2) each of them must have a leader. The leader in G_1 broadcasts the value $S_1 = |P_1| + |P_2| + \cdots + |P_{\frac{k}{2}}|$ on channel $C(1)$. Every station in G_2 monitors channel $C(1)$ and updates the value of its prefix sum by adding the value broadcast to the value it stores (i.e its local prefix sum within G_2). It is easy to see that the value obtained corresponds to the correct prefix sum within G. Finally, the leader in G_2 broadcasts the sum $S_2 = |P_{\frac{k}{2}+1}| + |P_{\frac{k}{2}+2}| + \cdots + |P_k|$ on channel $C(1)$. The leader in G_1 adds together the sums S_1 and S_2 and is elected the leader of G. Note that the stations in G_1 do not have to do any updating, for the prefix-sum within G_1 is also that within G. If no station belongs to G_1, every station in G_2 can detect it, because the status of $C(1)$ is NULL when the leader in G_1 broadcasts S_1. In this case, the prefix-sum within G_2 is also that within G. Further, the leader of G_2 becomes that of G. Thus, conditions (ps2) and (ps3) are satisfied.

It is clear that the prefix sums protocol we just described performs two broadcast rounds excluding the recursive execution. Since the depth of recursion is $\lceil \log k \rceil$, the protocol performs $O(\log k)$ broadcast rounds.

After executing the prefix sums protocol, each station $p_{i,1}$ broadcasts the value $|P_1|+|P_2|+\cdots+|P_i|$ on channel $C(i+1)$. Every station $p_{i+1,j}$ monitors channel $C(i+1)$ and computes $|P_1| + |P_2| + \cdots + |P_i| + j - 1$, which is its ID number. Consequently, Stage 2 terminates in $O(\log k)$ rounds. Thus, we have proved the following result.

Theorem 3.3 *Even if the number n of stations is not known beforehand, the k-channel PRN with CD can be initialized in $O(\frac{n}{k} + \log k)$ rounds with probability at least $1 - k \cdot e^{-\frac{2n}{3k}}$.*

Assuming $k \leq \frac{n}{3 \log n}$, we have

$$k \cdot e^{-\frac{2n}{3k}} = \frac{n}{3 \log n} \cdot e^{-2 \log n} < n \cdot \frac{1}{n^{2.88}} < \frac{1}{n},$$

and the number of rounds is $O(\frac{n}{k} + \log k) = O(\frac{n}{k})$. Consequently, we have the following result.

Corollary 3.4 *Even if the number n of stations is not known, the k-channel PRN with CD can be initialized in $O(\frac{n}{k})$ rounds with probability at least $1-\frac{1}{n}$, whenever $k \leq \frac{n}{3\log n}$.*

4 Initializing the PRN with no CD and unknown n

The main goal of this section is to address the problems of:

- leader election, and
- initialization

on the PRN with no collision resolution in case the number n of stations in not known beforehand.

4.1 Leader election on the PRN with no CD for unknown n

This subsection presents a leader election protocol on the PRN with no CD. The details of the protocol follow.

Protocol `Election-with-no-CD`
for $i \leftarrow 0$ to ∞ do
  for $j \leftarrow 0$ to i do
    each station broadcasts on channel C(1) with probability $\frac{1}{2^j}$;
    if the status of C(1) is SINGLE `then`
      the station broadcasting is declared the leader and
      the protocol terminates;
    endif
  end for
end for

It is clear that protocol `Election-with-no-CD` terminates with the correct election of a leader. Let s be the unique integer satisfying $2^s \leq n < 2^{s+1}$. We say that a broadcast round is *good* if $j = s$, that is, if $2^j \leq n < 2^{j+1}$. A good round succeeds in finding a leader with probability at least

$$\binom{n}{1}\left(\frac{1}{2^s}\right)^1\left(1-\frac{1}{2^s}\right)^{n-1} \geq 2^s\frac{1}{2^s}\left(1-\frac{1}{2^s}\right)^{n-1} \geq \left(1-\frac{1}{2^s}\right)^{2^{s+1}-1} > \frac{1}{e^2}.$$

Thus, the first t good rounds fail with probability at most

$$\left(1-\frac{1}{e^2}\right)^t \leq e^{-\frac{t}{e^2}}. \qquad \text{(since } 1+x \leq e^x\text{)}$$

In other words, with probability at least $1-e^{-\frac{t}{e^2}}$ a leader is elected in t good rounds. On the other hand, the first $s+t+1$ iterations of the outer `for`-loop, corresponding to $i = 0, 1, \ldots, s, s+1, \ldots, s+t$ contain t good rounds. It follows that the first $1 + 2 + \cdots + (s+t+1) = O((s+t)^2)$ broadcast rounds contain t good rounds. Since $s = \lfloor \log n \rfloor$ we have proved the following result.

Theorem 4.1 *Protocol* `Election-with-no-CD` *elects a leader in* $O(t^2 + (\log n)^2)$ *broadcast rounds with probability at least* $1 - e^{-\frac{t}{e^2}}$.

Choosing $t = e^2 \log n$ and $t = e^2\sqrt{n}$, we obtain the following important result.

Corollary 4.2 *Protocol* `Election-with-no-CD` *elects a leader in* $O((\log n)^2)$ *broadcast rounds with probability at least* $1 - \frac{1}{n}$ *or in* $O(n)$ *broadcast rounds with probability at least* $1 - \frac{1}{2\sqrt{n}}$.

4.2 Initializing the PRN with no CD and unknown n

Suppose that the stations in a subset $\mathcal{P}$ of $\mathcal{S}$ broadcast on channel C(1). If the PRN can detect collisions, then every station can determine whether $|\mathcal{P}| = 0$, $|\mathcal{P}| = 1$, or $|\mathcal{P}| \geq 2$. On the other hand, if the PRN does not have this capability, then one can only determine whether $|\mathcal{P}| = 1$ or $|\mathcal{P}| \neq 1$.

However, once a leader is elected in $\mathcal{S}$, the PRN with no CD can simulate that with CD in $O(1)$ rounds. In other words, the PRN with no CD can determine whether $|\mathcal{P}| = 0$, $|\mathcal{P}| = 1$, or $|\mathcal{P}| \geq 2$. The details follow.

First, the leader p broadcast to all stations whether $p \in \mathcal{P}$ or not. After that, the following protocol is executed:

Case 1: $p \in \mathcal{P}$. Since $|\mathcal{P}| \geq 1$, it is sufficient to check whether $|\mathcal{P}| = 1$ or $|\mathcal{P}| \geq 2$. The stations in $\mathcal{P}$ broadcast on channel C(1). If the status C(1) is SINGLE, then $|\mathcal{P}| = 1$, otherwise, it must be that $|\mathcal{P}| \geq 2$.

Case 2: $p \notin \mathcal{P}$. By mandating the stations in $\mathcal{P}$ to broadcast on channel C(1), we can determine if $|\mathcal{P}| = 1$ or $|\mathcal{P}| \neq 1$. Similarly, by mandating the stations in $\mathcal{P} \cup p$ to broadcast, we can determine if $|\mathcal{P}| = 0$ or $|\mathcal{P}| \neq 0$. These two results combined can determine whether $|\mathcal{P}| = 0$, $|\mathcal{P}| = 1$, or $|\mathcal{P}| \geq 2$ (i.e. $|\mathcal{P}| \neq 0$ and $|\mathcal{P}| \neq 1$).

Thus, three broadcast rounds are sufficient to simulate the PRN with CD. It follows that, if a leader is elected beforehand, a round of the single-channel PRN with CD can be simulated by three rounds of the PRN with no CD. By using the leader election protocol in the previous subsection, we have the following result.

Theorem 4.3 *If a protocol running* T *rounds on the single-channel PRN with CD, it can be simulated in* $O(T + t^2 + (\log n)^2)$ *rounds on the single-channel PRN with no CD with probability at least* $1 - O(e^{-\frac{t}{e^2}})$.

Theorems 3.1 and 4.3 combined imply that the single-channel n-station PRN with no CD can be initialized in $O(n + t^2 + (\log n)^2)$ rounds with probability at least $1 - O(\frac{1}{2^n} + e^{-\frac{t}{e^2}})$. Choosing $t = e^2\sqrt{n}$, we have the following

Corollary 4.4 *The single-channel, n-station PRN with no CD can be initialized in $O(n)$ broadcast rounds with probability at least $1 - \frac{1}{2\sqrt{n}}$.*

Next, we discuss an initialization protocol for the k-channel PRN with no CD. For this purpose, we will implement `Initialization-k-channel-PRN` on the k-channel PRN with no CD. The reader will recall that protocol `Initialization-k-channel-PRN` executes k instances of `Initialization-with-CD`, one in each of the k channels.

Instead, we execute the no CD version of `Initialization-with-CD` in Corollary 4.4. Thus, all the P_is, $(1 \leq i \leq k)$, can be initialized in $O(\frac{n}{k})$ broadcast rounds with probability at least $1 - \frac{k}{2\sqrt{\frac{n}{k}}}$. Further, we need to check whether the initialization of all the P_is is finished. Similarly to the implementation on the PRN with CD, this can be done by suspending the initialization in every, say, 10 rounds. Using one broadcast round in this suspended round, we will check whether all stations are locally initialized.

For this purpose, we need to find a leader of P in advance. By Corollary 4.2, this can be done in $O((\log n)^2)$ time with probability at least $1 - \frac{1}{n}$. Consequently, Stage 1 of `Initialization-k-channel-PRN` can be implemented on the PRN with no CD in $O(\frac{n}{k} + (\log n)^2)$ broadcast rounds with probability at least $1 - O(\frac{k}{2\sqrt{\frac{n}{k}}} + \frac{1}{n})$.

The prefix-sums computation executed in Stage 2 involves no broadcast causing collision, it can also executed on the PRN with no CD in $O(\log k)$ rounds.

As a consequence, Stages 1 and 2 can be implemented on the PRN with no CD to run in $O(\frac{n}{k} + (\log n)^2 + \log k)$ rounds with probability $1 - O(\frac{k}{2\sqrt{\frac{n}{k}}} + \frac{1}{n})$. Assuming $k \leq \frac{n}{4(\log n)^2}$, we have

Theorem 4.5 *The k-channel PRN with no CD can be initialized in $O(\frac{n}{k})$ rounds with probability $1 - O(\frac{1}{n})$, whenever $k \leq \frac{n}{4(\log n)^2}$.*

References

[1] N. Abramson, Ed., *Multiple Access Communications: Foundations for Emerging Technologies*, IEEE Press, New York, 1993.

[2] N. Abramson, Multiple access in wireless digital networks, *Proceedings of the IEEE*, 82, (1994), 1360–1370.

[3] N. Alon, A. Bar-Noy, N. Linial, and D. Peleg, Single-round simulation on radio networks, *Journal of Algorithms*, 13, (1992), 188–210.

[4] Attiya H., Distributed Computing Theory, in A. Y. Zomaya, ed., *Parallel and Distributed Computing Handbook*, McGraw-Hill, New York, 1996, 127–162.

[5] R. Bar-Yehuda, O. Goldreich, and A. Itai, Efficient emulation of single-hop radio network with collision detection on multi-hop radio network with no collision detection, *Distributed Computing*, 5, (1991), 67–71.

[6] R. Bar-Yehuda, O. Goldreich, and A. Itai, On the time-complexity of broadcast in multi-hop radio networks: An exponential gap between determinism and randomization, *Journal of Computer and Systems Sciences*, 45, (1992), 104–126.

[7] D. Bertzekas and R. Gallager, *Data Networks*, Second Edition, Prentice-Hall, 1992.

[8] R. Binder, N. Abramson, F. Kuo, A. Okinaka, and D. Wax, ALOHA packet broadcasting – a retrospect, *AFIPS Conf Proceedings*, May 1975, 203–216.

[9] U. Black, *Mobile and Wireless Networks*, Prentice-Hall, Upper Saddle River, NJ, 1996.

[10] I. Chlamtac and S. Kutten, Tree-based broadcasting in multihop radio networks, *IEEE Transaction on Computers*, C-36, (1987), 1209–1223.

[11] R. Dechter and L. Kleinrock, Broadcast communication and distributed algorithms, *IEEE Transactions on Computers*, C-35, (1986), 210–219.

[12] L. Hu, Distributed code assignments for CDMA packet radio networks, *IEEE/ACM Transactions on Networks*, 1, (1993), 668–677.

[13] J. JáJá. *An Introduction to Parallel Algorithms*. Addison-Wesley, 1992.

[14] R. M. Metcalfe and D. R. Boggs, Ethernet: distributed packet switching for local computer networks, *Communications of the ACM*, 19, (1976), 395–404.

[15] R. Motwani and P. Raghavan, *Randomized Algorithms*, Cambridge University Press, 1995.

[16] S. Murthy and J. J. Garcia-Luna-Aceves, A routing protocol for packet radio networks, *Proc. First Annual ACM Conference on Mobile Computing and Networking*, Berkeley, California, 1995, 86–95.

[17] K. Nakano, Optimal initializing algorithms for a reconfigurable mesh, *Journal of Parallel and Distributed Computing*, 24, (1995), 218-223.

[18] K. Nakano, Optimal sorting algorithms on bus-connected processor arrays, *IEICE Transactions Fundamentals*, E-76A, 11, (1994), 2008–2015.

[19] G. Singh, Leader election in the presence of link failures, *IEEE Transactions on Parallel and Distributed Systems*, 7, (1996), 231–236.

[20] D. E. Willard, Log-logarithmic selection resolution protocols in a multiple access channel, *SIAM Journal on Computing*, 15, (1986), 468–477.

DIMACS Series in Discrete Mathematics
and Theoretical Computer Science
Volume **52**, 2000

Energy-Conserving Software Design for Mobile Computers

Kshirasagar Naik and David S. L. Wei

ABSTRACT. In this paper we study how mobile computers can be made energy efficient. We investigate two orthogonal strategies for energy saving. First, a mobile is put into a *doze* mode when it is idling. We model the behavior of a mobile as an event driven system alternating between an *active* and a *dozing* state. The state transitions are defined on the basis of both internal events and messages coming in from a support station. Simulation study of the model suggests a wide range of energy saving depending on a mobile application's timing characteristics. Second, energy is saved by making a mobile send fewer messages while running a distributed algorithm. For this purpose, the processes of a distributed algorithm running on a set of mobile computers communicate through a *tuple space* managed on the fixed network. The tuple space mechanism allows mobiles to exchange fewer messages than if they were directly exchanging messages for the same problem. Thus, energy is saved in mobiles at the cost of managing a tuple space on the fixed network. We study the cost of managing a tuple space on the fixed network. It can be safely concluded that the impact of energy saving in a mobile on the fixed network is quite tolerable.

Key words: Distributed algorithm, dozing, energy efficient design, mobile computing, tuple space

1. Introduction

There is rapid development in high-speed computing and communication, miniaturization of computers, and mobile communication infrastructure. Users will not be confined to their offices and homes to access the Internet. Advanced communication style will put a significant impact on the way users do distributed computing. However, miniaturization of computers and the ability of people to move from place to place, while still being connected to the network, will offer mixed blessings. There will be both flexibility in accessing information anywhere and at any time and new challenges in designing distributed software systems.

Figure 1 shows an architecture for mobile networking [**21**]. A *mobile host* (MH) is one that can move while retaining its network connections. A static host with a wireless interface supporting a collection of mobile hosts is called a *mobile support station* (MSS). The static hosts are interconnected by a fixed network. A geographical area served by an MSS is called a *cell*. Each MSS and the collection of MHs belonging to its cell form a *wireless network*. All the MHs belonging to a cell rely on the MSS of the cell for availability to the rest of the network. An MH can directly communicate only with the MSS of the cell to which the MH belongs. A fixed host without a wireless interface can also be viewed as an MSS whose cell is never visited by any MH. Though there are some other mobile computing architectures [**15**], [**33**], we will use the one in Fig. 1 in our study.

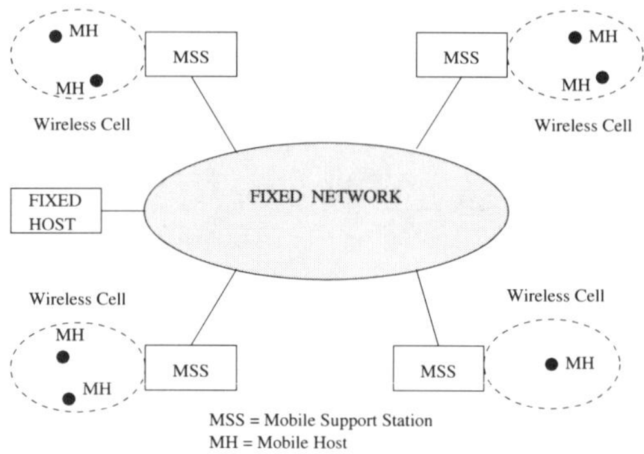

FIGURE 1. A Mobile Network

Mobile computers are characterized by their small size and finite energy supply for a couple of hours before recharging or replacing the batteries. Energy conservation is of paramount importance in the design of both hardware and software for such systems. In a mobile computer, energy can be saved in several ways, such as by:

- using low power electronics in the design of the hardware [**2**], [**3**],
- using power efficient data structures in softwares [**32**],
- allowing an MH to operate in *doze* mode, and
- reducing the number of messages sent from an MH to its MSS.

In this paper, we study the last two items above. Specifically, we explore how much energy can be saved by operating a mobile in a *doze* mode whenever possible. Clearly, a mobile can doze if it is neither computing nor sending/receiving any message. We also study how an *agent* based approach can be used in reducing the number of messages sent from a mobile process[1] in a distributed algorithm. In the agent based approach, mobile processes do not directly communicate among themselves; rather, they communicate through an agent system running on the fixed network.

We present a detailed model of state transitions between an *active* and a *dozing* state of a mobile. Intuitively, the proportion of time spent by a mobile in the dozing state represents the percentage of energy saved in the mobile. This model is a finite-state machine representing the behavior of a mobile in an abstract sense. Since the amount of dozing of a mobile highly depends on the applications running on it, obtaining an analytic model of energy saving from the abstract behavior of a mobile is an extremely difficult task. Therefore, we focus our attention on simulating the abstract behavior. A key idea in the simulation process is that a mobile application is represented as a trace of alternating durations of computation and communication. We simulate the interaction between a mobile and a support

[1]By a *mobile process* we mean a process on a mobile.

station for a wide range of computation and communication durations for both a mobile and a support station. The simulation results show a wide range of energy saving from almost 0% to 98%. The sensitivity of energy saving to different system and application parameters is interesting to observe.

Our second approach to energy saving is orthogonal to the first one. In this approach, the main idea is to reduce the number of messages required to be sent from a mobile while running a distributed algorithm. Our emphasis here is on designing distributed algorithms for mobile computers such that the total number of messages sent by a mobile process to other mobile processes in the algorithm is minimized. This reduction in the number of messages is achieved at the expense of extra messages on the fixed network. The extra messages on the fixed network may be acceptable in view of the fact that the fixed network is much faster than the wireless channel, and the support stations have much more computing power than the mobiles. Thus, energy efficiency in a mobile host is achieved at the expense of additional messages on the fixed network. The following is an outline of our approach to designing energy efficient distributed algorithms for mobile hosts.

- A mechanism for interprocess communication plays a major role in formulating an algorithm and, therefore, in deciding its message complexity. We show that a new kind of process communication, supported by the idea of *tuple space* [**17**], among mobile processes leads to less messages sent by a mobile process.
- A tuple space supports a two-tier computing environment, where the mobile processes are on one tier and the fixed network forms the second. The mobile processes need to send less messages if the tuple space is managed on the fixed network.

We demonstrate the effectiveness of the process communication mechanism of tuple space in saving energy in mobile computers by designing algorithms for three classical problems in distributed computing: *mutual exclusion*, *k-entry mutual exclusion*, and *multiway rendezvous*. For a fixed network, the traditional distributed algorithms for the multiway rendezvous problem, the message complexities for N processes are $O(N^2)$ [**16**] and $O(N\sqrt{N})$ [**38**]. However, for the same problem, an algorithm designed using a tuple space, the complexity of messages sent by the mobile processes is $O(N)$. The new algorithm for mutual exclusion requires a mobile process to send $O(1)$ messages, whereas traditional algorithms for mutual exclusion have complexities of $O(log_2 N)$ [**19**], $O(\sqrt{N})$ [**24**], and $O(N)$ [**30**].

In Section 2, we outline the general structure of traditionally designed distributed algorithms. Some important attributes of mobile computers are studied in Section 3. The idea of power computation in software systems is reviewed in Section 4. Energy saving due to the dozing of a mobile is studied in Section 5. In Section 6, we study how distributed algorithms can be designed for mobile hosts using the tuple space communication mechanism. Finally, some concluding remarks are presented in Section 7.

2. General Structure of Distributed Algorithms

In this section, we review the general structure of distributed algorithms and some common assumptions made in them. This review will help us in understanding the difference between a traditionally designed algorithm and an algorithm based on tuple space communication for an identical problem.

A distributed application consists of a collection of processes running on a network of computers. The speed of the processes are independent of each other. The processes do not share a common memory, but exchange messages from time to time to coordinate their activities [**11**], [**24**], [**8**], [**29**]. The two communication primitives used in the design of distributed algorithms are `SEND(to, msg)` and `RECEIVE(msg)`. The `SEND(to, msg)` primitive sends message `msg` to the process identified by `to`. The message arrives ate the destination process after an arbitrary, but finite delay. The `RECEIVE(msg)` primitive receives a message into the location `msg` from any process. A process may send various kinds of messages, such as `REQUEST`, `PRIVILEGE` [**29**], and state information, which can be encoded into `msg`.

The communication service supporting the `SEND` and `RECEIVE` primitives define their *blocking* semantics [**12**]. With a blocking `SEND`, the sender blocks until the message is received by the destination process; with a blocking `RECEIVE`, a process is blocked until a message arrives at the process. In the design of distributed algorithms, non-blocking and blocking semantics are assumed for `SEND` and `RECEIVE`, respectively.

A large number of distributed algorithms for some classical problems, such as *mutual exclusion*, *binary interprocess synchronization*, and *multiway rendezvous*, assume that the processes are arranged in a *communication structure*. A communication structure is a logical interconnection of processes. Two processes communicate between themselves using the `SEND` and `RECEIVE` primitives if there exists a link between them in the communication structure. Some examples of communication structure are *completely connected graph*, *unrooted tree*, *directed graph*, *acyclic graph*, *ring*, *undirected graph*, *open cube*, *maximal binomial structure*, and *finite projective plane*.

We define a *distributed algorithm* (DA) for a task T as a tuple $DA^T = < CS, CP, CB >$, where CS is a communication structure, CP is a set of communication primitives, and CB is a computational behavior that includes both computation on "local variables" and communications on CS using CP. For a distributed algorithm DA^T, the two broad kinds of *complexities* are *message complexity* and *delay complexity*.

Most of the distributed algorithms use the `SEND` and `RECEIVE` primitives. While designing a distributed algorithm, a designer first selects a communication structure and then defines a computational behavior. Thus, by using different communication structures and different computational behaviors, one may obtain different algorithms for the same task with different complexities. Of course, it is also possible to obtain two different algorithms for the same task by using the same CS, but using two different CBs. Selection of a communication structure is very important from the viewpoints of analyzing the complexity of an algorithm and obtaining an algorithm with low complexity. In Section 6.6, we explain how low-complexity distributed algorithms can be designed using the tuple space communication mechanism.

3. Attributes of a Mobile Computer

A. Mobility: Assume that mobile hosts MH.1 and MH.2 belong to two cells supported by MSS.1 and MSS.2, respectively, and there exists a network connection between the two mobile hosts. Let MH.1 disappear from the cell of MSS.1 and appear in the cell of a third station MSS.3 at a later time. First, MH.1 must inform its new support station MSS.3 that its previous support station was MSS.1; second, MSS.3 must communicate with MSS.1 to get the end point of the connection transferred to MSS.3; finally, MH.1 can begin communicating with MH.2 from its new host. Thus, a mobile host can move while retaining its network connections [**21**]. The process of being recognized and supported by a new MSS is called cellular *handoff*. It is possible to preserve the identity of a mobile host across cells and transparent to the user and its communication partners of the host [**4**]. Thus, existing distributed algorithms can correctly function on a mobile network, albeit slower than when run on a fixed network.

B. Low Power: The requirement of mobile computers being compact and light poses a direct constraint on the power supply. Because a mobile computer can run only for a couple hours before changing or recharging its batteries, both the users and designers of such computers may choose their own ways of conserving energy. The user can switch off the computer, whereas the designer can include a *doze* mode into the computer.

C. Disconnection: A logical connection, say TCP connection, can be disconnected in several ways, such as graceful disconnection, disconnected service provider, and timeout. Therefore, in case of a switch off of a mobile host, the peer process at a remote host will be disconnected after a timeout, and the connection cannot be used when the mobile host is powered on at a later time. Such a disconnection will lead to an abortion of the application. However, disconnection in a mobile environment is rather a requirement, than a fault. A mobile host can inform the local MSS, by sending a *disconnect(msg)* request where *msg* is a list of parameters, of an impending disconnection. This will allow the MSS to act on behalf of the mobile host in preserving the state of the connection. At a later time, the mobile host can restore the connection by getting the state information through a *reconnect(msg)* request [**7**].

D. Doze Mode: The *doze* mode is an innovative approach to conserving energy in a mobile host [**7**]. In this mode, the clock speed is reduced and no user process is executed. Rather, a mobile host simply waits for any incoming message. Upon receiving a message, the host resumes its normal mode of operation. There is a distinction between the doze mode and a disconnection in the sense that a mobile host is reachable in its doze mode. Because an incoming message forces a mobile host to make a transition to its normal mode, algorithms must be suitably designed to prevent such forced switch over if the host wants to control the dozing period.

E. Wireless Messages: Following are the three kinds of messages used in the communication between two mobile hosts:

- *wireless message:* This is a message sent from a mobile host to its MSS.
- *fixed message:* This is a message from one MSS to another.
- *broadcast message:* This is a message sent from an MSS to a mobile host in its cell.

The idea behind making a distinction between a wireless message and a broadcast message is that the energy required to send a wireless message is drawn from a mobile host, whereas the energy consumed for a broadcast message comes from a fixed host. Let MH.1 and MH.2 be two mobile hosts supported by stations MSS.1 and MSS.2, respectively. For MH.1 to send a message to MH.2, three messages are involved: (i) MH.1 sends a wireless message to MSS.1; (ii) MSS.1 sends a fixed message to MSS.2; and (iii) MSS.2 sends a broadcast message to MH.2.

We can fairly assume that a mobile host draws energy from a *finite* source, whereas a fixed host draws energy from an *infinite* source. Message traffics from a mobile host and a fixed host have different implications—a message from a mobile host to an MSS consumes energy from the mobile host, whereas a message from one fixed host to another contributes to network congestion. Thus, ideally, while designing a distributed algorithm for a mobile network, we should reduce all kinds of messages. However, conserving energy in a finite power source of a mobile host may take precedence over sending additional messages in the fixed network. Algorithm designers may even be forced to use more messages in the fixed network to reduce the number of wireless messages.

4. Review of Power Computation in Software

Traditionally, power computation and low power design have been studied at the hardware level [**2**]. This is because it is easy to talk about the fundamental elements of power consumption—voltage, current, and capacitance—at the hardware level. The idea of power computation at the software level is new. Power computation at the software design level is important due to the proliferation of portable computing systems and the severe limitations of the present-day battery technology [**27**].

Tiwari, Malik and Wolfe [**37**] propose a methodology to compute the power consumed by an embedded software. The methodology is based on the actual measurement of power consumed by each assembly-level instruction of a processor. Thus, the total power consumed by a program is obtained by adding the power requirement of each instruction in the program. For better accuracy, inter-instruction effects are also taken into account. From the viewpoint of power optimization, two considerations have been studied: *instruction reordering* and *generation of energy efficient code*. It was found that instruction reordering had no significant effect on power consumption. However, the actual choice of instructions in the generated code can significantly affect the power requirement. Specifically, for the 486DX2 processor, instructions with memory operands draw much more current than register operands. Thus, reduction in the number of memory operands can lead to a reduction in average current.

Wuytack, Francky and De Man [**32**] propose a method of implementing set data types with minimum power consumption. A set data type is an abstract data type widely used in communication systems [**23**] and database systems. In a programming language, one can implement the set data type using a variety of concrete data structures such as linked list (LL), binary tree (BT), array (AR), and pointer array (PA). Thus, to implement the set operations, such as locate, insert, and remove a record from a set, one has to manipulate the memory elements in a concrete data structure. It is the memory accesses in the process of set operations that actually consume power. Energy consumption at the memory level is generally computed by modeling a bit of memory location in terms of the capacitance of the associated logic gates[2]. Thus, the power consumption in set operations is a function of the number of memory elements used in implementing a set data type, the number of read and write operations done in the implementation, and some logic details such as capacitance of memory elements, voltage level, and frequency of operation. The concrete data structures are compared on the basis of a *filling factor*, which

[2]The energy required to charge a capacitor with capacitance C to a voltage level V is CV^2.

is the fraction of the locations that would be filled if implementation is in arrays. It has been shown that for different levels of filling factor, different concrete data structures lead to low values of the power cost function. For example, for filling factor greater than 60%, arrays are better than the LL, BT, and PA structures in implementing power efficient set data type.

Thus, to develop energy efficient softwares, we need to make a correspondence between the operations in a software and *hardware activities.* A hardware activity is any electrical activity drawing current, such as driving a disk, reading a memory location, refreshing a screen, a transmitter sending a wireless message, a receiver amplifying a received signal, etc. The *strength* of such a correspondence depends on the level of abstraction of the software representation. Specifically, if we represent a piece of software as a sequence of assembler instructions, we can make a very strong correspondence between an instruction and the energy consumed by that instruction [**37**]. This is because one knows from measurement how much current a processor draws to execute the instruction, the number of cycles required by the instruction, and the supply voltage. Power analysis of a program written in a high-level language such C can also be done by analyzing the sequence of assembly code generated by a C compiler.

Now consider a software specification manipulating an abstract data type such as set data type. There is a large *gap* between the operations on a set data type and the energy consumption at the hardware level. This is because there are several implementation alternatives to realize a set data type [**32**]. The actual energy consumption of the set data type is always tied to a particular implementation. Without an implementation assumption, one is faced with a large variance in power cost. For instance, for the set data type, the range of power costs for different implementations easily spans four orders of magnitude.

As a software specification becomes more and more abstract, we have more and more implementation alternatives for each concept in the specification. This will lead to a power cost function with too much variance. Also, it may not be possible to derive a sequence of assembly-level instructions from an operation at the abstract level. This problem makes the computation of the actual energy consumption due to an operation at the abstract level very difficult. However, if we are interested in comparing various strategies at the abstract level, we can do so by assuming the same implementation alternative for all strategies.

5. Energy Saving due to Dozing

In this section, we study the effect of dozing on energy saving in a mobile. Intuitively, when a mobile is neither computing nor sending/receiving any message it can go into a doze mode. This basic idea behind dozing is very broad. We need to define a precise model of interaction between a mobile and a support station. The motivation for such a model is that we can simulate it. Since analytic models of softwares are difficult to construct, we believe that simulating the interaction between a mobile and an MSS will give us insights into the energy saving characteristic of the dozing concept. By simulating the model for a number of parameters, we can observe the sensitivity of the energy saving characteristic to those parameters.

A wireless medium is generally unreliable with greater chance for message corruption. Thus, we need an acknowledgement (ACK) policy between a mobile and a support station. The acknowledgement mechanism is outlined in Fig. 2. Every message, except an ACK message, in each direction is acknowledged. Both a mobile and a support station retransmit a message if a timeout occurs before receiving an ACK. We have not shown the details of the retransmission in Fig. 2.

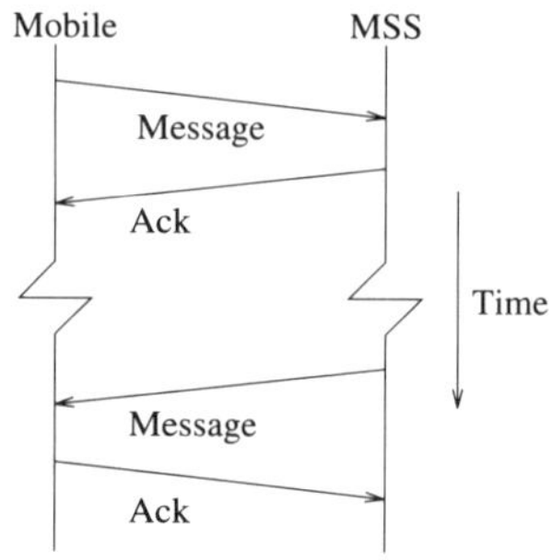

FIGURE 2. Communication between a mobile and a support station

From the viewpoint of energy saving, while a mobile is waiting for an acknowledgement message, it need not be consuming power. In the wait state of a process, if there are no other processes active, a mobile can easily go into a doze mode. In Fig. 3, we capture the state transitions of a process on a mobile. We define two major states for a process: *Active (A)* and *Dozing (D)*. If a process must wait either for a fixed time or indefinitely for a message to arrive from the support station, it waits in the *Dozing* state. When a process is not waiting for any message or timeout, it is said to be in an *Active* state. In the following, we explain the rationale behind the state transitions in Fig. 3.

Transition 1: A mobile process makes a transition from state A to D after sending a message to the support station. This is because the process expects an ACK in response to the message and, therefore, the mobile need not waste power while the process waits.

Transition 5: When a mobile process receives a message in its D state, it makes a transition to its A state so that normal activities of the process can resume. This ACK message is in response to the message sent just before making transition 1.

Transition 6: When a mobile process is interrupted by a timeout, it switches back to the A state to resume normal activities such as resending the message, etc. This transition is non-deterministic in the sense that it depends on the timeout event.

Transition 3: If a process must wait for a message from the message station to be able to continue, it goes to the D state. This transition is non-deterministic in the sense that the decision to wait for a message is internally taken by the process.

Transition 4: If a process receives a message, as explained in *Transitions 3*, it moves to the A state.

Transitions 2, 8: If a mobile process receives a message it must send an ACK to the support station. An ACK message is sent only in the A state. If the received message must be immediately acted upon, the process continues in its A state as denoted by *Transition 8*. However, if the process does not have to act upon the received message immediately, or it does not have to at all, then it may switch to the D state. The process can return to the A state through *Transition 4* or *6*.

Intuitively, the percentage of time spent by a mobile in state D represents energy saving in the mobile. A mobile alternates between the A and D states depending on the communication with a support station.

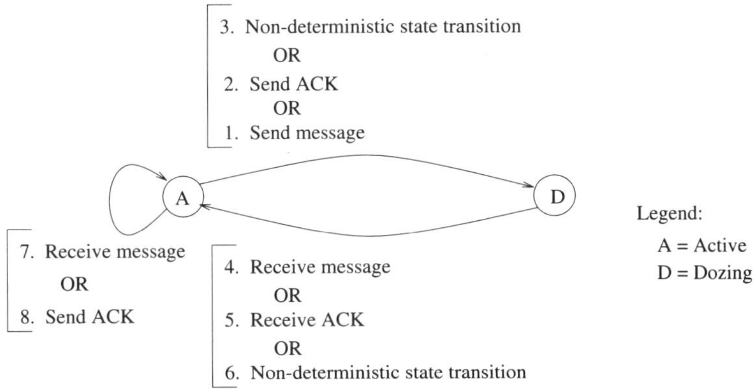

FIGURE 3. State transitions of a mobile process to save energy

REMARK 1. From a cost-effectiveness perspective, one may ask what is the difference between the power consumptions in the A and D states of a processor. If the difference is not significant, any effort to make a mobile doze may consume more energy than what we may save. Such a situation will have negative effects on energy saving. Thus, for the doze mode to be useful, the power consumption in the doze mode must be much less than what is consumed in the active mode. The Hobbit processor from $AT\&T$ consumes about 250 mW in the full operation mode (*active mode*), but consumes only about 50 μW in the doze mode. Thus, one can easily save energy by incorporating the idea of dozing into system design.

In order to have a deeper understanding of the energy saving ability of the state-transition model of a mobile, shown in Fig. 3, we carry out a simulation study as follows. As shown in Fig. 4(a), a mobile and a support station communicate through a wireless channel. We assume that to perform a certain job a mobile communicates with the support station from time to time. Thus, the behavior of a mobile alternates between *Computation* and *Communication*, as shown in Fig. 4(b). A similar behavior is assumed for the support station as well. Fig. 4(a) is useful in the sense that by simulating the interaction between a mobile and a support station, we compute the amount of time spent by a mobile in state D of Fig. 3. Similarly, Fig. 4(b) helps us in identifying the following simulation parameters:

- T_m is a random variable denoting the computation time in a mobile,
- T_b is a random variable denoting The computation time in a support station, and
- T_c is a random variable denoting the communication time between a mobile and an MSS.

We simulated Fig. 4(a) using the SES Workbench [**1**]. In our simulation we assume that T_m, T_b, and T_c are exponentially distributed. In fact, the SES Workbench supports a variety of distributions. The simulation length was set to one million time units. The average values of T_m and T_b were varied from 2 to 500 time units, whereas the average value of T_c was varied from 0 to 100 time units. The mobile in Fig. 4(a) was represented by the state machine of Fig. 3. For simplicity of the simulation model, we did not take *Transitions 3* and *6* of Fig. 3 into account. We monitored the amount of time T_D spent by a mobile in state D. Therefore, energy saving (E_s) is given by the expression T_D/T, where T is the simulation length. We show the simulation result, *i.e.* the percentage of energy saved, in Fig. 5. Fig. 5(a) shows energy saving for $T_c = 2$, Fig. 5(b) shows energy saving for $T_c = 5$, and so on.

The three-dimensional plots in Fig. 5 suggest a wide range of energy saving from almost 0% to 98% depending on the values of the parameters. The plots also reveal the sensitivity of energy saving to different parameters. For example, on the one hand, with increase in the average value of T_m the energy saving quickly drops down. On the other hand, with increase in the average value of T_b the energy saving increases. This is natural because a large value of T_m suggests that the CPU of the mobile is very busy and the mobile cannot doze for a long time. A large value of T_b suggests that the support station does not interrupt a mobile too often by sending messages. Thus, once a mobile goes into the D state, it can stay in that state for a long time. Hence, E_s increases with increase in T_b. For large values of T_c, the change in E_s is less drastic with respect to T_m and T_b. Thus, the actual amount of energy saving depends on the computation and communication parameters.

REMARK 2. The growth in battery technology in terms of *energy density*[3] and *power density*[4] [**27**] is expected to increase only 20% over the next decade [**31**]. Therefore, even 10–20% saving in energy due to dozing is quite significant.

REMARK 3. Lowering the power level of a mobile to switch to a doze mode is a system-wide activity. This switching is essentially the job of the operating system and hardware. The details of the switching is outside the scope of this paper.

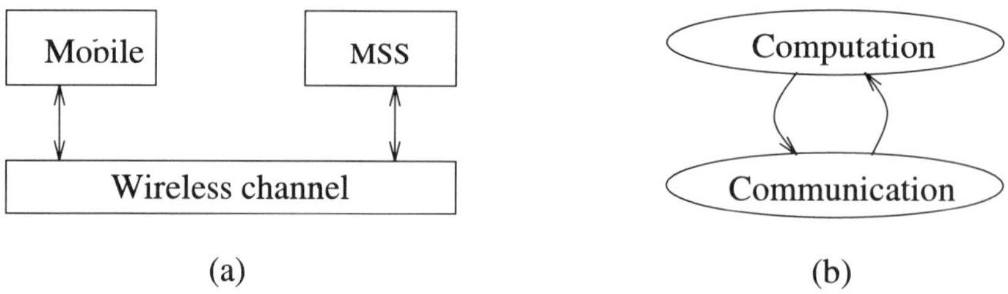

FIGURE 4. Simulation model

[3]The nominal energy content of a battery expressed in watt-hours per liter is referred to as energy density.

[4]The power delivered by a battery expressed in watts per liter is referred to as power density.

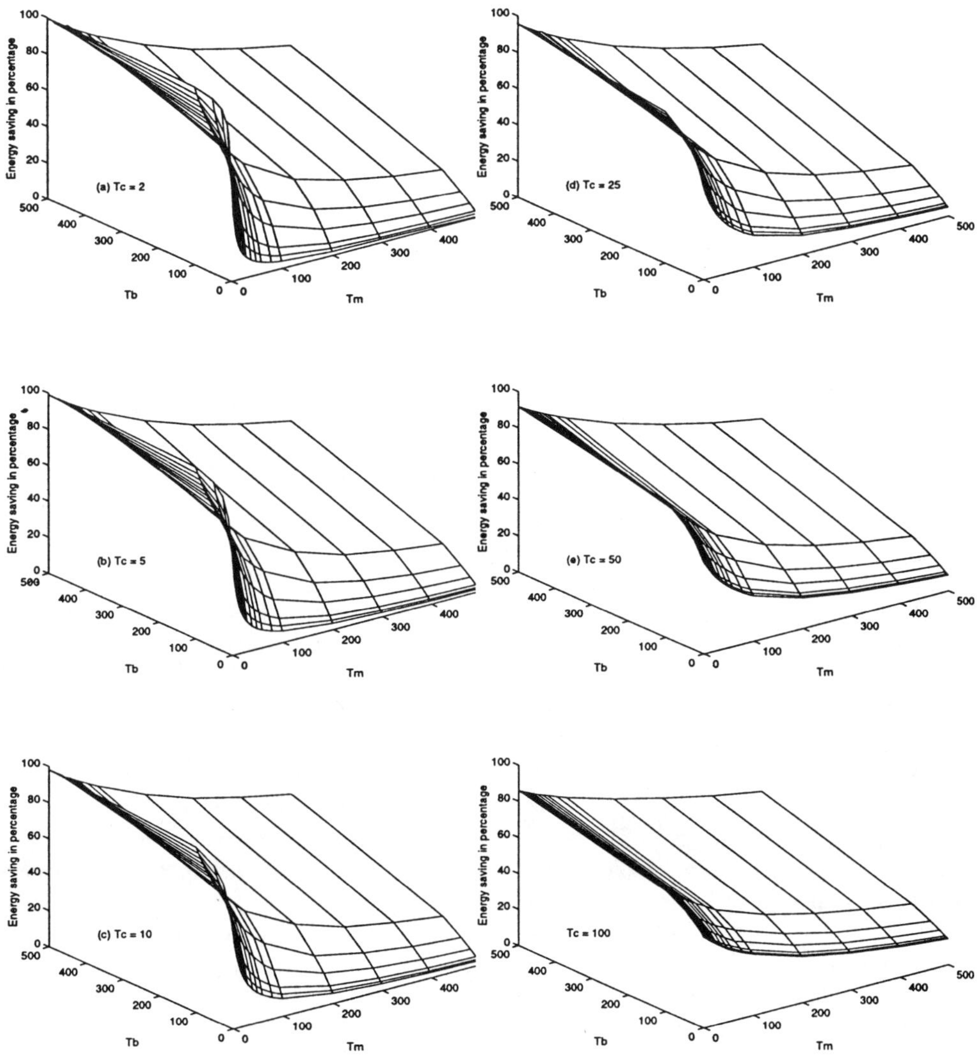

FIGURE 5. Simulation result

6. Energy Saving Using Agent Based Computation

In this section, we introduce the idea behind agent based computing, and establish a link between agent based computing and Linda's tuple space. We define a measure of energy saving due to the indirect communication supported by tuple space. We analyze the impact of using a tuple space on the fixed network for communications among mobile processes. Finally, we show how a few efficient distributed algorithms can be designed using tuple space.

6.1. Agent Based Computation. An agent is someone who manages the business affairs of another or others. Users obtain services from an agent at a high level of abstraction. A user deals at the *what* level, whereas an agent operates at the *how* level. Agent based computation is a relatively new idea in distributed computing [**20**], [**25**], [**14**], [**22**], [**36**]. Some of the advantages of agent based computing are

- masking of distribution aspect among application entities,
- flexible communication among application entities,
- on-line expansion and maintenance, and
- reliability.

Masking of the distribution aspect means the activities in a process are transparent to the locations of the cooperating processes. Flexible communication among processes means a process, say P_i, is not obliged to receive a message when another process, say P_j, sends a message to P_i. Rather, P_i autonomously decides when it actually needs a message from P_j, and waits for the message from P_j. On-line expansion means an entity is added to a distributed system without shutting down a functioning system. Similarly, on-line maintenance means maintenance and test procedures are carried out without suspending the system operation. Since a decentralized agent based system comprises of independent subsystem functions, the structure of such a system shields all the other parts from the influence of partial functional failure.

Ihara and Mori [**20**] and Mori [**25**] propose an agent based architecture for autonomous decentralized systems (ADS). In this architecture, processes are connected only through a mechanism called *Data Field*. Conceptually, a Data Field very closely resembles the *tuple space* of Linda [**17**], though Mori does not refer to Linda. Based on the idea of a *LAKE*, Sato et. al. [**14**] have built a distributed autonomous system called *Noah* (Network oriented application harmony) in their Mitsubishi laboratory. In fact, *Noah*'s LAKE is similar to Linda's tuple space.

Recently, Acharya and Badrinath [**5**] have proposed an agent based architecture for a multicast protocol in a mobile environment. In order to save power in a mobile host, a part of the protocol runs on a mobile host and another part runs on a support station. In this structure, the protocol part running on the support station can be considered an agent of the protocol process on a mobile host. The motivation behind such a structure is that the communication and computation load of a protocol execution is fulfilled within the static network to the largest extent possible. As a result, a mobile host saves energy in its CPU, memory, transmitter, and receiver systems.

6.2. A Measure of Energy Saving. In this section, we define a measure of energy saving in terms of messages sent out of a mobile. Transmitting a wireless message consumes energy in a mobile. Thus, any reduction in the number of wireless message directly contributes to energy saving. Reducing the number of wireless messages has other ramifications as well. If a distributed algorithm running on a mobile needs less wireless messages, it is very likely that the overall execution time of the algorithm is reduced. This situation also contributes to energy saving. Analyzing the effect of reduction of wireless message on the reduction in execution time of an algorithm is outside the scope of this paper. We analyze energy saving due to less wireless messages only.

In Section 2, we defined a distributed algorithm as a tuple $DA^T = < CS, CP, CB >$ for task T. Let $F^T(N)$ represent the message complexity of DA^T for N processes directly communicating with each other by using the SEND and RECEIVE primitives.

Definition: The *wireless complexity* of a distributed algorithm DA^T, running on a mobile network with N mobile hosts, is defined to be the number of wireless messages required to achieve the goal of the algorithm.

REMARK 4. Generally, the complexity of a distributed algorithm is expressed in terms of the *order* of messages without taking the constant factor into account. However, from the viewpoint of energy saving, even the constant factor is very important. This is because, a mobile process sending N wireless messages is obviously better than a process sending $4N$ wireless messages, though both send an identical *order* of messages.

Claim: If an algorithm DA^T designed using the SEND and RECEIVE primitives is run on a mobile network, the resulting wireless complexity is equal to the message complexity of DA^T. The proof follows from the fact that for each message sent by DA^T on a fixed network, there is a corresponding wireless message when we run DA^T on a mobile network.

Therefore, if we simply port a DA^T from a fixed network to a mobile network, we cannot reduce the wireless complexity. It seems that the only way to reduce the wireless complexity is to obtain a DA with a lower complexity. Let $F_m^T(N)$ denote the total number of messages exchanged when a DA^T is run on a mobile network. We can decompose $F_m^T(N)$ into two parts as follows:

$$F_m^T(N) = F_{fb}^T(N) + F_w^T(N), \tag{1}$$

where $F_{fb}^T(N)$ is the sum of fixed and broadcast messages, and $F_w^T(N)$ is the number of wireless messages. Thus, $F_w^T(N)$ is the *wireless complexity* of the algorithm. When a DA^T designed for a fixed network is run on a mobile network, Eq. (1) reduces to the following:

$$F_m^T(N) = 2.F^T(N) + F^T(N).$$

Now assume that we have a new kind of communication structure NCS and a new set of communication primitives NCP, and an appropriate computational behavior NCB for task T, such that for the resulting new distributed algorithm $NDA^T = < NCS, NCP, NCB >$, $F_{fb}^T(N)$ and $F_w^T(N)$ are *independent* of each other. That is, if we port NDA^T from a fixed network to a mobile network, each message in the fixed network *does not map* to *three* messages: one wireless message, one fixed message, and one broadcast message. Thus, we can reduce the wireless complexity if we can show that $F_w^T(N) < F^T(N)$.

Energy saving due to reduced wireless messages is expressed as follows. Let E_m be the amount of energy required to generate and transmit a wireless message of unit length. Then, the amount of energy saved by an algorithm running on a mobile is proportional to $(F^T(N) - F^T_w(N))E_m$.

In Section 6.3 we will present a *tuple space*, which provides a new kind of communication structure and a new set of communication primitives, as a first step toward designing energy efficient algorithms. In Section 6.6, we present some example algorithms for which $F^T_w(N) < F^T(N)$.

6.3. Process Communication Through a Tuple Space. In this section, we discuss a process communication mechanism very different from the traditionally used `SEND/RECEIVE` primitives. The main idea is that processes communicate in an *indirect* fashion as opposed to the direct message passing between processes supported by `SEND/RECEIVE`. The kind of communication supported by Linda [**17**] differs from the well-known mechanisms such as monitors, message passing, and remote operations. Linda was developed toward incorporating communication in a programming language. The abstract computation environment called *tuple space* (TS) is the basis of Linda's model of communication. In the Linda model, processes communicate by inserting and extracting messages from an information structure called tuple space. However many concurrent processes make up a distributed program, all are encompassed within one TS. Tuple space is a global, associative, object memory that holds *tuples.* A tuple is a sequence of typed fields, where each field is an expression that has a value or a *potential* value. Here, a *formal* is called a potential value. For instance,
$(\text{"}example1\text{"}, 1, ?x)$
is a tuple with three fields: a character string, an integer, and a potential value—the variable x. Linda operations place tuples into TS and read or remove them from TS. In case of reading and removing, the desired tuple is located by matching it with a *template.*

A tuple is said to match a template when all of their fields have the same type, length, and if applicable, literal value. For example, the template $(\text{"}example2\text{"}, i, ?j)$ would match the following tuples when i and j have been declared integers, and i has the value 10:
$(\text{"}example2\text{"}, 10, 200)$,
$(\text{"}example2\text{"}, i, 200)$.

In the template, the form $?j$ is called a formal. The formal says to assign the value in the corresponding tuple field in the matched tuple to the variable j. A value of 200 will be assigned to j. Linda provides the following four operators for accessing a tuple space.

- **(i) in(s)::** The *in* operation searches the TS for one matching the argument template s. If there is a matching tuple, the formals in the template are set to the values in its corresponding fields and the tuple is removed from the TS. If no matching tuple is found, the operation blocks until one becomes available.
- **(ii) rd(s)::** The *rd* operation is similar to *in* except that it does not remove the matching tuple from the TS. It merely *reads* the tuple.
- **(iii) out(t)::** The *out* operation adds tuple t to TS.
- **(iv) eval(t)::** An *eval* operation creates a new process to evaluate each of its fields. *evals* are most useful for tuples containing function calls. For example:
  $eval(\text{"}example3\text{"}, worker(count));$

creates a new process to run the routine *worker* with a parameter *count*. This operator is similar to the *out* except that field values are determined *after* adding the tuple to TS.

An $in()$ primitive extracts a tuple from the TS generated with an $out()$ if both the tuple and $in()$'s template satisfy the following conditions:

- the tuple and the template must have identical number of fields;
- corresponding fields must have the same type;
- a formal field can match only an actual field;
- corresponding actual field must have the same value.

In Linda, communication is *orthogonal* in the sense that, just as the receiver has no prior knowledge about the sender, the sender has no knowledge about the receiver. This property results in *space uncoupling (distributed naming)* and *time uncoupling.* Thus, many address-space-disjoint processes can share an object by depositing it in the TS. That is, Linda provides a mechanism for sharing distributed objects. A tuple added to TS by $out()$ remains in TS until it is removed by an $in()$. Nondeterminism is inherent in the definitions of the *in*, *out*, *rd*, and *eval* operators, and all these operations are atomic.

Linda is available as a commercial package [**34**]. While the emphasis has been focussed on the use of Linda to build efficient parallel programs [**10**], process communication through tuple space is an attractive paradigm for easily implementing a distributed application. A good survey of programming languages including the Linda environment for distributed computing systems can be found in [**18**].

6.4. A Tuple Space for Mobile Computers. Saving the number of wireless messages in an algorithm designed using the communication mechanism of tuple space largely depends on how the tuple space is managed. If a tuple space is managed by a collection of processes running on mobile hosts, then obviously the wireless complexity of an algorithm is the same as its fixed message complexity. To gain any benefit in terms of reduced wireless messages, we propose a two-layer structure as shown in Fig. 6. In this approach, the tuple space is managed by the fixed network and the interface layer manages the mobility of the processes in a distributed algorithm. The interface layer can be a part of the handoff mechanism [**6**]. The basic function of the interface layer can be summarized as follows. Let a process in mobile host MH make an $in(s)$ call to the local support station MSS.1. Before the tuple space finds a tuple matching with the template s, MH.1 moves to another cell supported by station MSS.2. The function of the interface layer is to forward the desired tuple to MH through its new support station MSS.2.

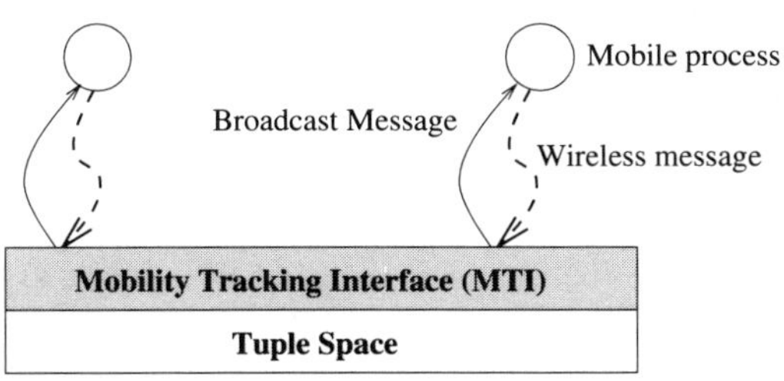

FIGURE 6. Interfacing a mobile process with Tuple Space

6.5. Impact of Tuple Space on the Fixed Network. We first review a few strategies to implement a tuple space. Next, we discuss how the mobility aspect affects the network load in terms of extra messages. The basic problems in implementing Linda on a network of computers are [**17**], [**13**]:

- how to locate the tuples;
- where to store the tuples.

There are two basic approaches to solving the above problems:

- the uniform distribution model;
- the hash model.

In the uniform distribution model, each node is assigned two predetermined *in_set* and *out_set* of nodes. Each *in_set* must have a non-empty intersection with every *out_set*. A tuple generated in a **out()** primitive in node i is broadcast to all members in the *out_set* of node i. A tuple requested in a **in()** primitive in node i is searched in the *in_set* of node i. In the S/Net implementation [**10**], the *out_set(i)* of node i is the entire network and the *in_set(i)* is the node i. In the POSYBL implementation [**35**], the *out_set(i)* is the node i and the *in_set(i)* is the entire network. In the SBN implementation [**17**], the network is a toroidal mesh and, for each i, the *out_set(i)* is the row of i and the *in_set(i)* is the column of i.

In the hash model, tuples are stored and searched in the node whose address is generated by the hash value of the tuple. The hash function is computed either over a fixed field of the tuple or over the structure of the tuple. In the iPSC implementation [**9**], the tuples are partitioned into two classes; tuples having a search key belong to one class and tuples not having a key belong to the other class; and, each class is stored in a different node.

Therefore, the hash model is more efficient than the uniform distribution model in terms of the number of messages required to store and search a tuple. In the hash model, since a tuple is stored in only one node and the address of the node is uniquely determined from the tuple, a **out()** operation requires one message, and the **rd()** and **in()** operations require two messages each.

Now we study the impact of mobility on a tuple space. We assume that the hash model is used. From the viewpoint of energy saving in a mobile, a tuple space must be managed on the fixed network. Since the MSS location to store a tuple is computed from the structure and value of the tuple, mobility does not affect the locations of the tuples. In fact, the location of a tuple is transparent to a mobile. When a mobile, say mh, requests its local support station, say MSS_i, using a **rd()** or **in()** operation, MSS_i computes the location of the support station, say MSS_j, storing the tuple. Next, MSS_i sends a message to MSS_j asking for the tuple. MSS_j sends a message back to MSS_i when the tuple is available. By the time MSS_i receives the tuple from MSS_j, if mh still belongs to the cell of MSS_i, then MSS_i gives the tuple to mh. However, before receiving the tuple from MSS_i mh may move to another support station MSS_k. We have to analyze this situation under two cases as follows:

- *Case 1:* The tuple arrives at MSS_i *before* the handoff procedure is initiated.
- *Case 2:* The tuple arrives at MSS_i *after* the handoff procedure is initiated.

In *Case 1*, the tuple can be sent to MSS_k by piggybacking the tuple on a handoff message from MSS_i to MSS_k. Thus, mh will receive the tuple at no extra message cost. In *Case 2*, an additional message is required to send the tuple from MSS_i to MSS_k. However, additional messages may be required to deliver the tuple to mh depending on when the tuple is available at MSS_j and how many cells mh has visited in the intervening period.

Let us define a term called *mobility factor*, denoted by λ, as the number of new cells visited by a mobile per unit time. During an interval of d time unit, it may be necessary to send a tuple to mh through $\lambda.d$ support stations from its original support station MSS_i to which a **rd()** or **in()** request was made. Thus, the extra load due to a tuple space on the fixed network largely depends on the interval between the **rd()** or **in()** request, the availability of the tuple at MSS_i, and the mobility factor of the mobiles.

6.6. Designing Distributed Algorithms Using Tuple Space. In this section, we explain how to design energy efficient distributed algorithms using the tuple space mechanism for three classical problems: *mutual exclusion*, *k-entry mutual exclusion*, and *multiway rendezvous*. We compare the wireless complexities of the new algorithms with the complexities of the traditionally designed algorithms.

A. Mutual Exclusion: The traditional mutual exclusion algorithms, in which processes directly communicate among themselves, have complexities of $O(log_2 N)$ [**19**], $O(\sqrt{N})$ [**24**], and $O(N)$ [**30**]. We can design one algorithm using the tuple space as the communication structure. As an initializing step we put the tuple

```
(``token'')
```

in the tuple space. Now a mobile process can execute the following piece of code to achieve mutual exclusion:

```
in(``token'');
  Execute critical section;
out(``token'');
```

It is easy to see that the above scheme is both starvation free and deadlock free. A mobile process needs only two wireless messages to achieve mutual exclusion. Thus, the wireless complexity is given by $O(1)$.

B. K-entry Mutual Exclusion: The K-entry mutual exclusion problem is a generalization of the mutual exclusion problem where K (≥ 1) processes are allowed to simultaneously execute in the critical section. Raymond's algorithm [**28**] requires $(2 * N - 1)$ messages, and the algorithm of Srimani and Reddy requires $(N + K - 1)$ messages in the worst case.

Now we give an outline of a K-entry mutual exclusion algorithm based on the tuple space. Define K initial tuples in the tuple space as follows:

```
(``token'', 1),
(``token'', 2),
:
(``token'', K).
```

A mobile process can execute the following piece of code to achieve K-entry mutual exclusion:

```
in(``token'', id);  /* id is of type integer. */
  Execute critical section;
out(``token'', id);
```

In the above scheme, at any time at most K processes can be in their critical sections. This is because we have only K tokens in the tuple space. It is easy to see that the above scheme is both starvation free and deadlock free. The wireless complexity of this algorithm is given by $O(1)$. Thus, our scheme of realizing mutual exclusion through a tuple space requires much less wireless messages than the algorithm in [**7**] which has a complexity of $O(N)$.

C. Multiway Rendezvous: The multiway rendezvous problem [**11**],[**8**] is more complex than the above two. We first informally explain the problem. Let $\mathbf{P} = \{p_1, p_2, \dots, p_N\}$ be a set of n processes, and $\mathbf{E} = \{e_1, e_2, \dots, e_m\}$ be a set of m events. The *event set* of a process p_j, denoted by $e(p_j)$, is the set of events in which p_j participates. Similarly, the *process set* of an event e_i, denoted by P_i, is the set of processes which participate in e_i. Two events e_i and e_j are said to be in *conflict* if and only if $P_i \cap P_j \neq \phi$.

A process has two states *idle* and *active*. The *active* state represents the scenario that the process is executing some event. Execution of an event is purely local. Upon completion of the event, the process goes to the *idle* state. Thus, the transition from *active* to *idle* is nondeterministic. However, the state transition from *idle* to *active* is nontrivial, and requires cooperation from other processes. A process $p_i \in P_k$ commits to an event e_k if and only if it is certain that all other members of P_k will commit to e_k. After committing to an event e_k, p_i moves to the *active* state.

Similar to a process, an event can be in one of its two states *enabled* and *disabled.* An event e_k is said to be *enabled* if all processes in P_k are *idle.* Thus, an *enabled* event can be "executed" if all the processes in P_k commit to e_k. Event e_k is said to be disabled if some member of P_k is *active.* Thus, a *disabled* event remains *disabled* until all its participating processes become *idle.*

Now we formulate the requirements of multiway rendezvous in terms of the following three properties:

- *Exclusion:* Two conflicting events never execute simultaneously.
- *Synchronization:* If p_k commits to e_k, then eventually P_k commits to e_k.
- *Progress:* When an event is enabled, either the event or some conflicting event will eventually be executed.

There are several algorithms to implement multiway rendezvous. Chandy and Misra [**11**] treat multiway rendezvous as a *committee coordination* problem. The algorithm of Bagrodia [**8**] is based on the idea of an *event manager* which uses a circulating token to guarantee the exclusion property. No complexity analysis of the algorithm is given. Analyzing the complexity of the algorithm may not be possible because of the continuously circulating token. The first algorithm for distributed implementation of multiway rendezvous in LOTOS is due to Gao and Bochmann [**16**]. This algorithm constructs a software structure, called *virtual ring,* for an event. For an N-way synchronization, its message complexity is $O(N^2)$. An algorithm based on *finite projective plane* [**24**] is given in [**38**]. In this scheme, synchronization conditions can be evaluated by making a process send requests only to a selected subset of processes of order $O(\sqrt{N})$. Thus, the message complexity of the algorithm reduces to $O(N\sqrt{N})$.

A multiway rendezvous algorithm using Linda primitives has been formulated in [**26**]. This algorithm is very easy to implement and has a message complexity of $O(Nf(N))$, where $f(N)$ is the complexity of a distributed mutual exclusion algorithm used in the implementation of Linda on a fixed network. There are several low complexity distributed mutual exclusion algorithms for the fixed network. Helary and Mostefaoui [**19**] propose a tree-based algorithm for mutual exclusion with a message complexity of $(log_2 N)$ for N nodes. Thus, the multiway rendezvous algorithm in [**26**] can run with message complexity $N.log_2 N$. Therefore, if mobile processes communicate through a tuple space, the multiway rendezvous algorithm in [**26**] is more efficient than the algorithms in [**16**], [**38**].

7. Concluding Remarks

In this paper, we studied how energy can be conserved in mobile computers. First, we modeled the behavior of a mobile as an event driven system. The abstract behavior of a mobile was specified as a finite-state machine with two states *active* and *dozing*. Messages received from a support station and internal events define state transitions. When and how long a mobile can doze largely depend on the characteristics of applications running on a mobile. In that sense energy saving depends on application softwares as well. Since the construction of an analytic model of an application program is very difficult, the finite-state model was very useful in simulating the energy saving characteristic of a mobile for a wide range of computation and communication parameters of the mobile.

Second, we studied how to reduce the number of messages sent by a mobile process while running a distributed algorithm. We argued how mobile process-communication through a tuple space managed on the fixed network can reduce the message complexity of distributed algorithms from a mobile's viewpoint. By studying various strategies to implement a tuple space on the fixed network, we concluded that the impact of energy saving on the fixed network is an inexpensive process.

Mobile computing and designing energy efficient software systems are in their nascent states. Primitives for process communication and communication structures largely outline an algorithm and its message complexity. Thus, to reduce the number of wireless messages from a mobile host, there is a need to identify new kinds of communication mechanisms. The fixed network is used not only for providing interprocess communication service, but can also provide computation service to the mobile processes.

References

[1] Ses workbench. *Scientific and Engineering Software, Inc., Austin*, 1996.

[2] Special issue. *Proceedings of the IEEE, Vol. 83, No. 4*, April, 1995.

[3] Special issue on low power design. *IEEE Trans. on Computer-Aided Design of Integrated Circuits and Systems, Vol. 15, No. 6*, June, 1996.

[4] D. B. Johnson A. Myles and C. Perkins. A mobile host protocol supporting route optimization and authentication. *IEEE JSAC, Vol. 13, No. 5*, pages 839–849, June 1995.

[5] A. Acharya and B. Badrinath. A framework for delivering multicast messages in networks with mobile hosts. *Submitted for publication.*

[6] A. Acharya and B. R. Badrinath. Delivering multicast messages in networks with mobile hosts. pages 292–299, 1993.

[7] A. Acharya B. R. Badrinath and T. Imielinski. Structuring distributed algorithms for mobile hosts. In *IEEE ICDCS'94*, pages 21–28, June, 1994.

[8] R. Bagrodia. Process synchronization: Design and performance evaluation of distributed algorithms. *IEEE Trans. on Software Eng., Vol. 15, No. 9*, pages 1053–1065, Sept. 1989.

[9] N. Carriero Bjornson R and D. Gelernter. The implementation and performance of hypercube linda. In *Proc. 4th Conf. on Hypercube Concurrent Comp. and Applications*, 1989.

[10] N. Carriero and D. Gelernter. The s/net's linda kernel. *ACM TOCS, Vol. 4, No. 2*, pages 111–129, May 1986.

[11] K. M. Chandy and J. Misra. *Parallel Program Design: A Foundation.* Reading, MA, Addison-Wesley, 1988.

[12] D. R. Cheriton. The v kernel: A software base for distributed systems. *IEEE Software 1, 2*, pages 19–42, (April, 1984).
[13] P. Ciancarini and N. Guerrini. Linda meets minix. *ACM Operating Systems Review, Vol. 27, No. 4*, pages 76–92, Oct. 1993.
[14] K. Miyazaki F. Sato, H. Kozuka and H. Fukuoka. Noah: An environment for distributed autonomous systems. In *2nd Intl. Symposium on Autonomous Decentralized Systems*, pages 412–418, 1995.
[15] A. Marlevi G. Y. Liu and G. Q. Maguire Jr. A mobile virtual-distributed system architecture for supporting wireless mobile computing and comm. *Wireless Networks 2*, pages 77–86, (1996).
[16] Q. Gao and G. v. Bochmann. Distributed implementation of lotos multi-rendezvous. In *Proc. of the 9th. IFIP Symp. on Protocol Spec., Testing and Verification*, June, 1989.
[17] D. Gelernter. Generative communication in linda. *ACM TOPLAS, Vol. 7, No. 1*, pages 80–112, Jan. 1985.
[18] J. G. Steiner H. E. Bal and A. S. Tanenbaum. Programming languages for distributed computing. *ACM Computing Surveys, Vol. 21, No. 3*, pages 261–322, Sept. 1989.
[19] J. M. Helary and A. Mostefaoui. A $(log_2 n)$ fault-tolerant distributed mutual exclusion algorithm based on open-cube structure. In *IEEE ICDCS'94*, June, 1994.
[20] H. Ihara and K. Mori. Autonomous decentralized computer control systems. *Computer*, pages 57–64, August, 1984.
[21] D. Duchamp J. Ioannidis and G. Q. Maguire. Ip-based protocols for mobile internetworking. In *ACM SIGCOMM Symp. on Communication, Architectures and Protocols*, 1991.
[22] et. al. K. Sugawara. Agent-oriented architecture for flexible networks. In *2nd Intl. Symposium on Autonomous Decentralized Systems*, pages 135–141, 1995.
[23] J.-Y. LeBoudec. The asynchronous transfer mode: A tutorial. *Computer Networks and ISDN Systems, Vol. 24*, pages 279–309, 1992.
[24] T. Maekawa. A $\sqrt{n}$ algorithm for mutual exclusion in decentralized systems. *ACM Trans. on Computer Systems, Vol. 3, No. 2*, pages 145–159, May, 1985.
[25] K. Mori. Autonomous decentralized systems: Concept, data field architecture and future trends. In *1st Intl. Symposium on Autonomous Decentralized Systems*, pages 28–34, 1993.
[26] K. Naik. Distributed implementation of multi-rendezvous in lotos using the orthogonal communication structure in linda. In *IEEE ICDCS'95*, June, 1995.
[27] R. A. Powers. Batteries for low power electronics. *Proc. of the IEEE, Vol. 83, No. 4*, pages 687–693, April, 1995.
[28] K. Raymond. A distributed algorithm for multiple entries to a critical section. *Information Processing Letters, Vol. 30, No. 4*, pages 189–193, 1989.
[29] K. Raymond. A tree based algorithm for mutual exclusion. *ACM Trans. on Computer Systems, Vol. 7, No. 1*, pages 61–77, January 1989.
[30] G. Ricart and A. Agrawala. An optimal algorithm for mutual exclusion in computer networks. *Communications of the ACM, Vol. 24, No. 1*, pages 9–17, January 1981.
[31] A. Chandrasekaran S. Sheng and R. B. Broderson. A portable multimedia terminal for personal communications. *IEEE Communications Magazine*, pages 64–75, Dec., 1992.
[32] F. V. M. Catthoor S. Wuytack and H. J. De Man. Transforming set data types to power optimal data structures. *IEEE Trans. on Computer-Aided Design of Integrated Circuits and Systems, Vol. 15, No. 6*, pages 619–629, June, 1996.
[33] K. Seal and S. Singh. Loss profiles: A quality of service measure in mobile computing. *Wireless Networks 2*, pages 45–61, (1996).
[34] A. H. Sherman. C-linda user's guide and reference manual. *Scientific Computing Associates Inc., New Haven, USA*, 1991.
[35] G. Shoinas. Issues on the implementation of a programming system for distributed applications. In *Tech. Report, Dept. of Computer Science, University of Crete, Greece*, 1991.
[36] M. Inoue T. Kondo and K. Nakai. Application of autonomous decentralized system to the steel production computer control. In *3rd International Workshop on Future Trends of Distributed Computing Systems*, pages 419–423, 1992.
[37] S. Malik V. Tiwari and A. Wolfe. Power analysis of embedded software: A first step towards software power estimation. *IEEE Trans. on VLSI Systems, Vol. 2, No. 4*, pages 437–445, Dec., 1994.

[38] T. Huang Z. Cheng and N. Shiratori. A new distributed algorithm for implementation of lotos multi-rendezvous. In *IFIP FORTE'94, North-Holland*, 1995.

DEPT. OF COMPUTER SOFTWARE, UNIVERSITY OF AIZU, AIZU-WAKAMATSU, 965 JAPAN
E-mail address: k-naik@u-aizu.ac.jp

DEPT. OF COMPUTER AND INFORMATION SCIENCES, FORDHAM UNIVERSITY, BRONX, NEW YORK 10458
E-mail address: wei@murray.fordham.edu

DIMACS Series in Discrete Mathematics
and Theoretical Computer Science
Volume **52**, 2000

Software Implementation Strategies for Power-Conscious Systems

Kshirasagar Naik and David S. L. Wei

ABSTRACT. A variety of systems with possibly embedded computing power, such as small portable robots, hand-held computers, and automated vehicles, have power supply constraints. Their batteries generally last only for a few hours before being replaced or recharged. It is important that all design efforts are made to conserve power in those systems. Energy consumption in a system can be reduced using a number of techniques, such as low-power electronics, architecture-level power reduction, compiler techniques, to name just a few. However, energy conservation at the application software-level has not yet been explored. In this paper, we show the impact of various software implementation techniques on energy saving. Based on the observation that different instructions of a processor cost different amount of energy, we propose three energy saving strategies, namely (i) assigning live variables to registers, (ii) avoiding repetitive address computations, and (iii) minimizing memory accesses. We also study how a variety of algorithm design and implementation techniques affect energy consumption. In particular, we focus on the following aspects: (i) recursive vs. iterative (with stacks and without stacks), (ii) different representations of the same algorithm, (iii) different algorithms—with identical asymptotic complexity—for the same problem, and (iv) different input representations. We demonstrate the energy saving capabilities of these approaches by studying a variety of applications related to power-conscious systems, such as sorting, pattern matching, matrix operations, depth-first search, and dynamic programming. From our experimental results, we conclude that by suitably choosing an algorithm for a problem and applying the energy saving techniques, energy savings in excess of 60% can be achieved.

Key Words: Low power system, energy saving, software design and implementation

1. Introduction

Miniaturization of computing systems is finding applications in special areas such as hand-held computation, tiny robots, guidance systems in automated vehicles, to name just a few. Also, these systems or their users move from place to place. Because of their small size and the mobility requirement, they are powered by a few batteries of low rating. In order to avoid frequent recharging and/or replacement of the batteries, there is significant interest in low-power system design.

Power consumption is increasingly becoming an area of growing concern in system design. It affects a variety of system concerns such as battery life, thermal limits, packaging constraints, and cooling options [**23**].

Let a program P run for T seconds to achieve its goal, V_{CC} be the supply voltage of the system, and I be the average current in ampere drawn from the power source for T seconds. We can rewrite T as $T = N \times \tau$, where N is the number of clock cycles and τ is the clock period. Then, the amount of energy consumed by P to achieve its goal is given by: $E = V_{CC} \times I \times N \times \tau$ *Joules*. Since for a given hardware, both V_{CC} and τ are fixed, $E \propto I \times N$. However, at the application level, it is more meaningful to talk about T than N, and therefore, we express energy as $E \propto I \times T$. Given the fact that power is the rate of energy consumption, in this paper, we refer to power and energy interchangeably.

Low power design is a complex endeavor requiring a broad range of strategies from floor planning on silicon substrate to design of application softwares. In Figure 1, we enumerate several strategies for achieving energy efficiency in a power-conscious system. In Section 2, we review some of these strategies. Though energy is actually consumed by the hardware, energy consumption can be reduced—apart from using low-power electronics—by suitably manipulating the software systems. This is because the hardware activities are controlled through the softwares. As the expression for energy suggests, the main idea in the design of energy-efficient softwares is to reduce both T and I.

From the running time (average case) of an algorithm we get a measure of T. However, to compute I, one must consider the current drawn during each clock cycle. Actual measurement of the current for different instructions shows that *different instructions lead to different amount of currents being drawn* [**26**], [**27**]. For example, in case of the Intel 486DX2 processor, moving data between two registers takes *one* clock cycle and draws 291.2 mA of current, whereas moving data from a register to a memory location can take up to *two* clock cycles drawing 451.7 mA during those cycles [**27**]. This exposes the potential for reducing the average current I by employing a number of implementation strategies which tend to use the instructions drawing less current and less number of instructions. This also leads to a reduced T.

In this paper, we study the impacts of the following three software implementation techniques on energy consumption:

EC1: coding a software by employing energy saving techniques,
EC2: choice of algorithms, and
EC3: general implementation strategies.

The first strategy **EC1** is based on the experimental result that different instructions draw different amount of current and take different number of clock cycles. We show a few programming techniques leading to a C compiler generating low-energy instructions and less number of total instructions for a program. The main ideas are to eliminate some redundant computations, reduce the number of memory accesses, assign temporary variables to registers, and avoid repetitive address computations.

Second, we study how a variety of algorithm design techniques affect energy consumption. Here, the main idea is to reduce the constant factor in the complexity of an algorithm. Finally, we study how different general implementation strategies affect the energy cost of a program. In particular, we study the following aspects of an implementation:

- recursive vs. iterative (with stacks and without stacks) techniques,
- different coding techniques, and
- different input data representations such as array, link-list, adjacency matrix, and adjacency list.

REMARK 1. Energy conservations due to **EC1**, **EC2**, and **EC3** are orthogonal and can independently be applied to a software design.

To study the impacts of **EC1**, **EC2**, and **EC3** on energy conservation, we studied algorithms for a number of problems, such as sorting, depth-first search, dynamic programming, matrix multiplication, Gauss elimination, and pattern matching. Sorting is a very general problem encountered in almost all applications. Users of the laptop computers generally do a lot of text editing, where pattern matching is an important function. Also, in mobile robots equipped with computer vision, pattern matching is essential to identifying objects. Other problems such as depth-first search, dynamic programming, matrix multiplication, and Gauss elimination are central to controlling the motion of robots. Though we have not identified all kinds of softwares used in power-conscious systems, we believe that our selection represents a cross-section of the kinds of problems encountered in those systems.

We study these impacts by implementing the algorithms with the energy saving programming techniques and computing their energy costs. The energy cost is computed by considering the patterns of code generated by a compiler for various control structures of the C language and the current drawn by each instruction of the Intel 486DX2 processor [**27**]. Similar analysis can also be done for other compilers and processors. Our finding is that by suitably implementing a solution to a problem, one can save in excess of 60% of energy.

REMARK 2. The growth in battery technology in terms of *energy density* and *power density*[1] [**17**] is expected to increase only 20% over the next decade [**21**]. Thus, energy saving in access of even 20% achieved through careful design of high-level softwares is of immense practical importance.

The paper is organized as follows. In Section 2, we review a number of techniques—both hardware and software—used in the design of low-power systems. Section 3 contains our main ideas and the experimental results. First, we identify three energy saving techniques that can be applied to any program in general. Second, we make a link between the constructs of the C language and the instructions of a processor from the viewpoint of computing the energy cost of a program. Third, we apply the energy saving techniques to a number of algorithms. Fourth, we study the impact of algorithm design techniques on energy cost, and then the impacts of implementation strategies on energy costs are explored. Finally, we give some concluding remarks in Section 4.

[1]The energy content of a battery expressed in watt-hours per liter is referred to as energy density, and the power delivered by a battery expressed in watts per liter is referred to as power density.

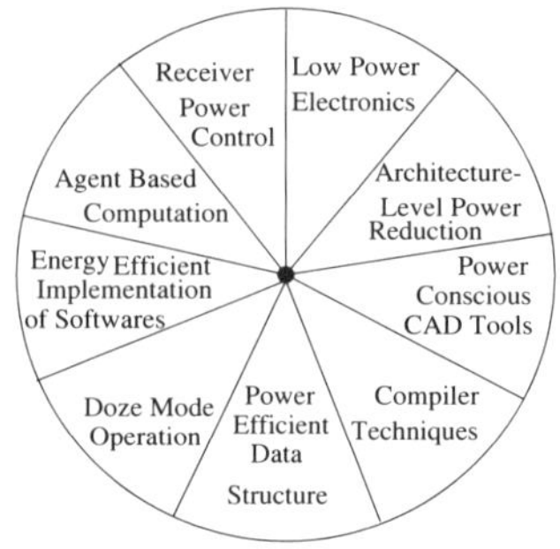

FIGURE 1. Various ways of reducing energy consumption in a system (not to scale).

2. Review of Energy Saving Strategies

We review a wide spectrum of strategies, shown in Fig. 1, ranging from the hardware fabrication process to receiver power control. Energy saving due to different approaches are, in the best case, multiplicative. For example, a 10% energy savings from low-power electronics together with a 15% savings from compiler techniques will yield a total energy saving of: (1 - (((1 - 0.1)* 0.85)) * 100 = 23.5%. However, generally the total energy savings is less, say in this example 20%, because the various energy saving strategies may adversely affect each other.

Low Power Electronics. The power equation in a CMOS digital circuit is expressed as [**4**]:

$$P = (C_L V_{DD}^2 f_p) + (I_{SC} V_{DD}) + (I_{leakage} V_{DD}), \quad (1)$$

where V_{DD} is the supply voltage, f_p is the output switching repetition frequency, C_L is the output capacitance load, I_{SC} is the short circuit current pulse generated when both n- and p-transistors are briefly turned on during output switching, and $I_{leakage}$ is the leakage current. The first term on the right-hand side of the power equation is the dominant factor. It is expected that power saving with two orders of magnitude can be achieved using low-power electronics. About half of the power reduction will come from architecture changes and management of switching activity. The other half of power reduction will come from using advanced materials technology to allow reduction of V_{DD} to 1 Volt or below from 5 or 3.5 Volts while also reducing C_L [**13**], [**1**].

Architecture-Level Power Reduction. The microprocessor can account for up to 33% of a notebook's power budget, which is around 15W. The processors designed for high-end desktops are not suitable for small mobile systems, because these processors dissipate up to 16W with a 5V power supply. Therefore, processor designers include a number of features to reduce power consumption. For the PowerPC 603^{TM} [**4**], some of the power reduction features are dynamic idle-time shutdown of separate execution units, low-power cache design, and power considerations for standard cells, data-path elements, and clocking. The processor also supports three static power management modes *doze, nap*, and *sleep*. These modes reduce power at a global level when the processor is idle for an extended period of time. Since CMOS circuits consume power during the charging and discharging of capacitances, reducing switching activity saves power. At the architecture-level, two strategies to reduce switching activities are Gray code addressing and cold scheduling of instructions [**24**]. Experimental results show that cold scheduling reduces switching by 20 to 30%. The Gray code's advantage over the binary code is that each memory access changes the address by only one bit. Thus, a significant number of bit switches can be eliminated using Gray code addressing. Also, by decomposing a finite-state machine into several submachines, it is possible to selectively turn off portions of a circuit, thereby reducing the switching activities [**3**].

Power-Conscious CAD Tools. The design of low-power systems cannot be achieved without good power-conscious CAD tools. CAD tools are used at all levels of hardware design: behavioral, architectural, logic and physical. For a detailed exposition of power-conscious CAD tools, the reader is referred to two tutorials by Singh, et. al [**23**] and Pedram [**16**].

Compiler Techniques. Compiler design techniques contribute to energy saving in several ways. Lee and Tiwari [**11**] address the problem of allocating memory to variables in embedded DSP (Digital Signal Processing) softwares. The goal is to maximize simultaneous data transfers from different memory banks to registers. In several DSP applications, two registers are loaded with the required data and an arithmetic operation is performed. Loading two registers with a single *double transfer* instruction draws a little more current than a move instruction. Both the instructions take one clock cycle each. However, energy is saved by using the double transfer, because the double transfer instruction loads the two registers in one clock cycle, whereas we need two clock cycles to sequentially load the registers. Experimental results for a few applications on a Fujitsu DSP processor show that up to 47% of energy can be saved by this approach.

Instructions with memory operands have much higher energy costs than instructions with register operands [**26**]. This suggests that energy can be saved by suitably assigning the live variables of a program to registers. But, a processor has only a small number of registers. When the number of simultaneously live variables is larger than the number of available registers, some of the variables must spill to memory. Register assignment for loop variables is important because loops are typically executed many times. Kolson, et. al. [**8**] present an algorithm for optimal register assignment to loop variables for embedded systems. This algorithm can be included in the code generation part of a compiler.

Power Efficient Data Structures. Wuytack, Francky and De Man [**28**] propose a method of implementing set data types with minimum power consumption. A set data type is an abstract data type widely used in communication systems [**10**] and database systems. In a programming language, one can implement the set data type using a variety of concrete data structures such as linked list (LL), binary tree (BT), array (AR), and pointer array (PA). Thus, to implement the set operations, such as locate, insert, and remove a record from a set, one has to manipulate the memory elements in a concrete data structure. It is the memory accesses in the process of set operations that actually consume power. Thus, the power consumption in set operations is a function of the number of memory elements used in implementing a set data type, the number of read and write operations done in the implementation, and some logic details such as capacitance of memory elements, voltage level, and frequency of operation. The concrete data structures are compared on the basis of a *filling factor*, which is the fraction of the locations that would be filled if implementation is in arrays. It has been shown that for different levels of filling factor, different concrete data structures lead to low values of the power cost function. For example, for filling factor greater than 60%, arrays are better than the LL, BT, and PA structures in implementing power efficient set data type.

Doze Mode of Operation. The *doze* mode is an innovative approach to conserving energy [**4**]. It is very attractive in a communication environment where a mobile system may occasionally send or receive messages. In the doze mode, the clock speed is reduced and no user process is executed. Rather, a mobile host simply waits for any incoming message. Upon receiving a message, the host resumes its normal mode of operation. The energy saving due to this mode depends on the local computations on a mobile and the pattern of communication between a mobile and a support station. Simulation studies show that energy saving due to this mode spreads over a wide range of 2–98% [**15**].

Agent Based Computation for Energy Saving. Agent based computation is a relatively new idea in distributed computing [**7**], [**14**], [**19**], [**25**], [**9**]. General agent-based distributed computing systems have been designed using the concept of Linda's *tuple space* [**5**], [**2**], [**22**]. Sato et. al. [**19**] have built a distributed autonomous system called *Noah* (Network oriented application harmony) in their Mitsubishi laboratory. Though the purpose of *Noah* is not to save energy, it demonstrates how agent based systems can be built using a tuple space as the medium for process communication. Naik and Wei [**15**] discuss how energy-efficient distributed algorithms in a mobile computing environment can be designed using a tuple space managed on the fixed network of a mobile system.

Receiver Power Control. The receiver subsystem of a mobile station need not be active all the time. Most digital cellular and cordless systems provide power cycling at the mobile units. Mobile stations can periodically relax (power cycle) their receivers as a means of conserving energy. Since the receiver of a mobile unit is not continuously ready to receive messages from the local support station (base station), some kind of coordination between a base station and a mobile unit is necessary. Salkintzis, et. al. [**18**] propose a *page-and-answer* protocol. Intuitively, the protocol works as follows. When a base station has a message for a mobile unit, the base station sends a small paging packet to the mobile unit. If the mobile unit receives the paging packet, that is if the mobile's receiver is up, the mobile sends an answer packet to the base station. Obviously, if the paging message is sent at a time when the receiver is powered off, no answer packet is generated by the mobile and the base station will once again page the mobile after some time. Upon receiving an answer packet, the base station sends the desired message to the mobile unit.

3. Energy Saving Potential of Implementation Techniques

To study the energy saving potential of algorithm design strategies and implementation techniques, we consider the C language. Though energy consumption of a software can precisely be computed by summing up the energy requirements of all the machine instructions executed, we are interested in computing the energy requirements of a software described in a high-level language. We selected

C because we can easily make a correspondence between a statement in C and a block of instructions by studying some example assembly codes produced by a C compiler. The energy consumption in such a block of instruction, in turn, is used in computing the energy consumption in a C program.

3.1. Energy Efficient Implementation Techniques. Measurement of energy costs of the instruction set of a few processors [**26**], [**12**] reveal that different instructions in the *same class* of instructions incur different energy costs. In Table 1, we show the energy cost of just a few instructions. By a class of instructions we mean a group of instructions with similar functionalities. For example, a conditional jump and an unconditional jump fall in the same class, whereas a jump and an add fall in different classes. Thus, we need to investigate the possibility of generating less number of instructions and low-cost instructions from a high level software. Toward this goal, we propose the following three strategies to reduce the energy cost of a program.

P1: Assign live variables to registers.
P2: Avoid repetitive computation of addresses.
P3: Minimize memory accesses.

To observe the impacts of the above strategies on energy saving, we apply those to the Bubblesort algorithm. We present the straightforward implementation of the algorithm in Table 2(a), and we apply the energy saving strategies to the algorithm in Table 2(b). The input to both the implementations is represented by an array called `list`. In the straightforward version, values of the live variables `n`, `i`, `j` and `temp` are stored in the memory, whereas in the second version they are stored in registers. Also, in the second version, we define a few extra variables to save some values to avoid repetitive computations.

TABLE 1. Energy cost of a subset of instructions of the Intel 486DX2 processor.

Instruction	**Current** (mA)	**Cycles**	**Instruction**	**Current** (mA)	**Cycles**
`MOV reg,imm`	299.2	1	`JCC imm - taken`	372.2	3
`MOV reg,reg`	291.2	1	`JCC imm - not taken`	356.8	1
`MOV reg,disp[base]`	434.7	1	`JMP imm`	370.1	3
`MOV reg,[base][index]`	409.0	2	`NOP`	275.7	1
`MOV disp[base],reg`	560.1	1	`ADD reg,imm`	315.6	1
`MOV disp[base],imm`	404.8	2	`ADD reg,reg`	309.0	1
`SAL reg,CL`	302.7	3	`ADD reg,dis[base]`	400.2	2
`CMP reg,imm`	296.0	1	`ADD disp[base],imm`	382.4	4
`CMP reg,reg`	288.0	1	`IMUL reg`	287.7	13
			`IMUL [base]`	305.0	13
			`IDIV [base]`	278.9	20
			`IDIV [base][index]`	281.8	21

In step 4 of Table 2(a), the `if` statement checks if the two array entries `list[j-1]` and `list[j]` array are to be swapped, and in steps 5 through 7, the swap is actually carried out. It may be noted that in case of a swap, the two array elements are read *twice*.

In Table 2(b), we implement the condition checking and swapping steps in an efficient manner. In steps 4 and 5, we first save the addresses of the two array elements in two register variables `p1` and `p2`, so that these addresses need not be recomputed while swapping the two array elements. However, it may be the case that no swapping is done in a certain iteration, in which case the two assignments become unnecessary. In the following section, experimental results show that, on the average, the two assignments contribute to energy saving. In step 6, we move the two array elements `list[j-1]` and `list[j]` into two registers denoted by `k1` and `k2`, respectively, and perform the condition checking. If the swap condition is satisfied, we carry out the swap by simply moving the contents of `k1` and `k2` into the appropriate locations, `p2` and `p1`, respectively, without reading `list[j-1]` and `list[j]` once again.

In Fig. 2, we illustrate the difference between the two versions of swap implemented in Table 2. A processor cannot directly move data from one memory location to another—the data must be read into a register and then written to the destination memory location. An assignment statement involving two memory locations needs a `load` and `store` instruction. Thus, in the straightforward implementation of the swap operation, shown in Fig. 2(a), *three* `load` and *three* `store` instructions are used. In the energy-efficient implementation of swap, illustrated in Fig. 2(b), we explicitly load the contents of the desired memory locations into two registers, and swap them through the registers needing only *two* `load` and *two* `store` instructions. This is verified by comparing their assembly codes given in Table 3.

Due to the high-level programming techniques **P1**, **P2**, and **P3**, less executable code is generated by a compiler from the source code. As an example, we apply these techniques to the bubblesort algorithm. The straightforward C representation of the bubblesort algorithm and the revised algorithm with the three techniques incorporated are shown in Table 2(a) and Table 2(b), respectively. For these two implementations of the bubblesort algorithm, the executable codes generated by a compiler are shown in Table 3. The left column of Table 3 corresponds to the straightforward implementation of Table 2(a), and the right column corresponds to Table 2(b). Comparing the two columns of code, it is clear that the high-level programming techniques **P1**, **P2**, and **P3** lead to less executable code.

REMARK 3. Though in general less code does not mean less energy cost, experimental results in Section 3.3 show that less code obtained using the strategies discussed in this section leads to much less energy cost.

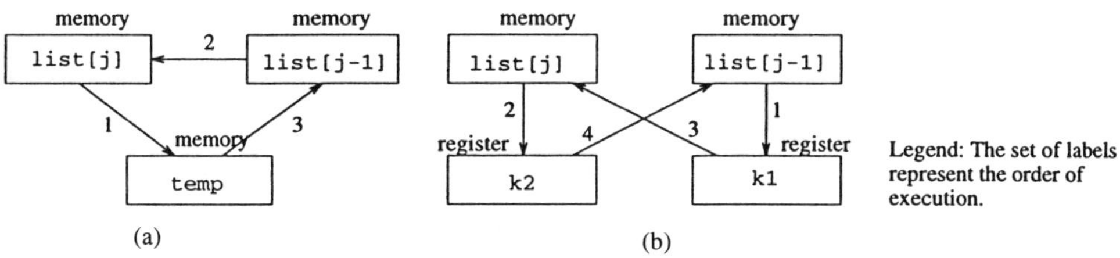

FIGURE 2. Swap Operation: (a) Straightforward, (b) Energy Efficient.

```
int list[10];
int i, j, n, temp;
for (i = n-1; i > 0; i--)
  for (j = 1; j <= i; j++){
    if (list[j-1] > list[j]){
      temp = list[j];
      list[j] = list[j-1];
      list[j-1] = temp;
    }
  }
```

(a)

```
int  list[10];
register int i, j, n, k1, k2, *p1, *p2;
for (i = n-1; i > 0; i--)
  for (j = 1; j <= i; j++){
    p1 = list+j-1;
    p2 = list+j;
    if ((k1 = *p1) > (k2 = *p2)){
       *p2 = k1; *p1 = k2;
    }
  }
```

(b)

TABLE 2. Bubble Sort: (a) Straightforward implementation, (b) Application of P1, P2, and P3.

```
----- Straightforward Implementation -----
.LL12:
        ld [%fp-60],%o0
        add %o0,-1,%o1
        st %o1,[%fp-424]
.LL20:
        ld [%fp-424],%o0
        cmp %o0,0
        bg .LL23
        nop
        b .LL21
        nop
.LL23:
        mov 1,%o0
        st %o0,[%fp-428]
.LL24:
        ld [%fp-428],%o0
        ld [%fp-424],%o1
        cmp %o0,%o1
        ble .LL27
        nop
        b .LL25
        nop
.LL27:
        ld [%fp-428],%o0
        mov %o0,%o1
        sll %o1,2,%o0
        add %fp,-16,%o1
        add %o1,%o0,%o0
        ld [%fp-428],%o1
        mov %o1,%o2
        sll %o2,2,%o1
        add %fp,-16,%o2
        add %o1,%o2,%o1
        ld [%o0-404],%o0
        ld [%o1-400],%o1
        cmp %o0,%o1
        ble .LL28
        nop
        ld [%fp-428],%o0
        mov %o0,%o1
        sll %o1,2,%o0
        add %fp,-16,%o1
        add %o0,%o1,%o0
        ld [%o0-400],%o1
        st %o1,[%fp-432]
        ld [%fp-428],%o0
        mov %o0,%o1
        sll %o1,2,%o0
        add %fp,-16,%o1
        add %o0,%o1,%o0
        ld [%fp-428],%o1
        mov %o1,%o2
        sll %o2,2,%o1
        add %fp,-16,%o2
        add %o2,%o1,%o1
        ld [%o1-404],%o2
        st %o2,[%o0-400]
        ld [%fp-428],%o0
        mov %o0,%o1
        sll %o1,2,%o0
        add %fp,-16,%o1
        add %o1,%o0,%o0
        ld [%fp-432],%o1
        st %o1,[%o0-404]
.LL28:
.LL26:
        ld [%fp-428],%o1
        add %o1,1,%o0
        mov %o0,%o1
        st %o1,[%fp-428]
        b .LL24
        nop
.LL25:
.LL22:
        ld [%fp-424],%o1
        add %o1,-1,%o0
        mov %o0,%o1
        st %o1,[%fp-424]
        b .LL20
.L21:   :
-----Energy Efficient Implementation-----
        add %l2,-1,%l0
.LL2:
        cmp %l0,0
        bg .LL5
        nop
        b .LL3
        nop
.LL5:
        mov 1,%l1
.LL6
        cmp %l1,%l0
        ble .LL9
        nop
        b .LL7
        nop
.LL9:
        add %fp,-416,%l7
        mov %l1,%o0
        sll %o0,2,%o1
        add %l7,%o1,%l7
        add %l7,-4,%l7
        add %fp,-416,%i0
        mov %l1,%o0
        sll %o0,2,%o1
        add %i0,%o1,%i0
        ld [%l7],%l4
        mov %l4,%o0
        ld [%i0],%l5
        mov %l5,%o1
        cmp %o0,%o1
        ble .LL10
        nop
        st %l4,[%i0]
        st %l5,[%l7]
.LL10:
.LL8:
        add %l1,1,%l1
        b .LL6
        nop
.LL7:
.LL4:
        add %l0,-1,%l0
        b .LL2
.LL3:   :
```

TABLE 3. Assembly codes from the two versions of Bubble sort.

3.2. Computation of Energy Cost. We explain how to compute the energy cost of a program in C. The basic idea is to make a link between a sequential block of C statements and the corresponding block of instructions produced by a C compiler. Intuitively, by summing up the energy costs of the instructions in a sequential block, we obtain the energy cost of the sequential block of C statements. The instructions generated from the bubblesort algorithm of Table 2(b) are presented in a block format in Table 4(a). There are nine blocks of instructions `BLOCK0` through `BLOCK8`. Precise computation of energy costs becomes more involved for a control statement, because the instructions corresponding to such a statement are not found as a contiguous block. For example, corresponding to the `for` statement in line 2 of Table 2(b), there are three blocks of instructions `BLOCK0`, `BLOCK1`, and `BLOCK7`. Also, these three blocks of instructions do not execute for an equal number of times—`BLOCK0` executes only once, whereas `BLOCK1` executes for $n - 1$ times.

From the instructions generated by a C compiler, we can identify a block of instructions corresponding to a sequence of C statements, and several blocks for one C statement. For example, `BLOCK4` corresponds to the C statements in lines 4, 5, and 6 of Table 2(b). Generation of multiple blocks of instructions for one C statement has already been explained in the context of the `for` statement above. From the knowledge of the energy cost of an instruction, we can compute the energy cost of a block of instructions. Thus, the total energy cost of an algorithm can be computed by suitably placing counting statements using the energy costs of blocks of instructions in the algorithm. We compute the energy cost of the bubblesort algorithm of Table 2(b) in Table 4(b). In Table 4(b), the blocks `BLOCK0` through `BLOCK8` represent the energy costs of the corresponding blocks of instructions in Table 4(a).

Thus, we compute the average energy cost of an algorithm in three steps as outlined below.

- Given an algorithm in C, identify a correspondence between the C statements and the blocks of instructions produced by a compiler.
- Using the block structures identified above, insert energy counting statements into the algorithm.
- Run the algorithm in a loop for many times using a variety of inputs, and compute the average energy cost.

In this approach, identification of instruction blocks and insertion of statements to accumulate the energy cost of an algorithm is done by hand. This approach is useful to study the energy costs of small programs. However, in our opinion, this needs to be done by an energy cost-conscious compiler which is outside the scope of this paper.

In the computation of energy cost of a program, the basic cost of an instruction is represented by the product of the *number of clock cycles* taken by the instruction and the *average current* drawn during that period. Thus, the energy cost of an instruction is represented in *ampere-cycles* unit. By dividing the *ampere-cycle* cost of a program for a given input by the clock frequency in *cycles/second* of the processor, we obtain the energy cost of the program in *ampere-seconds*.

3.3. Impact of Energy Saving Techniques. In this section, we apply the energy saving techniques **P1, P2**, and **P3** to a number of algorithms and study their impacts on energy saving. As examples, we apply these techniques to the bubblesort algorithm of Table 2(a), Table 2(b), Table 5(a), and Table 5(b). A

```
        add %12,-1,%10         } BLOCK0
.LL2:
        cmp %10,0              } BLOCK1
        bg .LL5
        nop                    } BLOCK8
        b .LL3
        nop
.LL5:
        mov 1,%11              } BLOCK2
.LL6:
        cmp %l1,%10            } BLOCK3
        ble .LL9
        nop                    }
        b .LL7
        nop
.LL9:
        add %fp,-416,%l7       }
        mov %l1,%o0
        sll %o0,2,%o0
        add %l7,%o1,%l7
        add %l7,-4,%l7
        add %fp,-416,%i0
        mov %l1,%o0
        sll %o0,2,%o1          } BLOCK4
        add %i0,%o1,%i0
        ld [%l7],%l4
        mov %l4,%o0
        ld [%i0],%l5
        mov %l5,%o1
        cmp %o0,%o1
        ble .LL10
        nop                    }
        st %l4,[%i0]           } BLOCK5
        st %l5,[%l7]
.LL10:
.LL8:
        add %l1,1,%l1          } BLOCK6
        b .LL6
        nop
.LL7:
.LL4:
        add %l0,-1,%l0         } BLOCK7
        b .LL2
.LL3:
        .
        .
        .
```

(a)

```
int list[10];
register int i, j, n, k1, k2, *p1, *p2;
double energy = 0;

energy += BLOCK0;
for (i = n-1; i > 0; i--) {
   energy += (BLOCK1 + BLOCK2);
   for (j = 1; j <= i; j++) {
      energy += (BLOCK3 + BLOCK4);
      p1 = list+j-1;
      p2 = list+j;
      if ((k1 = *p1) > (k2 = *p2)) {
           energy += BLOCK5;
           *p2 = k1;
           *p1 = k2;
      }
      energy += BLOCK6;
   }
   energy += BLOCK7;
}
energy += BLOCK8;
```

(b)

TABLE 4. (a) Blocks of Instructions generated from Bubblesort, (b) Computing Energy Cost.

program's energy cost is computed using the strategy explained in Section 3.2. We experiment algorithms for three different classes of problems, namely sorting, dynamic programming, and matrix operations. In particular, we choose bubblesort, matrix chain product, matrix multiplication, and Gauss elimination algorithms.

```
int list[10];                                  |  int list[10];
register int i, j, n, k1, k2;                  |  register int i, j, n, temp;
                                               |
for (i = n-1; i > 0; i--){                     | for (i = n-1; i > 0; i--){
  for (j = 1; j <= i; j++){                    |  for (j = 1; j <= i; j++){
    if ((k1 = list[j-1]) > (k2 = list[j])){    |    if (list[j-1] > list[j]){
      list[j] = k1;                            |      temp= list[j];
      list[j-1] = k2;                          |      list[j]=list[j-1];list[j-1]=temp;
    }                                          |    }
  }                                            |  }
}                                              | }
```

TABLE 5. Bubble Sort: (a) Using techs. P1 and P3, and (b) Using P1.

For the bubblesort algorithm, we show its energy cost in Fig. 3. Fig. 3 shows that technique **P1** alone saves 46% of energy. Both **P1** and **P3** together save 54% of energy, and all three techniques applied together save up to 59% of energy. In case of the matrix chain product, matrix multiplication, and Gauss elimination algorithms, Figures 4, 5, and 6 show energy savings up to 27%, 46%, and 48%, respectively. The suitability of a particular technique depends on the algorithms. For example, we cannot apply **P2** to the matrix multiplication algorithm.

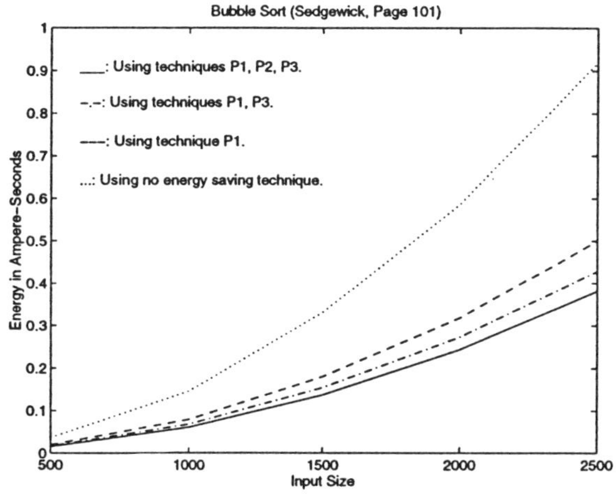

FIGURE 3. Impact of energy saving techniques demonstrated using Bubblesort.

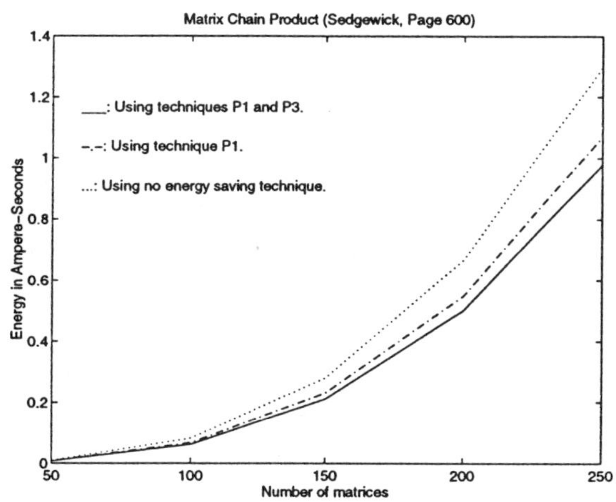

FIGURE 4. Impact of energy saving techniques demonstrated using Matrix Chain Product.

3.4. Impact of Algorithm Design Techniques on Energy Cost. In general, any strategy that can reduce the running time of an algorithm is also useful in saving energy. Algorithms, for the same problem, with identical asymptotic complexity differ in their energy costs depending on the constant factor behind the big-O. Thus, to design an energy efficient algorithm, we need to reduce not only its asymptotic complexity, but also the constant factor. So far the idea of reducing the asymptotic complexity has received a great amount of attention and this area is quite mature. However, reducing the constant factor side has been largely ignored,

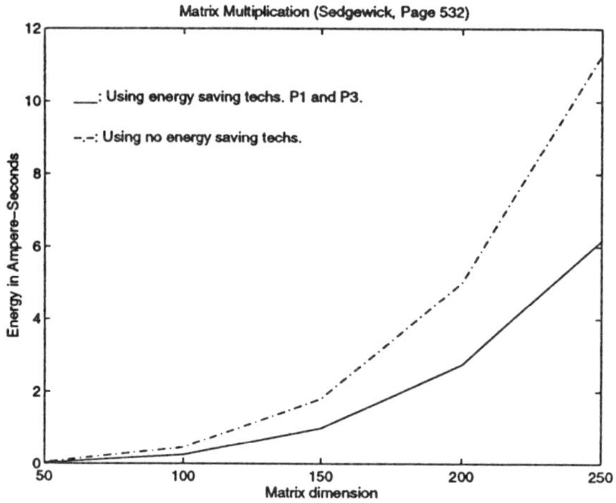

FIGURE 5. Impact of energy saving techniques demonstrated using Matrix Multiplication.

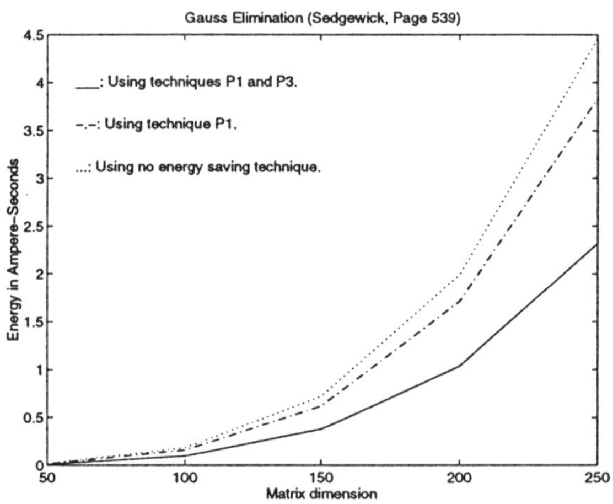

FIGURE 6. Impact of energy saving techniques demonstrated using Gauss Elimination.

which is vital to saving energy. Reducing the constant factor is important in the sense that bringing a constant factor from say 4 down to 2 leads to a 50% energy saving.

Intuitively, loop control contributes to the asymptotic complexity, whereas the number of statements in a loop affects the constant factor. Therefore, reducing the number of statements in each loop is likely to reduce the constant factor. Let us consider the sorting problem. Assuming that the input is an array, reducing the number of array accesses, that is the read and write operations, can potentially reduce the constant factor. In Fig. 7, we show the energy costs of three sorting algorithms, namely quicksort[2], mergesort, and heapsort, with identical asymptotic

[2]In our experiment, we use randomly generated input data leading to an average behavior of the algorithm.

complexity, say $O(n \lg n)$. It is apparent that the constant factor of quicksort is the lowest among the three. Precisely, it is about half of that of the heapsort. Also, the constant factor of the mergesort is about 0.85 times that of the heapsort. According to our observation, this is due to quicksort's relatively less access, that is read and write operations, of the input array. To confirm our observation, we count the number of array accesses in the three algorithms and show the result in Fig. 8. This follows from the fact that the merge and heapsort algorithms have more read/write operations per iteration than those in the quicksort.

As another example, we study the impact of algorithm design techniques on energy saving using the pattern matching problem. We consider four algorithms, namely the brute-force approach, Rabin Karp (RK) algorithm, Boyre Moore (BM) algorithm, and the Knuth, Morris and Pratt (KMP) algorithm, whose energy costs are shown in Fig. 9. It is clear that the RK algorithm is the most expensive one and the BM algorithm is least expensive. This is because of two reasons: (i) the RK algorithm uses the modulo operation, which costs more than 20 times than a simple operation such as an addition, in the loop; and (ii) the BM algorithm uses an elegant way of skipping while comparing the pattern against the input string. Due to the efficient skipping, the algorithm accesses the input fewer number of times.

We also show the impact of the energy saving techniques on the pattern matching algorithms in Fig. 10. Since the RK algorithm is anyway a lot more expensive than the others, we do not show it in Fig. 10. The figure shows that energy saving in the range of 18 – 29% can be achieved. In case of the BM algorithm, the energy saving techniques do not save significant amount of energy. In spite of this, the BM algorithm is the best choice because overall it is the least expensive one.

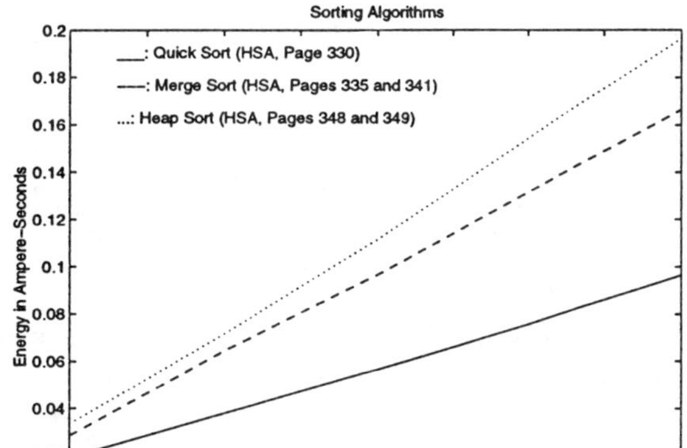

FIGURE 7. Energy cost of different algorithms for the same problem (sorting).

3.5. Impact of Implementation Strategies on Energy Cost. Usually, after an algorithm is designed, not much attention is paid to the style of implementations. In terms of energy saving, the constant factor behind the big-O of an algorithm is important, and different implementation techniques may lead to different constant factors. We investigate several factors influencing the constant factors in what follows.

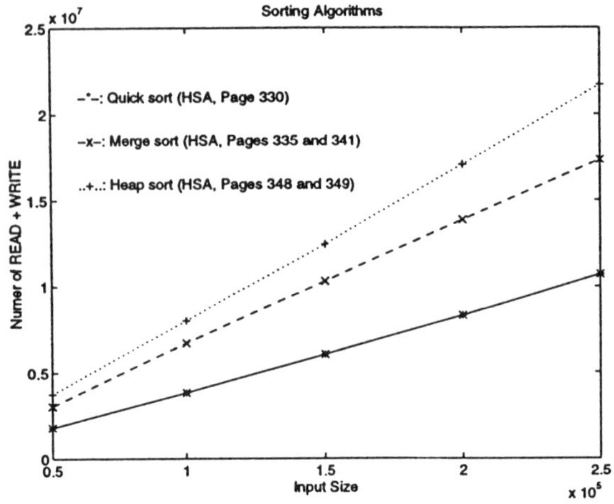

FIGURE 8. Read and Write Cost of a few sorting algorithms.

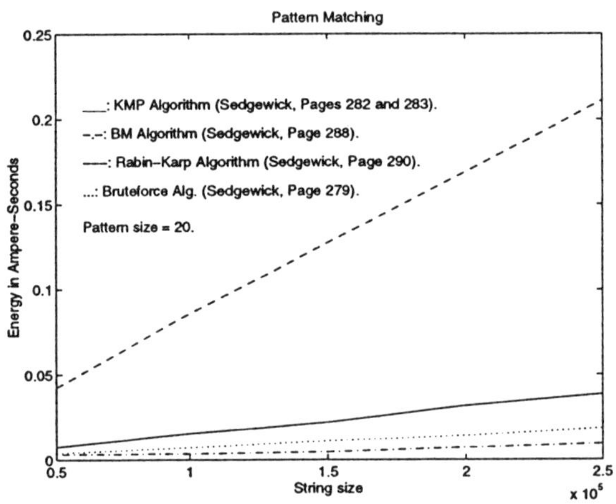

FIGURE 9. Energy cost of different algorithms for the same problem (pattern matching).

3.5.1. *Recursion vs. Iteration.* For some problems, algorithms are designed using both recursive and iterative styles. An advantage of recursive representation is that it is compact. However, due to the necessity of a stack managed by the system, it is often the case that a recursive algorithm takes longer time than its iterative counterpart. In the following, we compare the energy costs of recursive algorithms with those of iterative algorithms—with and without stacks.

The energy costs of two versions of quicksort—a recursive one and an iterative one using a stack at the user level—are compared in Fig. 11. The iterative one is beaten by the recursive one because of using a stack at the user level. One the one hand, managing a stack at the user level requires implementing the *push* and *pop* functions. On the other hand, the idea of stack involved in a recursive algorithm is efficiently managed by the compiler using static information about a

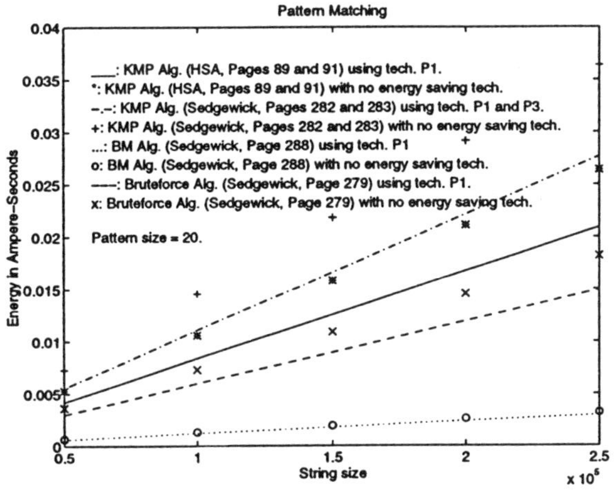

FIGURE 10. Energy cost of different algorithms for the same problem (pattern matching).

procedure, such that the stack is basically implemented as a simple procedure call at the instruction level.

The energy costs of two versions of mergesort—a recursive one and an iterative one *without* using a stack—are compared in Fig. 12. Because the iterative one does not need a stack, it is less expensive.

This suggests that an iterative algorithm without using a stack tends to be better than its recursive counterpart. Comparing Figures 11 and 12, we see that the recursive version of the quicksort costs less energy than the iterative version of the mergesort. Thus, one might ask if one can design an iterative version of the quicksort without using a stack. The answer is that one cannot avoid using a stack, because the loop control parameters (the boundaries of each sub-list) must be remembered, rather than be computed as in the case of the mergesort.

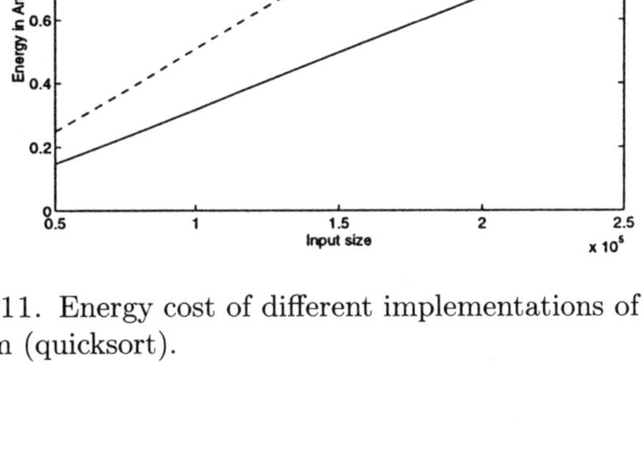

FIGURE 11. Energy cost of different implementations of the same algorithm (quicksort).

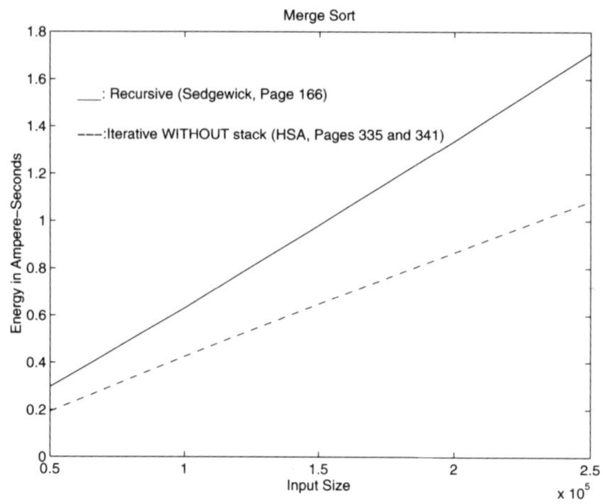

FIGURE 12. Energy cost of different implementations of the same algorithm (mergesort).

3.5.2. *Various Coding Techniques.* We show how coding the same algorithm even in slightly different ways can lead to widely varying energy costs. Those slightly varying codes can easily be overlooked, but their energy costs may significantly be different. We use the KMP pattern matching and quicksort algorithms to explore this aspect. In Table 6, we show two different codings of the KMP algorithm whose energy costs are shown in Fig. 13. It is apparent that the code in Horowitz, Sahni and Anderson (HSA) [**6**] consumes 31% less energy than the code in Sedgewick [**20**]. The reason behind this is that in the former case each iteration executes about one or two statements less than the latter case out of about five or six statements depending on the loop condition.

```
int pmatch(char *string, char *pat) {
/* KMP algorithm              */
  int i = 0, j = 0;
  int lens = strlen(string);
  int lenp = strlen(pat);
  while (i < lens && j < lenp) {
    if (string[i] == pat[j]) {
          i++; j++;
    }
    else if (j == 0) i++;
        else j = failure[j-1]+1;
  }
  return ((j == lenp) ? (i-lenp) : -1);
}
                  (a)
```

```
int kmpsearch (char *p, char *a){
  int i, j;
  int M = strlen(p), N = strlen(a);
  initnext(p);
  for (i=0, j=0; j<M && i<N; i++, j++)
    while ((j>=0)&&(a[i]!=p[j])) j=next[j];
    if (j == M) return i-M;else return i;
}
                  (b)
```

TABLE 6. KMP Algorithm: (a) Horowitz, Sahni and Anderson, P. 89, (b) Sedgewick, P. 282.

Fig. 14 shows the energy costs of two different codings of quicksort. The code in Sedgewick [**20**] consumes about 14% less energy than that in HSA [**6**]. This is because of two reasons. First, the `for` loop in the Sedgewick code allows one less

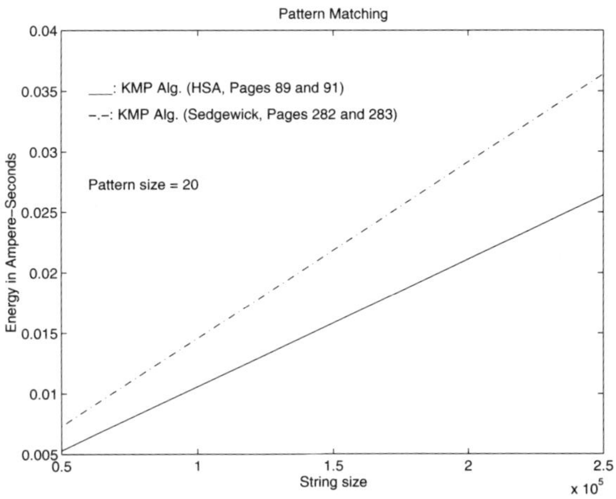

FIGURE 13. Energy cost of different coding techniques applied to the KMP algorithm.

conditional check to be done than the HSA code using the `while` construct. Second, the `while` statements in the Sedgewick code generate a segment of more efficient instructions than the HSA's `do-while` statements. More precisely, the former generates one less `load` instruction. It is our belief that by carefully designing the control flow in a program, less number of statements are executed.

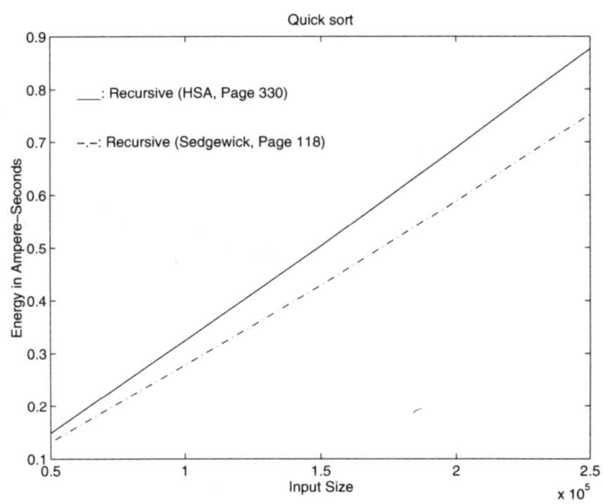

FIGURE 14. Energy cost of different coding techniques applied to the same algorithm (quicksort).

3.5.3. *Different Input Representations.* In this section, we show that even input formats can also significantly affect the energy cost of a program. We choose the depth-first search and mergesort algorithms for this study. In case of the depth-first search problem, an input graph can be represented by its adjacency matrix or adjacency list. Because of the different input formats, the algorithms have to be designed in different ways, and it turns out that the one with adjacency list input performs better in cases of sparse graphs. Figure 15 demonstrates this aspect in terms of energy cost. The suitability of an algorithm depends on the *sparsity factor*, where sparsity factor is intuitively defined as the ratio of the number of edges in an N-node graph to $N(N-1)/2$.

It is apparent from Fig. 16 that in the mergesort, algorithms with array inputs outperform those with linked list input. In our experiment we used only integer data. However, for large record data type, we believe that the algorithms with linked list input will perform better. This is because swapping the records, which is a lot more expensive than manipulating the pointers of the list data, will be avoided.

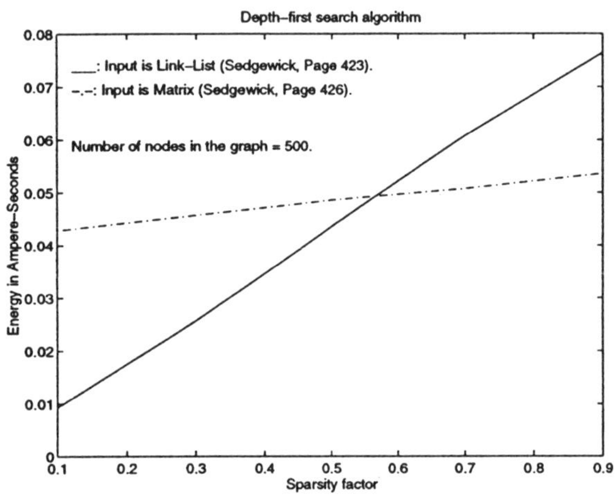

FIGURE 15. Energy cost of algorithms with different input formats.

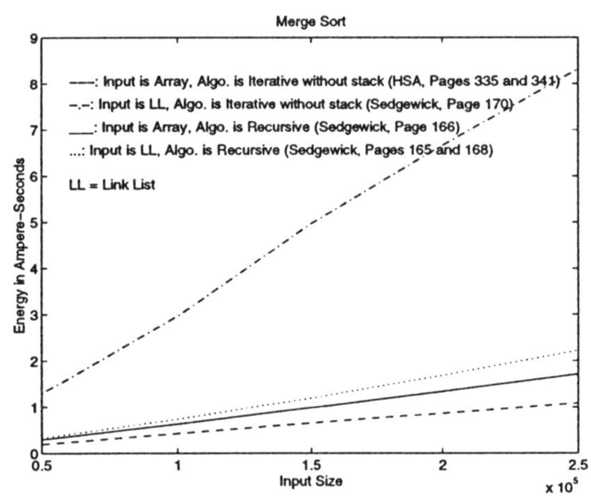

FIGURE 16. Energy cost of algorithms with different input formats.

4. Concluding Remarks

In this paper we addressed how by employing various techniques at the software implementation level, we can save energy in power-conscious systems. In particular, we presented three energy saving techniques, namely assigning live variables to registers, avoiding repetitive address computations, and minimizing memory accesses. Using the energy cost of individual instructions, we proposed a way to compute the energy cost of a high level program in C. We applied these techniques to algorithms for sorting, matrix chain product, matrix multiplication, Gauss elimination, and pattern matching. These techniques lead to energy savings from 18% to 60%.

We also studied the impact of algorithm design and implementation techniques with various formats of inputs on energy saving. From the viewpoint of energy saving, we need to reduce the constant factor in the complexity of an algorithm. Using a few sorting and pattern matching algorithms with identical asymptotic behavior, we showed how different constant factors lead to different energy costs. In case of sorting, one can save up to 50% of energy by choosing the quicksort over the heapsort, and in case of pattern matching, one can save up to 95% of energy by using the Boyre-Moore algorithm instead of the Rabin-Karp algorithm. Among the many possible ways of implementing an algorithm, we selected three general aspects, namely recursion versus iteration, different ways of coding an algorithm, and input formats. Experimental results show that an iterative implementation without using a stack performs better than a recursive one which in turn performs better than an iterative implementation using a stack. Using this knowledge, one can save energy in excess of 35%. By carefully designing the control flow in a program, it is possible to save energy. In the two versions of the quicksort and pattern matching algorithms, energy savings up to 14% and 31%, respectively, can be achieved. We also showed how different input representations, such as array, link-list, adjacency matrix, and adjacency list, affect the energy costs.

Minimizing the energy cost due to a high-level program is a new research idea. In this paper, we studied how different approaches to software design can be exploited in reducing the energy costs.

References

[1] Special issue on low power electronics. *Proceedings of the IEEE, Vol. 83, No. 4*, April, 1995.

[2] H. E. Bal, J. G. Steiner, and A. S. Tanenbaum. Programming languages for distributed computing. *ACM Computing Surveys, Vol. 21, No. 3*, pages 261–322, Sept. 1989.

[3] S-H. Chow, Y-C. Ho, and T. Hwang. Low power realization of finite state machines–a decompostion approach. *ACM Trans. on Design Automation of Electronic Systems, Vol. 1, No. 3*, pages 315–340, July, 1996.

[4] S. Gary, P. Ippolito, G. Gerosa, C. Dietz, J. Eno, and H. Sanchez. Powerpc 603, a microprocessor for portable computers. *IEEE Design and Test of Computers*, pages 14–23, Winter, 1994.

[5] D. Gelernter. Generative communication in linda. *ACM TOPLAS, Vol. 7, No. 1*, pages 80–112, Jan. 1985.

[6] E. Horowitz, S. Sahni, and S. Anderson-Freed. *Fundamentals of Data Structures in C.* Computer Science Press, 1993.

[7] H. Ihara and K. Mori. Autonomous decentralized computer control systems. *Computer*, pages 57–64, August, 1984.

[8] D. J. Kolson, A. Nicolau, and N. Dutt. Optimal register assignment to loops for embedded code generation. *ACM Trans. on Design Automation of Electronic Systems, Vol. 1, No. 2*, pages 251–279, April, 1996.

[9] T. Kondo, M. Inoue, and K. Nakai. Application of autonomous decentralized system to the steel production computer control. In *3rd International Workshop on Future Trends of Distributed Computing Systems*, pages 419–423, 1992.

[10] J.-Y. LeBoudec. The asynchronous transfer mode: A tutorial. *Computer Networks and ISDN Systems, Vol. 24*, pages 279–309, 1992.

[11] M. T. Lee and V. Tiwari. A memory allocation technique for low-energy embedded dsp software. In *Symposium on Low Power Electronics*, pages 24–25, 1995.
[12] M. T. Lee, V. Tiwari, S. Malik, and M. Fujita. Power analysis and minimization techniques for embedded dsp software. *IEEE Trans. on VLSI Systems, Vol. 5, No. 1*, pages 123–135, March, 1997.
[13] Z. J. Lemnios and K. J. Gabriel. Low-power electronics. *IEEE Design and Test of Computers*, pages 8–13, Winter, 1994.
[14] K. Mori. Autonomous decentralized systems: Concept, data field architecture and future trends. In *1st Intl. Symposium on Autonomous Decentralized Systems*, pages 28–34, 1993.
[15] K. Naik and D. S. L. Wei. Energy-conserving software design for mobile computers. *submitted for publication*, Dec., 1996.
[16] M. Pedram. Power minimization in ic design: Principles and applications. *ACM Trans. on Design Automation of Electronic Systems, Vol. 1, No. 1*, pages 3–56, Jan. 1996.
[17] R. A. Powers. Batteries for low power electronics. *Proc. of the IEEE, Vol. 83, No. 4*, pages 687–693, April, 1995.
[18] A. K. Salkintzis, C. Chamzas, and C. Koukourlis. An energy saving protocol for mobile data networks. *International Conference on Advances in Communications and Control (COMCON 5)*, pages 26–30, June, 1995.
[19] F. Sato, H. Kozuka, K. Miyazaki, and H. Fukuoka. Noah: An environment for distributed autonomous systems. In *2nd Intl. Symposium on Autonomous Decentralized Systems*, pages 412–418, 1995.
[20] R. Sedgewick. *Algorithms in C*. Addison-Wesley Publishing Co., 1990.
[21] S. Sheng, A. Chandrasekaran, and R. B. Broderson. A portable multimedia terminal for personal communications. *IEEE Comm. Magazine*, pages 64–75, Dec., 1992.
[22] A. H. Sherman. C-linda user's guide and reference manual. *Scientific Computing Associates Inc., New Haven, USA*, 1991.
[23] D. Singh, J. M. Rabaey, M. Pedram, F. Catthoor, S. Rajgopal, N. Sehgal, and T. J. Mozdzen. Power conscious cad tools and methodologies: A perspective. *Proc. of the IEEE, Vol. 83, No. 4*, pages 570–594, April, 1995.
[24] C-L. Su, C-Y. Tsui, and A. M. Despain. Saving power in the control path of embedded processors. *IEEE Design and Test of Computers*, pages 24–30, Winter, 1994.
[25] K. Sugawara. Agent-oriented architecture for flexible networks. In *2nd Intl. Symposium on Autonomous Decentralized Systems*, pages 135–141, 1995.
[26] V. Tiwari, S. Malik, and A. Wolfe. Power analysis of embedded software: A first step towards software power estimation. *IEEE Trans. on VLSI Systems, Vol. 2, No. 4*, pages 437–445, Dec., 1994.
[27] V. Tiwari, S. Malik, and A. Wolfe. Power analysis of the intel 486dx2. *Tech. Report No. CE-M94-5*, Jan., 1994.
[28] S. Wuytack, F. V. M. Catthoor, and H. J. De Man. Transforming set data types to power optimal data structures. *IEEE Trans. on Computer-Aided Design of Integrated Circuits and Systems, Vol. 15, No. 6*, pages 619–629, June, 1996.

DEPT. OF COMPUTER SOFTWARE, UNIVERSITY OF AIZU, AIZU-WAKAMATSU, 965-8580 JAPAN
E-mail address: `k-naik@u-aizu.ac.jp`

DEPT. OF COMPUTER AND INFORMATION SCIENCES, FORDHAM UNIVERSITY, BRONX, NEW YORK 10458
E-mail address: `wei@murray.fordham.edu`

DIMACS Series in Discrete Mathematics
and Theoretical Computer Science
Volume **52**, 2000

Impact of Unidirectional Links in Wireless Ad-Hoc Networks

Ravi Prakash and Mukesh Singhal

ABSTRACT. Ad-hoc networks consist of a set of mobile nodes with similar capabilities, communicating with each other using wireless links. As a node may not be able to directly reach every other node, a packet may have to traverse multiple wireless links from its source to its destination. Extensive research has been conducted to develop routing, transport layer communication, and resource allocation algorithms for ad-hoc networks. Most of the solutions assume that all wireless links are bidirectional. In reality, several links may be unidirectional. The presence of such links can jeopardize the performance of the existing solutions. In this paper we discuss the impact of such links. We also propose data structures and an information dissemination strategy to detect the presence of unidirectional links. This information is then used to modify a routing scheme, DSDV, to operate in ad-hoc networks with unidirectional links.

1. Introduction

An ad-hoc network is composed of a set of mobile nodes communicating with each other over wireless links. All the nodes have similar capabilities. The communication range of each node is finite. Therefore, two non-neighbor nodes may communicate with each other over a path of multiple wireless hops. Each node can be the source/destination of packets, and is also responsible for forwarding packets meant for other nodes. Often, the mobility pattern of the nodes is non-deterministic. Hence, the network topology is always in a flux. Paths between node pairs may change significantly over time. Two nodes may also be temporarily unreachable from each other if they fall in different network partitions.

Over the years there has been a significant amount of effort towards developing routing algorithms for ad-hoc networks. Routing algorithms for ad-hoc networks can be classified into (a) *cluster-based* algorithms, and (b) *flat* algorithms. In cluster-based algorithms, at regular intervals, a subset of nodes is elected as *cluster-heads*. A node is either a cluster-head or one wireless hop away from a cluster-head. Nodes that are not cluster-heads will, henceforth, be referred to as *ordinary* nodes. Only cluster-heads maintain information about network topology and routing. When an ordinary node has to send a packet, the node can send

---

1991 *Mathematics Subject Classification.* Primary 68Q22, 68M10; Secondary 68-06, 68-02.
The first author was supported in part by the National Science Foundation grant number CCR-9796331.

the packet to the cluster-head which routes that packet towards the destination. Alternatively, the source can obtain routing information from its cluster-head and forward the packet based on this information. Some examples of cluster-based routing algorithms are the *Linked Cluster Algorithm (LCA)* [**A, B**] and *Minimum Connected Dominating Set (MCDS)* [**E, F**].

There are several flat routing schemes. The *Destination Sequenced Distance Vector (DSDV)* [**O**] approach is a modification of the distance vector routing algorithm used earlier in ARPANET. In DSDV, each node maintains a distance vector that contains entries for each destination. The entry indicates the distance estimate and the next hop to be taken by a packet to reach a destination. Each entry has a sequence number associated with it indicating its *freshness*. If a destination is unreachable, the distance metric is set to infinity. Periodically a node's distance estimates are diffused to neighbors. When a node p loses a link that it was using to forward packets meant for destination q, p sets its distance metric for q to infinity and propagates this information with a higher sequence number. Such updates are diffused immediately, without waiting for the next update time. Similarly, when a path is found to a hitherto unreachable node the finite distance metric to that destination is propagated immediately through the network.

Dynamic Source Routing (DSR) [**I**] uses a diffusion based mechanism to find a route to the destination. Instead of periodically exchanging routing information between nodes, route(s) are discovered when a node has to send packets to some destination node. During this process intermediate nodes can use the discovered routes to update their own routing information. Caching of recently discovered routing information is employed to speed up the routing process. The route maintenance mechanism (i) sends a *route error* packet to the source if it detects that the route to the destination is broken, and (ii) either tries to use any other cached route to the destination or invokes route discovery once again. In order to route packets, the source completely specifies the path the packet should take.

In the *Ad-hoc On-Demand Distance Vector (AODV)* scheme [**N**], route discovery and maintenance are performed on demand, as in DSR, along with hop-based routing as in DSDV. In order to reduce communication overheads, as compared to DSDV, updates are propagated only along *active* routes, *i.e.*, routes that have seen some traffic in the recent past.

The *Temporally Ordered Routing Algorithm (TORA)* [**L**] is based on the notion of edge-reversal [**G**]. One instance of the algorithm is executed for each destination and a directed graph is maintained with respect to each destination. Only bidirectional links are considered, and a direction is associate with each link. Directed paths between every pair of nodes are initially determined through a sequence of edge reversals. When any node detects that it has lost the path to a destination (all edges incident on the node are directed towards it, in the graph for that destination) it performs full edge reversal so that it has only outgoing links to all its neighbors, and initiates route rediscovery for that destination. If a network partition is detected, the source is informed about the same.

The routing algorithms mentioned above have contributed significantly towards the understanding of the problem and the feasible solution approaches. Now that wireless communication and the technology to manufacture lightweight personal communication devices have matured, the time is ripe for the deployment of ad-hoc networks for various civilian applications.

However, to successfully deploy these systems we need to make a transition to more realistic system models! By realistic system models we mean propagation models that are not limited to simple circular propagation of signals. The IEEE 802.11 physical and medium-access control layer protocol [**H**] considers the propagation issues. We will not go into these issues in detail. Instead, we will concentrate on a manifestation of the realistic propagation models, namely presence of some *unidirectional links* in the network.

Some links may be unidirectional due to the *hidden terminal problem* [**Q**]. Node A may be able to receive messages from node B as there may not be any interfering node in A's vicinity. However, B may be in the vicinity of an interfering node and, therefore, be unable to receive A's messages. So, the link between A and B is directed from B to A.

The intent of this paper is twofold:

1. Evaluate the impact of unidirectional links on some of the existing algorithms proposed for routing, distributed resource allocation, and transport layer connections in ad-hoc networks. Almost all existing algorithms tend to assume that all links are bidirectional.
2. Propose preliminary strategies to capture information about unidirectional links, and guidelines to use this information to modify existing algorithms. The ultimate goal is to modify existing algorithms so that they can work correctly in an ad-hoc network that has a combination of unidirectional and bidirectional links.

In Section 2 we discuss the impact of unidirectional links on some of the existing algorithms for ad-hoc networks. In Section 3 we propose modification strategies. Finally, we present the conclusions and directions for future work in Section 4.

2. Problem Description

As mentioned earlier, we will consider three issues for ad-hoc networks with unidirectional links:

1. *Routing:* a network layer issue.
2. *TCP and UDP:* transport layer issues.
3. *Mutual exclusion:* an issue that arises at various layers, right from the MAC layer (avoidance of co-channel interference) to application layer (allocation of shared resources).

2.1. Routing. Several flat routing protocols [**O, N, L**] and hierarchical routing protocols [**A, B, E, F**] assume that all wireless links are bidirectional.[1] In the presence of unidirectional links several problems arise. For the purpose of illustration, let us consider DSDV [**O**]. AODV [**N**] has similar behavior. Other routing protocols may also exhibit similar problems.

Consider the network configuration shown in Figure 1. One can observe three interesting departures from accepted notions of routing:

1. *Knowledge Asymmetry:* There is a path from j to a of two hops: jia. In DSDV and AODV j would come to know of the path on being informed by i that i is one hop away from a. However, due to the link being unidirectional, i cannot directly inform j about the path. *Just because i knows that j is its neighbor, i cannot assume that j also knows that i is its neighbor.*

[1]DSR [**I**] does not explicitly assume the presence of only bidirectional links.

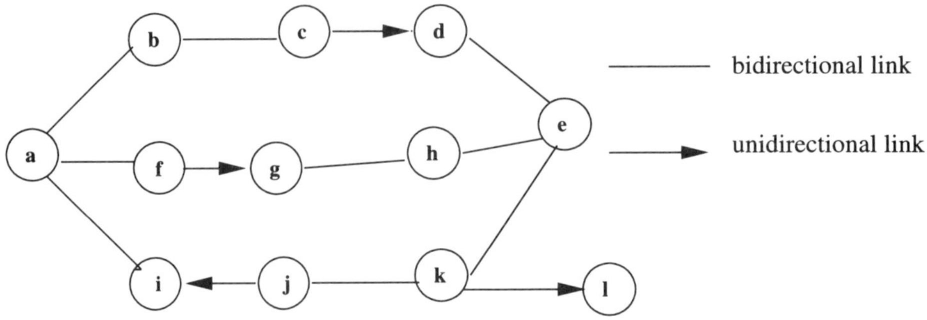

FIGURE 1. Ad-hoc network with unidirectional and bidirectional links.

Simple diffusion strategy may not be sufficient to propagate information about network topology. More sophisticated strategies are needed.

2. Routing Asymmetry: In DSDV and AODV, assuming all links are bidirectional, the path from node i to j would consist of one wireless hop. However, as the link $\vec{ji}$ is unidirectional, the shortest path from i to j consists of seven hops and the path from j to i consists of one hop: *a routing asymmetry.*
3. Sink Unreachability: A more complicated situation arises in the context of node l. In DSDV path updates are initiated by the destination node. In AODV a source node finds a route to the destination only when a sequence of *route replies* flows back on the path from the destination to the source. In Figure 1, there exists a path to node l. So, it could be the destination of packets. However, there is no way node l can inform k that the latter can reach the former in one hop. So, reachability information about l cannot propagate to other nodes. Node l is a *sink* node as all its incident links are directed towards it. *The network topology may indicate that a sink is reachable from other nodes. But due to the limitations of the routing algorithm no node knows of the existence of the sink, making it effectively unreachable.*

In fact, the problem with DSDV and AODV in the scenario shown in Figure 1 is quite serious. As they can only use bidirectional links for routing purposes, they will ignore links $\vec{cd}$, $\vec{fg}$, $\vec{ji}$, and $\vec{kl}$. As a result, even though nodes a and e are reachable from each other, DSDV and AODV will perceive a and c to be in different network partitions.

Similarly, TORA and LCA operate on bidirectional links only. In TORA, even if there exists an out-bound unidirectional link from a node, that node fails to utilize the information and may initiate a new route discovery. This can contribute significantly to communication overheads incurred due to route rediscovery.

In the context of LCA, in addition to paths between cluster-heads that are composed entirely of bidirectional links, there may be other paths in the network using some unidirectional links, too. However, LCA will fail to detect and use these paths, potentially causing congestion along the subset of paths traversed by packets between cluster-heads.

In DSR, let there be a unidirectional link from j to i along which i receives a path discovery message. When i has to send an acknowledgment to j it may need to initiate a new path discovery to find a route to j. The acknowledgment should then be sent along this route. Thus, while DSR does not ignore the possiblity of unidirectional links, it makes an implicit assumption that routes in both directions always exist between a pair of nodes. Such an assumption *may not always be valid* in a network with a combination of bidirectional and unidirectional links.

2.2. Transport Layer. In several implementations of TCP there is an implicit assumption that data and acknowledgment packets pass through the same set of routers, in opposite directions. Lakshman, Madhow and Suter have studied asymmetric networks where the links are bidirectional, but the bandwidth in one direction is significantly greater than the bandwidth in the other direction [**J**]. So, the data and acknowledgment packets traveling along the same set of links may experience different degrees of congestion.

However, due to the existence of unidirectional links in wireless ad-hoc networks, coupled with the tendency of some routing schemes to introduce asymmetry, the path from a TCP source to a destination (downstream path) may not be the same as the path from the destination to the source (upstream path). Therefore, data and acknowledgment packets may take entirely different routes. For example, in Figure 1 there are two paths from a TCP source a to the destination e: $ab\overrightarrow{cd}e$ and $a\overrightarrow{fg}he$, along which data packets can travel. However, the only path that acknowledgment packets can take from e to a is $e\overrightarrow{kj}ia$.

In such a case, if the upstream path gets disrupted, acknowledgments do not reach a as there is no other path from e to a. The underlying routing schemes may inform e of path disruption. However, there is no way to inform a about the upstream path disruption. What is needed is an explicit notification scheme, as in [**C**], urging node a to suspend transmission until a new upstream route is discovered. In the absence of such a notification, node a will continue to send packets and expect acknowledgments, and ultimately timeout. As a result, node a will enter into the congestion control mode even though there may be no congestion along the downstream path. Paxson [**M**] observed this problem in the context of a wide-area network and stated that *routing asymmetry complicates network troubleshooting, because it increases the likelihood that a network problem apparent in one direction along a virtual path cannot be detected in the other direction.*

An even severe problem arises when sinks are involved. Let us consider the node pair a and l. It is not possible to establish a TCP connection between the two nodes. Even though the connection establishment request from a reaches l, the response in the opposite direction cannot be sent. However, a UDP stream can be sent from a to l. This creates an interesting predicament. An application that only requires unidirectional data transfer may choose to have more reliable packet delivery from source a to destination l. So, it may go for a TCP connection and fail to communicate at all. Had the application not cared about reliability, and chosen to send UDP packets, it would have been successful. If transport layer reliability is not required by the application (but is desirable), how is the application to choose between TCP and UDP? A simple solution could be to always try to set up a TCP connection first, and if failure is encountered, to go with UDP. However, this would be an expensive solution.

2.3. Resource Allocation and Mutual Exclusion. Resource allocation and mutual exclusion have been extensively studied in distributed computing. Recently, there has also been an interest in these problems in the context of mobile systems [**R**]. Several resource allocation and mutual exclusion algorithms involve token passing or message exchange to request and grant permissions. The holder of the token can access the shared resource. In essence they are variations of the directed acyclic graph (DAG) resource allocation algorithm.

In DAG based algorithms it is assumed that all edges are capable of carrying messages in both directions. A logical direction is associated with each edge. An edge directed from a to b indicates that b holds the token shared by a and b. Node b can enter its critical section when all the edges incident on it are directed towards it. On exiting the critical section, the node reverses the direction of edges incident on it. This grants tokens to the neighbors and ensures starvation freedom and fairness. The solution to the *drinking philosophers problem* proposed by Chandy and Misra [**D**], Singhal's dynamic information structure mutual exclusion algorithm [**P**], Walter, Welch and Vaidya's algorithm [**R**], etc. fall into this class of solutions.

If some wireless links are unidirectional, edge reversal is not possible as message/token can be passed in one direction only. A pertinent question to ask at this juncture is: *Are there distributed mobile applications where links may be unidirectional and mutual exclusion is required between neighbors at either ends of such links?* If there are no such applications then considering unidirectional links would be an exercise in futility.

However, at various layers of the protocol stack there are relevant applications. For example, let us consider the distributed wireless channel allocation problem at the MAC layer [**K**]. Node a may not be able to inform a neighbor b about its channel usage if only unidirectional communication is possible from b to a. In such a situation, unless channel usage information is propagated through the entire network to reach b, node b may end up selecting the same channel as a, thus causing interference. At higher layers of the protocol stack a similar situation may arise if *disconnected mode* of operation is supported. A mobile process may acquire a lock on a shared resource and may get partitioned from the rest of the network before releasing the lock. Thus, when the lock is to be released, the node may not be able to send the corresponding release message. Once again the culprit is unidirectional flow of information. Here, neighborhood does not indicate physical proximity. Instead, it indicates logical proximity.

The preceding discussion illustrates the fact that presence of unidirectional links can jeopardize the performance of various algorithms that assume all links to be bidirectional. In the next section we will give an overview of our solution approach.

3. Solution Approach

Each node needs to maintain enough information to distinguish between bidirectional and unidirectional links to its neighbors. A node may not be able to directly send information to a neighbor if there is no link from the node to the neighbor. Therefore, each node also has to forward topology information being exchanged among other nodes. Once knowledge of link orientations is available, appropriate routing decisions can be made.

Hence, we will concentrate on modifications to protocols like DSDV and AODV to cope with the presence of unidirectional links.

It is assumed that each node emits a beacon at regular intervals. A node can hear beacons transmitted by a neighboring node provided the link between them is bidirectional, or directed from the neighbor to itself. The transmission of beacons by different nodes is not synchronized as there is no global clock in the system.

3.1. Data Structures. Each node p maintains the following data structures:

- *Nodesheard:* set of nodes whose beacons have been heard by node p within the last t time units. $Nodesheard_p$ indicates the *Nodesheard* set for node p. However, the subscript will be dropped when the node's identity is obvious from the context. If $q \in Nodesheard_p$ and $p \in Nodesheard_q$, then there exists a bidirectional link between p and q. However, if $q \in Nodesheard_p$ and $p \notin Nodesheard_q$, then there is a unidirectional link from q to p. This data structure is modeled after the one by the same name used in the Linked Cluster Algorithm [**A, B**].
- *D:* an $n \times n$ matrix of 2-tuples, where n is the number of nodes in the network. If $D[i,j] = (seq,\ dist)$, it means node p knows that the path from node i to node j if of length $dist$, and the sequence number associated with this information is seq. Due to the possibility of unidirectional links, $D[i,j].dist$ may not be equal to $D[j,i].dist$.
- *To* and *From:* vectors of length n, where each entry is a 3-tuple of the form $(seq,\ dist,\ node\_id)$. The To vector is similar to the distance vector of DSDV as it maintains information about the path length from a node to all other nodes, and the next hop on the path to those nodes. $From_p$ vector contains information about paths from other nodes to p. Due to the presence of unidirectional links in the network, and the resultant routing asymmetry, the corresponding $dist$ values in the To and $From$ vectors may be different from each other. When routing information stabilizes, To_p should have the same $dist$ and seq values as the corresponding entries in the pth row of D_p. There should be a similar match between $From_p$ and the pth column of D_p.

3.2. Determination of Link Orientation. We employ the Nodesheard set in a manner similar to [**A**] to determine network adjacency. As stated earlier, each node periodically transmits its Nodesheard set with its beacon. It also continuously listens for similar transmissions from other nodes. If node p hears node q's transmission and finds that $p \in Nodesheard_q$, node p knows that there exists a bidirectional link between p and q. The next time p broadcasts its beacon it includes q in its Nodesheard set. When q hears this beacon, it, too, gets to know of the presence of the bidirectional link.

If node p hears q's transmission and $p \notin Nodesheard_q$, p concludes that there exists a unidirectional link from q to p. However, how does q get to know of the presence of this link? For this purpose we employ the matrix D.

Construction of D matrix: Initially, the D matrix at each process p only contains its adjacency information. If there is a bidirectional link between p and q, then $D_p[p,q].dist = D_p[q,p].dist = 1$. If p knows that there exists a unidirectional link from q to p, then $D_p[q,p].dist = 1$. The sequence number associated with each entry is analogous to the sequence number associated with routing table entries in DSDV and AODV.

Each node periodically transmits its D matrix. The time between successive transmissions of D is a multiple of the time between successive transmissions of the $Nodesheard$ set. This is so for two reasons:

1. Transmission of D consumes much more bandwidth than the transmission of $Nodesheard$.
2. Transient noise that may interfere with the reception of a few successive $Nodesheard$ messages from a neighbor does not lead a node into erroneously concluding that its path to/from that neighbor is broken.

Let node p receive matrix D_{recv} from node q. If $\overline{pq}$ is a bidirectional link or a unidirectional link from q to p, p modifies its D matrix in the following manner on receiving the matrix:

- For any arbitrary node r that is different from p and q:
  - If $D_{recv}[r,q].seq$ is less than $D[r,p].seq$ then leave $D[r,q]$ unchanged.
  - If $D_{recv}[r,q].seq$ is equal to $D[r,p].seq$ then:
    - $D[r,p].dist = min(D_{recv}[r,q].dist + 1,\ D[r,p].dist)$
    - $D[r,q].dist = min(D_{recv}[r,q].dist,\ D[r,q].dist)$
  - If $D_{recv}[r,q].seq$ is greater than $D[r,p].seq$ then:
    - $D[r,p].dist = D_{recv}[r,q].dist + 1$
    - $D[r,p].seq = D_{recv}[r,q].seq$, and
    - $D[r,q].dist = D_{recv}[r,q].dist$.

  These operations enable node p to determine its distance from other nodes.
- For any arbitrary pair of nodes r and s, different from p and q: if $D_{recv}[r,s].seq$ is greater than $D[r,s].seq$, or $D_{recv}[r,s].seq = D[r,s].seq$ and $D_{recv}[r,s].dist < D[r,s].dist$, replace $D[r,s]$ with $D_{recv}[r,s]$.

If link $\bar{pq}$ is a bidirectional link, node p performs the following operations as well:

- If $D_{recv}[q,r].seq$ is less than $D[p,r].seq$ then leave $D[p,r]$ unchanged.
- If $D_{recv}[q,r].seq$ is equal to $D[p,r].seq$ then:
  - $D[p,r].dist = min(D_{recv}[q,r].dist + 1,\ D[p,r].dist)$,
  - $D[q,r].dist = min(D_{recv}[q,r].dist,\ D[q,r].dist)$.
- If $D_{recv}[q,r].seq$ is greater than $D[q,r].seq$ then:
  - $D[p,r].dist = D_{recv}[q,r].dist + 1$
  - $D[p,r].seq = D_{recv}[q,r].seq$, and
  - $D[q,r].dist = D_{recv}[q,r].dist$.

The preceding operations are similar to the updates performed by DSDV and AODV. They enable node p to determine its distance to other nodes.

If distance estimates of p are revised as a result of receiving D_{recv} from q, then the corresponding entries of To_p and $From_p$ are also updated and the $node\_id$ values of those entries are set to q.

If the received D matrix from node q is such that $D_{recv}[p,r].dist = 1$ and $r \notin Nodesheard_p$, node p concludes that there exists a unidirectional link from p to r. Therefore, for every arbitrary node s that is different from p and r, p updates its D matrix as follows: $D[p,s].dist = mind(D_{recv}[r,s].dist) + 1$ over all r such that $r \notin Nodesheard_p$ and $p \in Nodesheard_r$.

Thus, each node updates its reacability information and propagates this information to other nodes.

Example: Let us refer back to Figure 1. Node i knows that there is a path of length one from j to i. This information is forwarded by i, through a to the rest of the network. Later, when node j receives D_{recv} matrix from k, j finds that $D_{recv}[j,i] = 1$. It is at this point that j realizes that it has an outgoing link to i.

Using this information, along with distance estimates from i to other nodes, j can revise its estimate of its distance to other nodes.

Also, when node b sends its D matrix to node c, c realizes that b is two hops away from i. Therefore, c concludes that it must be three hops away from i.

Also, node l can gather information about the distance from all other nodes to itself. However, as there is no outgoing path from l, no other node knows of l's existence. Hence, l may receive broadcast messages in the network, but no message can be explicitly sent to it, *i.e.*, sink nodes know of other nodes while other nodes are oblivious of its existence. An interesting question to consider at this point is if the presence of sinks can be a security loophold for the network?

3.3. Storage and Communication Overheads. The storage requirement at each node is $O(n^2)$, where n is the number of nodes in the system. This is significantly greater than DSDV which only requires $O(n)$ data items to be stored by each mobile node. The increased storage complexity of the proposed scheme is due to the topology matrix D maintained by each node. The other data structures: the $Nodesheard$ set, and the To and $From$ vectors require $O(n)$ storage. Similarly, the largest message is of size $O(n^2)$ data items, once again greater than the communication overheads of DSDV.

Considering that memory and bandwidth are scarce commodities in a mobile computing environment, the high storage and communication complexity can severely impact the performance of the system. A pertinent question to ask at this juncture is: *Is it possible to reduce the storage and communication complexity of the algorithm for a network with potentially unidirectional links?*

At this point we do not have a conclusive proof about the impossibility of a lower communication and storage complexity. However, the following scenario leads us to believe that $O(n^2)$ is the best we can do. Consider a node p which has m outgoing links to different nodes, one incoming link from node q, and no bidirectional link. The outgoing links terminate at nodes $o_1, o_2, \ldots, o_m$. Node p can reach all nodes in the network through the outgoing links. Also, let there be multiple paths from p, along different outgoing links, to other nodes in the network. For such a system, node q would have to inform p about the distance from o_i to all other nodes in the network, where $1 \leq i \leq m$. Therefore, amount of information that q would need to convey to p for nodes reachable through o_i would be $O(n)$. Therefore, total amount of data that q would need to gather and forward to p would be equal to $O(m \times n)$. As $m = O(n)$, we feel that the storage and communication requirements cannot be better than $O(n^2)$.

4. Conclusion and Future Work

Most of the research in mobile computing tends to assume that all links are bidirectional. However, due to a variety of reasons, for example propagation characteristics and interference, only unidirectional communication may be possible between some pairs of adjacent nodes. Existing algorithms will fail in the presence of such links.

We described the adverse impact of unidirectional links on existing algorithms for routing, transport layer communication, and mutual exclusion. We also described simple data structures and proposed a strategy to propagate routing information in networks with a combination of unidirectional and bidirectional links.

The proposed strategy is a modification of DSDV and AODV: well known routing algorithms for wireless ad-hoc networks. It incurs higher communication and storage overheads of $O(n^2)$. However, such overheads seem unavoidable.

While it may not be possible to reduce the storage and communication complexity, we will work on efficient storage and information propagation strategies to reduce the absolute size of messages exchanged between neighboring nodes. This is of significance due to the low bandwidth of wireless links. We also intend to work towards a proof that communication and storage overheads cannot be improved beyond $O(n^2)$. If it turns out that our initial belief is wrong, we will work towards developing alternative solutions that incur lower overheads. We will also try to gain a better understanding of the role of sink nodes in a network.

References

[A] D. J. Baker and A. Ephremides. The Architectural Organization of a Mobile Radio Network via a Distributed Algorithm. *IEEE Transactions on Communications*, COM-29(11):1694–1701, November 1981.

[B] D.J. Baker, A. Ephremides, and J. A. Flynn. The Design and Simulation of a Mobile Radio Network with Distributed Control. *IEEE Journal on Selected Areas in Communications*, pages 226–237, 1984.

[C] K. Chandran, S. Raghunathan, S. Venkatesan, and R. Prakash. A Feedback Based Scheme for Improving TCP Performance in Ad-Hoc Wireless Networks. In *Proceedings of the* 18th *International Conference on Distributed Computing Systems (ICDCS'98)*, pages 472–479. IEEE, May 1998.

[D] K. M. Chandy and J. Misra. The Drinking Philosophers Problem. *ACM Transactions on Programming Languages and Systems*, 6(4):632–646, October 1984.

[E] B. Das and V. Bharghavan. Routing in Ad-Hoc Networks Using Minimum Connected Dominating Sets. In *Proceedings of ICC*, 1997.

[F] B. Das, R. Sivakumar, and V. Bharghavan. Routing in Ad-Hoc Networks Using a Spine. In *Proceedings of IEEE IC3N*, 1997.

[G] E. Gafni and D. Bertsekas. Distributed Algorithms for Generating Loop-free Routes in Networks with Frequently Changing Topology. *IEEE Transactions on Communications*, pages 11–18, January 1981.

[H] IEEE. *P802.11, IEEE Draft Standard for Wireless LAN Medium Access Control (MAC) and Physical Layer (PHY) Specification, D2.0*, July 1995.

[I] D. B. Johnson and D. A. Maltz. Dynamic Source Routing in Ad-Hoc Wireless Networks. In T. Imielinski and H. Korth, editors, *Mobile Computing*. Kluwer Academic Publishers, 1996.

[J] T.V. Lakshman, U. Madhow, and B. Suter. Window-based error recovery and flow control with a slow acknowledgment channel: a study of TCP/IP performance. In *Proceedings of INFOCOM*. IEEE, 1997.

[K] S. Nesargi and R. Prakash. Distributed Wireless Channel Allocation in Networks with Mobile Base Stations. In *Proceedings of IEEE INFOCOM'99*, New York, N.Y., March 1999.

[L] V. D. Park and M. S. Corson. A Highly Adaptive Distributed Routing Algorithm for Mobile Wireless Networks. In *Proceedings of IEEE INFOCOM*, April 1997.

[M] V. Paxson. End-to-End Routing Behavior in the Internet. *IEEE/ACM Transactions on Networking*, 5(5):601–615, October 1997.

[N] C. Perkins. Ad Hoc On Demand Distance Vector (AODV) Routing. Internet-Draft, draft-ietf-manet-aodv-02.txt, November 1998.

[O] C.E. Perkins and P. Bhagwat. Highly Dynamic Destination-Sequenced Distance-Vector Routing (DSDV) for Mobile Computers. In *Proceedings of ACM SIGCOMM Conference on Communication Architectures, Protocols and Applications*, pages 234–244, August 1994.

[P] M. Singhal. A Dynamic Information-Structure Mutual Exclusion Algorithm for Distributed Systems. *IEEE Transactions on Parallel and Distributed Systems*, 3(1):121–125, January 1992.

[Q] A. Tanenbaum. *Computer Networks(*3rd *Edition)*. Prentice Hall, Upper Saddle River, N.J., 1996.

[R] J. Walter, J.L. Welch, and N. Vaidya. A Mutual Exclusion Algorithm for Ad Hoc Mobile Networks. In *Proceedings of the* 2nd *International Workshop on Discrete Algorithms and Methods for Mobile Computing and Communications (DIAL M for Mobility)*, Dallas, October 1998.

Computer Science Program, University of Texas at Dallas, Richardson, TX 75083-0688.

E-mail address: ravip@utdallas.edu

National Science Foundation, 4201 Wilson Boulevard, Room 1145, Arlington, VA 22230.

E-mail address: msinghal@nsf.gov

DIMACS Series in Discrete Mathematics
and Theoretical Computer Science
Volume **52**, 2000

On Frequency Assignment in Cellular Networks

Sanguthevar Rajasekaran, K. Naik, and David Wei

ABSTRACT. In this paper we consider the problem of frequency assignment in cellular networks. The model we consider is general. We present an algorithm for frequency assignment and analyze its performance.

1. Introduction

Frequency assignment is an important problem in the operation of mobile networks. The area available for the network is typically partitioned into hexagonal *cells* (see Figure 1 for an example). A base station is located at the center of each cell. This station is in-charge of handling calls arising from mobile hosts situated in its cell. The problem of frequency assignment is to allocate a frequency to each call for communication such that the interference among different calls is minimized and the total number of distinct frequencies used is minimized.

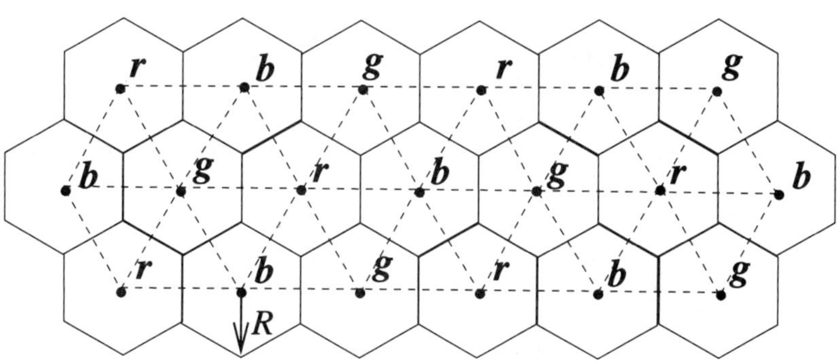

FIGURE 1. Configuration of cells

Cellular networks are usually represented as planar graphs where the nodes correspond to cell centers. There is an edge between two nodes if the corresponding hexagons have at least one common side. The graph G corresponding to the cells of Figure 1 is shown with dotted lines.

This work was supported in part by an NSF Award CCR-95-03-007 and an EPA Grant R-825-293-01-0.

Mobile hosts in any cell communicate with the base station in the cell to serve their various needs. Each node of the graph G can be weighted with the number of hosts in the corresponding cell. Each host should be assigned a frequency for communication in such a way that interference with the other hosts in the network is minimized. The hosts in the same cell should be assigned different frequencies. Also, any two hosts that are at a distance of d or less should be assigned different frequencies (where d is chosen appropriately to minimize the interference). A fourth power attenuation law is typically assumed [**1**].

It is conceivable that a host can dynamically migrate from one cell to another. Thus the weighted graph G also changes dynamically. If one assumes that possible interferences are restricted to adjacent cells, then the frequency assignment problem reduces to weighted graph coloring. In this case, the graph G is refered to as the *interference graph*.

The *static frequency assignment* problem refers to the problem of assigning frequencies to the hosts given a snapshot of the network at a specified time. The *online (or dynamic) frequency assignment* refers to the problem of continuous frequency assignment taking into account the changes that might occur in the graph G. In this case, the state of the network can be modeled with an ordered sequence of graphs $G_t, t \geq 0$. The problem corresponding to G_t should be solved before considering G_{t+1}.

The static frequency assignment problem has been studied extensively. Even for the case when the interference is restricted to adjacent cells, it has been shown that the problem of optimal coloring of the interference graph G is NP-complete [**6**]. Several approximation algorithms have been devised for this version. For example, a simple algorithm called *Fixed Allocation (FA)* uses no more than three times the optimal number of frequencies (or colors). Janssen et. al.'s algorithm [**2**] uses no more than 1.5 times the optimal number of colors. The number of colors used by Narayanan and Shende's algorithm [**5**] is no more than $\frac{4}{3}$ times the optimal. Janssen et. al. [**3**] obtain the same performance for the dynamic version of the problem also.

Several dynamic algorithms are known for the general version of the problem as well. Four of these algorithms are: the geometric strategy; Borrowing with Directional Channel-Locking (BDCL) [**7**]; Nanda-Goodman [**4**]; and the Two-Step Dynamic-Priority (TSDP) [**1**]. These algorithms have been demonstrated to do well in practice. However no bounds have been proven on their performances.

In this paper we consider a general version of the frequency assignment problem and present an algorithm that uses no more than three times the optimal number of colors. This is the first algorithm with a proven bound on the number of colors. We also illustrate how to use this algorithm to develop dynamic frequency assignment algorithms.

The rest of the paper is organized as follows. In Section 2 we provide a summary of known techniques. In Section 3 we present a coloring algorithm and analyze its performance. In this section we also extend our coloring algorithm to obtain dynamic frequency assignment algorithms. Section 4 concludes the paper.

2. Literature Survey

Static frequency assignment problem where interference is restricted to adjacent cells (call this version *unit-distance frequency assignment problem*) can be stated as follows. Input is a graph $G(V, E, w)$ where V is a set of nodes (cell centers), E is a set of edges (there is an edge from node u to node v if they are adjacent cells), and w is the weight function. In particular, $w(u)$ stands for the number of hosts that the cell u has to handle. The problem is to *multicolor* the graph with as few colors as possible such that 1) for any $u \in V$, $w(u)$ distinct colors are assigned to node u, and 2) for any edge $(u, v) \in E$, the colors assigned to the nodes u and v are disjoint. Let $\chi(G)$ be the minimum number of colors needed to color G. In the *general frequency assignment* problem, there cannot be a common frequency between two cells if they are separated by a distance of $\leq d$ (for some specified d).

We can think of the space available for the mobile network as consisting of rows and columns of cells where the leftmost bottom cell is denoted as $(0, 0)$ whose center is located at $(0, 0)$. Let R be the radius of each cell (see Figure 1). The cell in row i and column j (denoted as (i, j)) has its center at $\left(\sqrt{3}Rj + \frac{\sqrt{3}}{2}Ri, \frac{3}{2}Ri\right)$.

Thus in the case of unit-distance frequency assignment, the value of d is $\sqrt{3}R$. For the general case, d could take on any value. We can also define an interference graph corresponding to the general frequency assignment problem. This graph has the cell centers as its nodes. There is an edge from node u to node v if these nodes are at a distance of $\leq d$.

The performance of any coloring algorithm is measured using its *competitive ratio* which is the ratio of the number of colors used to $\chi(G)$. An algorithm is said to be *c-competitive* if its competitive ratio is c or better.

2.1. Unit-distance Frequency Assignment. Clearly, the total weight on any maximal clique of the interference graph is a lower bound on $\chi(G)$. The only maximal cliques in the interference graph are edges and triangles. If D_2^G and D_3^G denote the maximum weight of any edge or triangle, respectively, then $D^G = \max\{D_2^G, D_3^G\}$ is a lower bound on $\chi(G)$.

A simple coloring algorithm that is 3-competitive can be devised using the fact that the interference graph (where the weight of each node is 1) is 3-colorable. Figure 1 shows this coloring. The three colors used are red, blue, and green. Let the colors to be used be called $0, 1, 2, \ldots$. To color new calls, the red, blue, and green nodes use the smallest integers available from the sets of integers that are mod $0, 1$, and 2, respectively. For example, the sequence of colors used by any red node will be $0, 3, 6, \ldots$. Clearly, there cannot be any interference between adjacent cells under this coloring. This algorithm is known as the *Fixed Allocation (FA)* algorithm.

The competitive ratio of the above algorithm can be improved to $\frac{3}{2}$ as has been shown by Janssen et. al. [**2**]. For any node v consider all the triangles in which v is a vertex. Let $D_1(v)$ denote the maximum total weight of any of these triangles. $D_1(v)$ is also known as 1-local maximum clique weight [**3**].

The colors are divided into three palettes. Red colors are integers that are mod 0. Blue and green colors are integers that are mod 1 and 2, respectively. Let v be any node. The node v employs three local spectra consisting of red, blue, and green colors each spectrum being of size $\lceil D_1(v)/2 \rceil$. If v's base color is red it uses the first $\lceil w(v)/2 \rceil$ colors from its red spectrum and the last $\lfloor w(v)/2 \rfloor$ colors from its blue spectrum. If v's color is blue it uses the first $\lceil w(v)/2 \rceil$ colors from its blue spectrum and the last $\lfloor w(v)/2 \rfloor$ colors from its green spectrum. Finally, if v's color is green it uses the first $\lceil w(v)/2 \rceil$ colors from its green spectrum and the last $\lfloor w(v)/2 \rfloor$ colors from its red spectrum.

Clearly, the number of colors used by any node is no more than $\frac{3}{2}\chi(G)$. It is easy to verify that no two adjacent cells will access the same color.

Narayanan and Shende [**5**] have given an intricate algorithm for coloring that is $\frac{4}{3}$-competitive. Based on this algorithm, Janssen et. al. [**3**] have offered a dynamic frequency assignment algorithm that is $\frac{4}{3}$-competitive. In this algorithm, at any given time, a node has to have a view of only a small neighborhood around it.

2.2. General Frequency Assignment. Several works have been done that address the problem of dynamic general frequency assignment under both TDMA and FDMA multiplexing schemes. The problem of frequency assignment is also refered to as the *carrier allocation problem* (see e.g., [**1**]). The bandwidth available for communication is partitioned into *carriers* by frequency division. Each carrier is divided into certain number of *channels* by time division. A channel is used to support a call.

In this section we give a brief summary of four techniques for carrier allocation, namely, the geometric strategy, borrowing with directional channel-locking (BDCL), Nanda-Goodman strategy, and the two-step dynamic-priority (TSDP). Let D_{min} be the minimum distance between two cells in order for them not to interfere with each other. If c is any cell, let $IN(c)$ stand for the interference region of c, i.e., $IN(c)$ is the set of cells that are at a distance of less than D_{min} from c. The status of a carrier r in a cell c, denoted as $status(r, c)$, is defined as follows [**1**]. 1) We say r is a used carrier for c if at least one channel in r is used by some user in c. In this case $status(r, c) = UC$. 2) We say r is an interfered carrier for c if there is a cell $c' \in IN(c)$ such that $status(r, c') = UC$. In this case, $status(r, c) = IC$. 3) Otherwise, we say r is available for c. In this case $status(r, c) = AC$.

Geometric Strategy. Here the cells are partitioned into k sets $S_0, S_1, \ldots, S_{k-1}$ as evenly as possible. These sets are such that no two cells in the same set are D_{min} or less apart, i.e., no two cells interfere with each other. The set of available carriers is also divided into k sets $P_0, P_1, \ldots, P_{k-1}$ as evenly as possible. Carriers in each P_i are ordered. When a cell in S_i needs a carrier, it chooses the first available carrier in the sequence $P_i, P_{i+1}, \ldots, P_{k-1}, P_0, P_1, \ldots, P_{i-1}$. Thus the cells in S_i prefer the carriers in P_i the most and hence P_i are called the *first-choice* or *nominal* carriers for S_i.

Borrowing with Directional Channel-Locking (BDCL). This technique [**7**] is closely related to the geometric strategy. Here also the cells are partitioned into $S_0, S_1, \ldots, S_{k-1}$ and the carriers are partitioned into $P_0, P_1, \ldots, P_{k-1}$. Carriers in P_i are the nominal carriers for cells in S_i. When a cell c in S_i needs a carrier, it chooses the first available carrier in the sequence P_i. (Note that P_i is a list ordered according to priority, with the first carrier having the highest priority.) If no nominal carrier is available, the cell borrows the carrier of lowest priority from the *richest cell* in $IN(c)$, i.e., the cell with the largest number of unassigned nominal carriers.

Nanda-Goodman Technique. In this technique [**4**] each carrier is assigned an *acquisition priority* and a *release priority.* These priorities are assigned dynamically. For any cell c and a carrier r available for c, c's acquisition priority for r is defined as the number of cells in $IN(c)$ in which r is an interfered carrier. In other words,

$$acquisition - priority(r, c) = |\{c' \in IN(c) : status(r, c') = IC\}|.$$

Let r be used in c. Then the release priority of r is defined as the number of cells in $IN(c)$ in which r will become available if r is released by c. In other words,

$$release-priority(r,c) = |\{c' \in IN(c) : \zeta(r,c') = 1\}|$$

where $\zeta(r,c')$ is the number of cells in $IN(c')$ that are currently using r.

Two-Step Dynamic Priority (TSDP) Technique. This algorithm [**1**] is based on an *optimal reuse pattern* of the carriers. Let the *reuse pattern* $S(r)$ of a carrier r be defined as $S(r) = \{c|status(r,c) = UC\}$. By *optimal reuse pattern* of a carrier we mean the largest reuse pattern possible for that carrier. Any carrier that is used in a cell c can be reused in another cell c' provided that these two cells are at least D_{min} apart. Dong and Lai [**1**] show that an optimal way of using a carrier is to employ it in cells that are regularly placed in the plane such that the distance between adjacent cells in any column (or row) is exactly equal to D_{min}. This notion can be formalized as follows.

Define an equivalence relation Γ on the collection C of cells. For any $c_1, c_2 \in C$, $(c_1, c_2) \in \Gamma$ if and only if one of the following is true: 1) $c_1 = c_2$; 2) $dist(c_1, c_2) = D_{min}$; or 3) There exists a c such that $(c_1, c) \in \Gamma$ and $(c, c_2) \in \Gamma$.

Dong and Lai [**1**] have shown that an optimal reuse pattern for any carrier is an equivalence class of Γ. When $D_{min} = 3\sqrt{3}R$, there arise nine equivalence classes as shown in Figure 2.

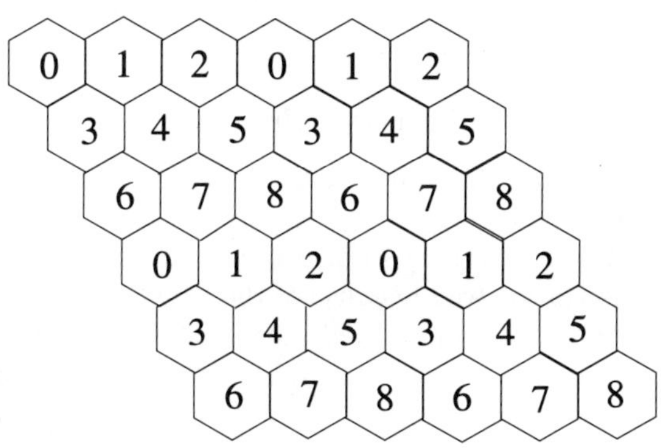

FIGURE 2. Equivalence classes when $D_{min} = 3\sqrt{3}R$

The TSDP algorithm partitions the cells into optimal reuse patterns $G_0, G_1, \ldots, G_{k-1}$ using the equivalence relation Γ. Consider only the case $D_{min} = 3\sqrt{3}R$. For any cell $c \in G_i$, define $LOP(c) = \{c' : dist(c,c') = D_{min}\}$. When $D_{min} = 3\sqrt{3}R$, there are six cells in $LOP(c)$. The algorithm also has an adjustable parameter λ which is an integer in the range $[0, 7]$.

Any carrier r is called a *primary carrier* for cell c if r is currently used by at least λ cells in $LOP(c)$. Otherwise, r is said to be a *secondary carrier* for c. By appropriately choosing λ, one can make sure that the carriers are used according to maximal reuse patterns as much as possible.

The algorithm description will be complete if we specify the acquisition and release priorities for the various carriers.

For any cell c and a primary carrier r for c, the acquisition priority is defined as follows.

$$acquisition - priority(r, c) = \sum_{c' \in LOP(c)} weight(r, c')$$

where $weight(r, c')$ is N if $status(r, c') = UC$; 1 if $status(r, c') = AC$; and 0 if $status(r, c') = IC$. Here N is a positive number greater than the number of cells in $IN(c)$. Also,

$$release - priority(r, c) = -acquisition - priority(r, c)$$

Dong and Lai [**1**] show that this priority scheme results in the primary carriers being used in close approximation to their optimal reuse patterns.

For the secondary carriers they employ the priority schemes of Nanda-Goodman in order to use these carriers in a compact manner. If r is a secondary carrier for c, then

$$acquisition - priority(r, c) = |\{c' \in IN(c) : status(r, c') = IC\}|.$$

Also,

$$release - priority(r, c) = |\{c' \in IN(c) : \zeta(c', r) = 1\}|$$

where $\zeta(r, c')$ is the number of cells in $IN(c')$ in which the carrier r is currently used.

The experimental results of [**1**] indicate that the TSDP algorithm performs better than the other three techniques namely, geometric, BDCL, and Nanda-Goodman. In [**1**] they also present some variants of the algorithm TSDP.

3. The New Algorithm

All the carrier allocation algorithms we have seen thus far have the following common theme: Partition the cells as well as carriers into groups. For each group of cells and each groups of carriers assign acquisition as well as release priorities. The TSDP algorithm performs well because of the way the cells are partitioned and also because of the priority schemes. Cell partition corresponds to optimal reuse patterns of the carriers and the priority scheme for primary carriers is such that they are used according to their optimal reuse patterns as much as possible.

Consider the case $D_{min} = k\sqrt{3}R$, for some positive integer k. For this case, there will be k^2 equivalence classes in the relation Γ. Call the intersection of k successive columns and k successive rows of cells a *$k \times k$-rectangle.* How many colors are needed to multicolor the interference graph? The optimal reuse pattern suggests that a different palette be used for each cell in a $k \times k$-rectangle. A lower bound Δ on the number of colors needed is the maximum of total weight on any circle of diameter $(k-1)\sqrt{3}R$. Figure 3 shows the case when $k = 3$. The total weight of all the cells whose centers lie within a circle of diameter $2\sqrt{3}R$ is a lower bound.

Thus the TSDP algorithm uses no more than k^2 times the optimal number of colors.

Example 1. Consider the following instance. Similar to the cells, name the rectangles also as $(0,0), (0,1)$, etc. Rectangle $(0,0)$ has q requests in its cell $(0,0)$. The rest of the cells in the rectangle do not have any calls. Rectangle $(0,1)$ has q calls in its cell $(0,1)$ and the rest of the cells do not have any calls, and so on. Also assume

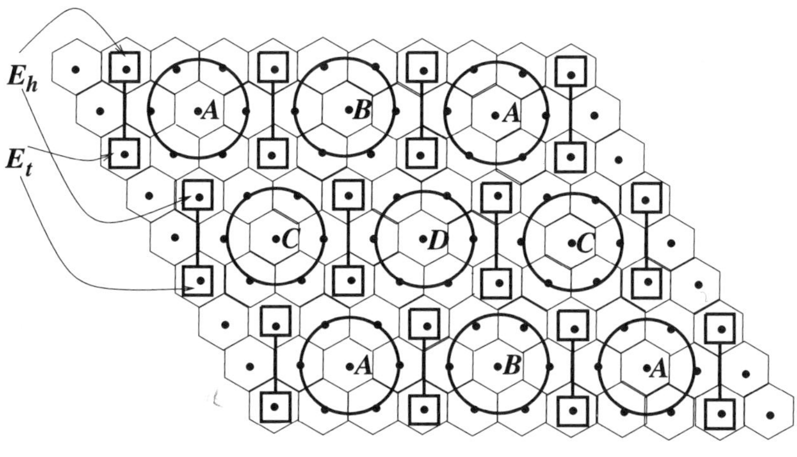

FIGURE 3. An efficient coloring of the graph

that this pattern persists for a long time. For this instance, the TSDP algorithm will use $k^2\Delta$ colors. Thus the TSDP algorithm is k^2-competitive. □

Our algorithm is based on the fact that the interference graph can be colored with no more than 3Δ colors. We begin with a 5-competitive scheme. Figure 3 shows this coloring for the case $k = 3$. But the result holds for any value of k.

LEMMA 3.1. *There exists a coloring scheme that is 5-competitive.*

Proof. The idea is to partition the cells as illustrated in Figure 3. In particular we draw circles of diameter $(k-1)\sqrt{3}R$. All the cells whose centers lie on or inside a circle constitute a class. We label these classes as follows. The classes that are in the same rows will be labeled $A, B, A, \ldots$. The classes immediately below will be labeled $C, D, C, \ldots$. There are cells that are not covered by these circles. These are grouped such that all the cells in between two successive circles belong to the same class. Label these classes E. We use five palettes. Four palettes are used to color the classes labeled A, B, C, and D. The fifth palette is used to color the class E. Let each of the five spectra be of size Δ.

Clearly, there won't be any conflicts among the classes A, B, C, and D. Two classes that are labeled E might have some conflicts, since there are nodes from two adjacent classes labeled E that are at a distance of $< D_{min}$. In order to avoid these conflicts, for any class labeled E, the cells in the top half (E_h) will use the spectrum from one end and the cells in the bottom half (E_t) will use the spectrum from the other end. This scheme reverses for the E classes that are in rows below.

It is easy to verify now that there won't be any conflicts among the E classes also. □

Our algorithm will partition the cells according to the coloring imposed by Lemma 3.1. There are many possibilities. If we know in advance the value of Δ, then we can use five spectra each of size Δ and there won't be any interferences or failures. Another possibility is that the number of colors (i.e., carriers) available may be fixed. In this case we partition the available carriers into five: $P_0, P_1, P_2, P_3,$

and P_4. Call the cell partitions S_0, S_1, S_2, S_3, and S_4 (corresponding to A, B, C, D, and E, respectively). We say P_i are the primary carriers for S_i, $0 \leq i \leq 4$. Depending on the priority schemes to be used, we can derive many dynamic carrier allocation strategies. We list below some of them.

Algorithm1. The priority scheme is similar to the geometric strategy. If a cell $c \in S_i$ requires a channel it picks the first available channel from the list $P_i, P_{i+1}, \ldots, P_4, P_0, \ldots, P_{i-1}$, for $0 \leq i \leq 4$.

Algorithm2. Here the priority scheme corresponds to that of BDCL.

Algorithm3. For each group S_i, primary carriers are P_i, $0 \leq i \leq 4$. For the secondary carriers, use the priority scheme of Nanda-Goodman.

A Useful Heuristic for Algorithm1. Algorithm1 is based on the geometric strategy. Here we mention a heuristic that has the potential of improving the performance. According to the geometric strategy, a cell m labelled B will look for a carrier in the ordered list B, C, A. If there is no available carrier in B, it will look for one from the palettes C and A. Say a carrier q is available in C. If q is assigned to m, q cannot then be assigned to any cell in the interference region of m. This will affect the reusability of the carrier. Instead we can do the following. When the cell m does not find a carrier in B, it first checks the carriers in C and A that are currently used by cells that are outside m's interference region. This way we can make sure that we are not using any "new" carriers from either C or A.

LEMMA 3.2. *There exists a coloring scheme that is 3-competitive.*

Proof. The cells are partitioned into three classes as shown in Figure 4. The labeling of these classes correspond to the palettes that will be used to color them. It is easy to see that there won't be any conflicts among the three classes. □

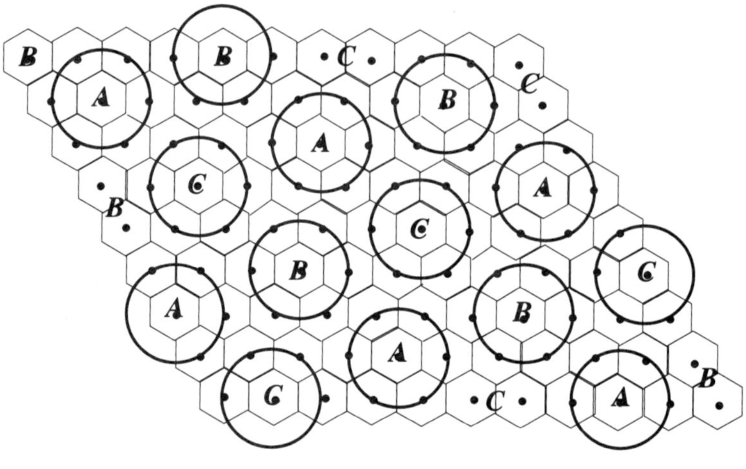

FIGURE 4. A 3-competitive coloring of the graph

4. Conclusions

In this paper we have provided a summary of known algorithms for the frequency assignment problem. We have considered both unit-distance and general versions. A coloring algorithm that is three-competitive for the general interference graph has been given. Based on this coloring scheme we have developed many dynamic carrier allocation algorithms.

References

[1] X. Dong and T. H. Lai, Dynamic Carrier Allocation Strategies for Mobile Cellular Networks, Technical Report OSU-CISRC-10/96-TR48, Dept. of Computer and Information Science, The Ohio State University, 1997.

[2] J. Janssen, K. Kilakos, and O. Marcotte, Fixed Preference Frequency Allocation for Cellular Telephone Systems, unpublished manuscript, April 1995.

[3] J. Janssen, D. Krizanc, L. Narayanan, and S. Shende, Distributed Online Frequency Assignment in Cellular Networks, Manuscript, 1997.

[4] S. Nanda and D. Goodman, Dynamic Resource Acquisition: Distributed Carrier Allocation for TDMA Cellular Systems, *IEEE Globecom*, 1991, pp. 883-889.

[5] L. Narayanan and S. Shende, Static Frequency Assignment in Cellular Networks, Technical Report, Dept. of Computer Science, Concordia University, 1997.

[6] C. McDiarmid and B. Reed, Channel Assignment and Weighted Colouring, manuscript, 1997.

[7] M. Zhang and T.-S. Yum, Comparisons of Channel-Assignment Strategies in Cellular Mobile Telephone Systems, *IEEE Transactions on Vehicular Technology* 38(4), 1989, pp. 211-215.

DEPT. OF CISE, UNIV. OF FLORIDA
E-mail address: raj@cise.ufl.edu

DEPT. OF CS, UNIV. OF AIZU

DEPT. OF CS, FORDHAM UNIVERSITY, NEW YORK, NY

DIMACS Series in Discrete Mathematics
and Theoretical Computer Science
Volume **52**, 2000

An Agent-Based Architecture for Securing Mobile IP

X. Yi, S. Kitazawa, H.Sakazaki, E. Okamoto and D.Frank Hsu

ABSTRACT. Mobile IP has been designed within the IETF to server the needs of rapid growth of mobile computer users who wish to connect the Internet and maintain communications as they move from place to place. Mobile IP has many security issues, which are authentication, privacy, replay detection and key management. In this paper, by application of intelligent agent technology, we present a new architecture for securing mobile IP, which can provide (1) the mutual authentication among mobile user, home agent and foreign agent; (2) key distribution; (3) privacy protection for mobile user and home agent; (4) protection against replaying attack.

1. Introduction

The proliferation of powerful notebook computers and wireless communications promises to provide users with network access at any time and in any location. This continuous connectivity will allow users to be quickly notified of changing events and provide them with the resources necessary to respond to them even when in transit. Laptop computer should no longer be considered the poor cousins of workstations or even mainframe, but should be thought of instead as another choice in wide spectrum of available computer resource.

Unfortunately, present internetworking protocols such as TCP/IP, IPX, and Appletalk behave awkwardly when dealing with host migration between networks. Current versions of the Internet Protocol, or IP, make an implicit assumption that the point at which a computer attaches to the Internet is fixed and its IP address identifies the network to which it is attached. Datagrams are sent to a computer based on the location information contained in its IP address.

If a mobile computer, or mobile host, moves to a new network while keeping its IP address unchanged, its address will not reflect the new point of attachment. Consequently, existing routing protocols will be unable to route datagrams to it correctly. In this situation, the mobile host must be reconfigured with a different IP address representative of its new location. Not only is this process cumbersome for ordinary users, but it also presents the problem of informing potential correspondents of the new address. Furthermore, changing the IP address will cause already-established transport layer connections to be lost. Simply, under the current Internet Protocol, if the mobile host moves without changing its address, it will lose routing; but if it does change its address, it will lose connections.

Mobile IP [**1**] has been designed within the Internet Engineering Task Force (IETF) to server the needs of rapid growth of mobile computer users who wish

2000 *Mathematics Subject Classification.* Primary 68-06.

to connect the Internet and maintain communications as they move from place to place. Mobile IP enables a mobile nodes to change its attachment point on the Internet which maintaining it IP address as well as its network connectivity using this IP address. The protocol permits mobile internetworking to be done on the network layer, however, it also introduces new vulnerabilities to the global Internet.

The IETF's IP Security (IPsec) Working Group is developing standards for IP-layer security mechanisms ([**2**] [**3**] [**4**] [**5**] [**6**]). The group is also developing generic key management protocols for use in the Internet. The current IPSec standards include 3 algorithm-independent base specifications which are currently standards-track RFCs. These 3 RFCs are in the process of being revised (per usual IETF procedures) and the revisions will take into account a number of security issues.

In this paper, we propose a new solution for securing mobile IP. This solution introduces intelligent agent technology to design an architecture for securing mobile IP. The proposed agent-based architecture can provide (1) the mutual authentication among mobile user, home agent and foreign agent; (2) key distribution; (3) privacy protection for mobile user and home agent; (4) protection against replaying attack.

2. Overview of Mobile IP and Security Requirements of Mobile IP

2.1. Overview of Mobile IP. Mobile IP is an enhancement to IP which allows a computer to roam freely on the Internet while still maintaining the same IP address. IETF is currently developing a Mobile-IP standard which, at the time of this writing, is in its fifteenth revision. The Mobile-IP architecture, as proposed by the IETF, defines special entities called the Home Agent (HA) and Foreign Agent (FA) which cooperate to allow a Mobile Host (MH) to move without changing its IP address.

Each Mobile Host is associated with a unique home network as indicated by its permanent IP address. Normal IP routing always delivers packets meant for the MH to this network. When a MH is away, a specially designated computer on this network, its Home Agent, is responsible for intercepting and forwarding its packets.

The MH uses a special registration protocol to keep its HA informed about its current location. Whenever a MH moves from its home network to a foreign network, or from one foreign network to another, it chooses a Foreign Agent on the new network and uses it to forward a registration message to its HA.

After a successful registration, packets arriving for the MH on its home network are encapsulated by its HA and sent to its FA. Encapsulation refers to the process of enclosing the original datagram as data inside another datagram with a new IP header. This is similar to the post office affixing a new address label over an older label when forwarding mail for a recipient who has moved. The source and destination address fields in the outer header correspond to the HA and FA, respectively. This mechanism is also called tunneling since intermediate routers remain oblivious of the original inner IP header. In the absence of this encapsulation, intermediate routers will simply return packets back to the home network. On receiving the encapsulated datagram, the FA strips off the outer header and delivers the newly exposed datagram to the appropriate visiting MH on its local network.

As a consequence, mobile IP can be thought of as the three major cooperative subsystems. Firstly, there is a discovery mechanism defined so that the mobile computer can determine the new attachment points (new IP address) as they move from place to place within the Internet. Secondly, once the mobile computer knows the IP address as its new attachment point, it registers with the home agent which represents it at its home network. Lastly, mobile IP defines simple mechanism to deliver datagrams to the mobile host when it is away from its home network.

2.2. Security Requirements of Mobile IP. Security requirements of mobile IP should be considered from two perspectives: (1) the expectation of the foreign networks to prevent from the illegal mobile hosts while they are visited by the mobile hosts; (2) the expectation of the mobile hosts to protect their communication when they visit the foreign networks.

The strong authentication of registration messages in both basic and route optimized Mobile IP is a crucial step to ensure correct and persistent IP connectivity for the mobile host. From the economic benefit point of view, the foreign agent should authenticate the mobile host by inquiring the home agent of the mobile host during mobile IP registration of the mobile host. In addition, the home agent also should verify whether the registration request comes from its mobile host in case of providing free service to an illegal user while undertaking its network usage expense.

Once mobile IP registration of a mobile host is successful, the home agent and the foreign agent should consider the following confidentiality of traffic:

1. Confidentiality of traffic between the mobile host and its home agent.
2. Confidentiality of traffic between the mobile host and the foreign agent.
3. Confidentiality of traffic between the home agent and the foreign agent.

3. Description of the Agent-Based Architecture for Securing Mobile IP

3.1. Overview of the Proposed Architecture. The agent-based architecture intends to deal with security issues in the following situation:

A mobile host has moved from its home network to a foreign network and wants to register the foreign network. It has got access to the foreign network and have received broadcast information about a foreign agent in the foreign network. The mobile host firstly generates an intelligent agent with its registration request and then motivates it.

The intelligent agent automatically roams from the mobile host into the foreign agent. In the foreign agent, the intelligent agent submits the registration request to the foreign agent and asks for the foreign agent's information, then carries this information and further roams from the foreign agent into its home agent.

After the intelligent agent enters the home agent, submits the registration request and information of the foreign agent to the home agent and asks for the verification reply from the home agent. Based on the request, the home agent will generate another new intelligent agent. The new intelligent agent brings the verification reply and randomly selected shared secret keys among the mobile host, the home agent and the foreign agent back to the foreign agent.

The new intelligent agent submits the verification reply and shared secret key between the foreign agent and the home agent to the foreign agent and asks for registration reply and the shared secret key between the foreign agent and the mobile host from the foreign agent. Finally, the new intelligent agent brings this information back to the mobile host.

The whole architecture for securing mobile IP can be illustrated in the following figure:

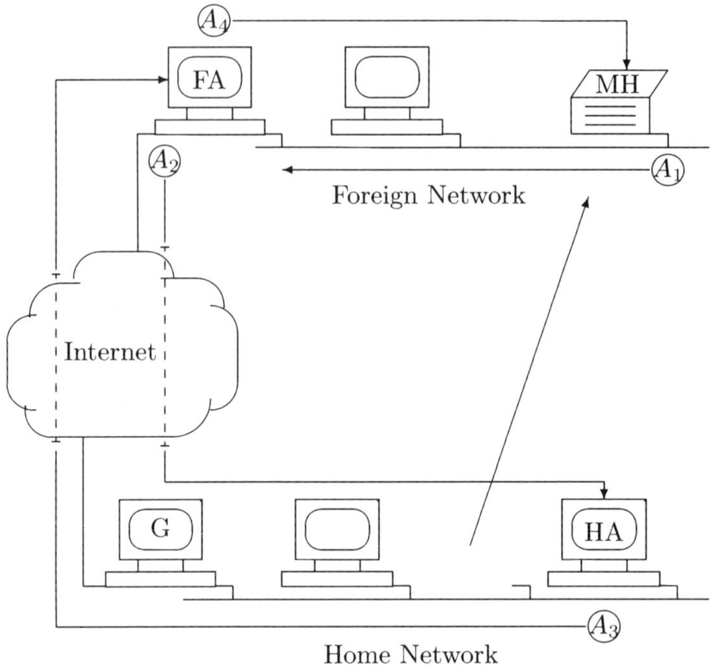

Fig.1 The agent-based architecture for securing mobile IP

In the following section, the procedures of securing mobile IP will be dealt with in detail.

3.2. Certificate and Signature Verification. We assume that the network involves a trusted certificate authority (CA) which provides participants of the network, including mobile hosts, home agents and foreign agents, with public key certificate service as follows:

1. As the Digital Signature Standard (DSS) [**7**], the certificate authority chooses three parameters (p, q, g), where p is a large prime, q is a large prime factor of $p-1$, $g = h^{(p-1)/q}\ (mod\ p)$ with h being an integer satisfying $1 < h < p-1$ and $h^{(p-1)/q}\ (mod\ p) > 1$.
2. Each participant in the network is required to generate a pair of signature public-secret keys. The pair of signature public-secret keys of the certificate authority is (y_{ca}, x_{ca}), where $y_{ca} = g^{x_{ca}}$ (mod p) and x_{ca} is a secret key chosen randomly from $GF(q)^*$. In the similar way, the pair of signature public-secret keys of a mobile host mh, denoted by (y_{mh}, x_{mh}), that of a home agent ha, denoted by (y_{ha}, x_{ha}), and that of a foreign agent fa, denoted by (y_{fa}, x_{fa}),
3. The certificate message is hashed. For example, as to a mobile host, we adopt the certificate format of X.509 [**8**] to construct C_{mh} which may contain such information as certificate serial number, validity period, the ID of mh, the public key (y_{mh}) of mh, the ID of CA, the public key (y_{ca}) of CA, etc, and then compute out the hash value $h(C_{mh})$ of C_{mh} by use of an one-way 2n-bit hash function which is based on a n-bit block cipher with 2n-bit key (such as IDEA [**9**]) and shown in the following figure:

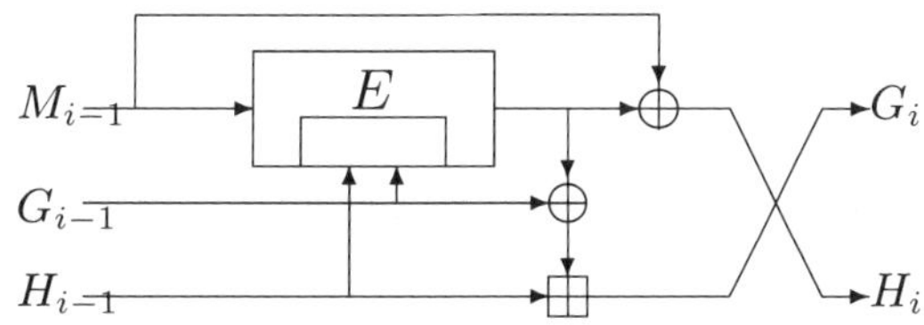

Fig.2 Computational graph for the hash round function h

In Fig.2, M_i, G_i and H_i are n-bit integers; E denotes a n-bit block cipher using 2n-bit key; $\bigoplus$ presents bitwise exclusive-or of n-bit blocks while $\boxed{+}$ indicates addition modulo 2^n of n-bit integers. The detail hash process can be seen in literature [**10**].

4. The certificate authority creates certificates for all participants. The digital signature of the certificate authority on a message C is composed of two numbers s and t which are defined as

$$s = g^r \ (mod\ p)\ (mod\ q) \tag{3.1}$$

$$t = -s \cdot x - h(C) \cdot r \ (mod\ q) \tag{3.2}$$

where r is a random number chosen from $GF(q)^*$.

Given (C^*, s^*, t^*), one can verify whether (s^*, t^*) is indeed a genuine signature of the certificate authority on C^* only by checking the following equation:

$$g^{t^*} \cdot y_{ca}^{s^*} \cdot s^{*h(C^*)} \ (mod\ p) = 1 \tag{3.3}$$

One can accept the digital signature of the certificate authority on the message C^* if the above equation holds.

3.3. Intelligent Agent Applications and its Structure. An intelligent agent can be defined as a software element (program, procedure, object, thread, etc.), owned by a user or another software element, capable of migrating, from one computer to another, to execute a set of tasks on behalf of its owner. Intelligent agents are said to be autonomous, in the sense that they can take their own decisions while away from their host. This implies that an intelligent agent is not just a piece of data being transferred between systems, but may also carry some code and state, which enables it to perform parts of its tasks in one system, migrate to another and continue its work there.

Intelligent agent technology has received growing interest from the research community and has matured significantly in the last few years [**11**] [**12**], however, the number of application using this technology is still scarce. In this paper, we try to apply the intelligent agent technology to fulfil the tasks of mobile IP registration and key distribution in a mobile IP protocol.

In the proposed architecture, the basic structure of an intelligent agent through Internet can be divided into seven distinct portions as follows:

1. Descriptor – the illustration of the intelligent agent's structure and format.
2. Certificate – the certificate of the intelligent agent's owner.
3. Code and state – the program executed in an agent execution environment to fulfil certain mission and the executed state.
4. Datagram – the information for receiver.
5. Time stamp – the time when the intelligent agent launches from the sender.
6. Signature – the signature of the intelligent agent's sender.
7. Attachment – the information attached by the foreign agent.

The structure of an intelligent agent can be illustrated in the following figure:

Descriptor	Certificate	Code / State	Datagram	Time Stamp	Signature	Attachment

Fig.3 The structure of an intelligent agent

3.4. Flow of the Proposed Architecture.

3.4.1. *Foreign agent discovery.* Foreign agent advertises their availability on each link for which they provide service. The process of detecting a foreign agent is quite similar to that used by Internet nodes to detect routers running Internet Control Message Protocol (ICMP) Router Discovery (RFC 1256). The basic operation involved periodic broadcasts of advertisements by the router onto their directly attached network.

We assume that a mobile host has got access to the foreign network and has received the advertisement of a foreign agent which indicates the information of the foreign agent, such as location of the foreign agent. Now the mobile host wants to perform mobile IP registration.

3.4.2. *Intelligent agent generation.* The mobile host generates an intelligent agent on basis of the structure of intelligent agent as shown in Fig.3. In this time, the certificate portion is occupied by the mobile host's certificate issued by the certificate authority. As described in section 3.2, the certificate is composed of certificate message C_{mh} (based on the certificate format of X.509) and a pair of parameter (s_{mh}, t_{mh}) which satisfy equation (3.1) and (3.2). It should be noticed that the signature public key of the mobile host (y_{mh}) is included in the certificate message.

The code consists of two parts, the first part will be executed in the foreign agent, the second part will run in the home agent.

The datagram is filled by the registration request which at least indicates the locations of the foreign agent and the home agent.

The signature portion is put in by the signature of the mobile host on the portion $(M_{mh_{req}})$ from descriptor to time stamp of the intelligent agent. The procedure of signature is as follows (same as in the issue of certificate):

$$s_{mh_{req}} = g^r \ (mod\ p)\ (mod\ q) \tag{3.4}$$

$$t_{mh_{req}} = -s_{mh_{req}} \cdot x_{mh} - h(M_{mh_{req}}) \cdot r \cdot \ (mod\ q) \tag{3.5}$$

where r is a random number chosen from $GF(q)^*$.

Given $(M^*_{mh_{req}}, s^*_{mh_{req}}, t^*_{mh_{req}})$, one can verify whether $(s^*_{mh_{req}}, t^*_{mh_{req}})$ is indeed a genuine signature of the mobile host on $M^*_{mh_{req}}$ only by checking the following equation:

$$g^{t^*_{mh_{req}}} \cdot y_{mh}^{s^*_{mh_{req}}} \cdot s^{*\ h(M^*_{mh_{req}})}_{mh_{req}} \ (mod\ p) = 1 \tag{3.6}$$

3.4.3. *Procedure of the Intelligent Agent Roaming Network.* The procedure of the intelligent agent roaming network can be illustrated in the following four distinct stages:

1. Mobile host $\Longrightarrow$ Foreign agent (it denotes the intelligent agent roams from the mobile host to the foreign agent)

Based on the location of the foreign agent, the intelligent agent roams from the mobile host into the foreign agent.

As soon as the intelligent agent enters the foreign agent, the foreign agent firstly performs verification procedure based on the certificate and signature of the intelligent agent by equations (3.3) and (3.6). If successful, the foreign agent supplies an agent execution environment for the intelligent agent to run.

In the environment, the intelligent agent submits the registration request to the foreign agent and asks it to provide its certificate, verification request to the home agent ($M_{fa_{req}}$) which at least involves the mobile host, the foreign agent and the current time, and its signature on ($M_{fa_{req}}$), i.e.,

$$s_{fa_{req}} = g^r \ (mod\ p)\ (mod\ q) \tag{3.7}$$

$$t_{fa_{req}} = -s_{fa_{req}} \cdot x_{fa} - h(M_{fa_{req}}) \cdot r \ (mod\ q) \tag{3.8}$$

where r is a random number chosen from $GF(q)^*$.

One can verify whether $(s^*_{fa_{req}}, t^*_{fa_{req}})$ is indeed a genuine signature of the foreign agent on $M^*_{fa_{req}}$ only by checking the following equation:

$$g^{t^*_{fa_{req}}} \cdot y_{fa}^{s^*_{fa_{req}}} \cdot s^*_{fa_{req}}{}^{h(M^*_{fa_{req}})} \ (mod\ p) = 1 \tag{3.9}$$

Then the above information from the foreign agent is attached behind the intelligent agent and brought into the home agent by the intelligent agent.

2. Foreign agent $\Longrightarrow$ Home agent

Before the home agent supplies an agent execution environment for the intelligent agent to run, it will verifies two signatures in the intelligent agent. One is signed by the mobile host. It can be checked by equation (3.6). Another is signed by the foreign agent. It can be verified by equation (3.9). If successful, the intelligent agent can reside at the agent execution environment and submits the registration request and the verification request to the home agent and asks it to form a new intelligent agent with the verification reply to the foreign agent.

According to the requirements from the intelligent agent and the structure of intelligent agent in Fig.3, the home agent generates another new intelligent agent. Now the certificate portion is replaced by the certificate of the home agent; the code is updated and specified to run at the foreign agent; the datagram is put in with the verification reply and two encapsulated secret keys which take the following forms respectively:

$$y_{fa}^{-t \cdot x_{ha}} \cdot K_1 \ (mod\ p) \tag{3.10}$$

$$y_{mh}^{-t \cdot x_{ha}} \cdot K_2 \ (mod\ p) \tag{3.11}$$

where K_1 and K_2 are 2n-bit random numbers which will act as the shared secret key of block cipher between the home agent and the foreign agent and the shared secret key of block cipher between the home agent and the mobile host respectively, t is the current time (i.e., time stamp of the new intelligent agent).

Finally, the home agent signs the portion ($M_{ha_{rep}}$) from the descriptor, to the time stamp of the new intelligent agent in the following way:

$$s_{ha_{rep}} = g^r \ (mod\ p)\ (mod\ q) \quad (3.12)$$

$$t_{ha_{rep}} = -s_{ha_{rep}} \cdot x_{ha} - h(M_{ha_{rep}}) \cdot r\ (mod\ q) \quad (3.13)$$

where r is a random number chosen from $GF(q)^*$.

One can verify whether $(s^*_{ha_{rep}}, t^*_{ha_{rep}})$ is indeed a genuine signature of the home agent on $M^*_{ha_{rep}}$ only by checking the following equation:

$$g^{t^*_{ha_{req}}} \cdot y_{ha}^{s^*_{ha_{rep}}} \cdot s^*_{ha_{rep}}{}^{h(M^*_{ha_{rep}})}\ (mod\ p) = 1 \quad (3.14)$$

3. Home agent $\Longrightarrow$ Foreign agent

   The new intelligent agent roams back to the foreign agent again. After verifying the intelligent agent by equation (3.14), the foreign agent supplies an agent execution environment for the new intelligent agent to run again. In the environment, it submits the verification reply and the encapsulated shared secret key (3.10) to the foreign agent. From (3.10), the foreign can elicit K_1 in the following way:

$$(y_{ha})^{t \cdot x_{fa}} \cdot (y_{fa}^{-t \cdot x_{ha}} \cdot K_1)\ (mod\ p) = K_1 \quad (3.15)$$

   where t is the time stamp in the intelligent agent.

   Then the new intelligent agent asks the foreign agent to provide its certificate, registration reply, the encapsulated shared secret key between the mobile host and the foreign agent, and signature on the above information (denoted as $M_{fa_{rep}}$).

   The registration reply should specify a care-of address in the foreign network to the mobile host.

   The encapsulated secret key takes the following form:

$$y_{mh}^{-t \cdot x_{fa}} \cdot K_3\ (mod\ p) \quad (3.16)$$

   where K_3 is a 2n-bit random number which will act as the shared secret key of block cipher between the foreign agent and the mobile host, t is the time stamp of the intelligent agent.

   The signature of the foreign agent on $M_{fa_{rep}}$ is:

$$s_{fa_{rep}} = g^r \ (mod\ p)\ (mod\ q) \quad (3.17)$$

$$t_{fa_{rep}} = -s_{fa_{rep}} \cdot x_{fa} - h(M_{fa_{rep}}) \cdot r\ (mod\ q) \quad (3.18)$$

   where r is a random number chosen from $GF(q)^*$.

   One can verify whether $(s^*_{fa_{rep}}, t^*_{fa_{rep}})$ is indeed a genuine signature of the foreign agent on $M^*_{fa_{rep}}$ only by checking the following equation:

$$g^{t^*_{fa_{req}}} \cdot y_{fa}^{s^*_{fa_{rep}}} \cdot s^*_{fa_{rep}}{}^{h(M^*_{fa_{rep}})}\ (mod\ p) = 1 \quad (3.19)$$

   The new intelligent agent put all the information provided by the foreign agent in its attachment and roams back to the mobile host.

4. Foreign agent $\Longrightarrow$ Mobile host

   After receiving the new intelligent agent, the mobile host will verify the signature of the home agent on the new intelligent agent by equation (3.14) and the signature of the foreign agent on the registration reply by equation (3.19). Then it elicits the secret keys K_2, K_3 in the following way:

$$(y_{ha})^{t \cdot x_{mh}} \cdot (y_{mh}^{-t \cdot x_{ha}} \cdot K_2) \ (mod\ p) \quad = \quad K_2 \tag{3.20}$$

$$(y_{fa})^{t \cdot x_{mh}} \cdot (y_{mh}^{-t \cdot x_{fa}} \cdot K_3) \ (mod\ p) \quad = \quad K_3 \tag{3.21}$$

Finally, The secret tunnels between the home agent and the foreign agent, between the home agent and the mobile host, and between the foreign agent and the mobile host are established with the shared secret keys K_1, K_2, K_3 of block cipher respectively. It is noticed that we suggest to adopt the same block cipher to encrypt information as the underlying block cipher in Fig.2 to save storage space.

4. Security Analysis of the Architecture for Securing Mobile IP

4.1. Security of the Signature Scheme in the Architecture. We now consider that a signer signs a message by adopting the signature scheme in the proposed architecture. For example, the mobile host signs $M_{mh_{req}}$ by generating a pair $(s_{mh_{req}}, t_{mh_{req}})$ according to equation (3.4) and (3.5). Because of the participation of the mobile host's secret key in the formation of the digital signature $(s_{mh_{req}}, t_{mh_{req}})$ of the mobile host on $M_{mh_{req}}$, an attacker can not directly use the equations (3.4) and (3.5) to forge a valid signature pair $(s_{mh_{req}}, t_{mh_{req}})$.

A direct method by which an attacker may try is choosing a random number $s_{mh_{req}}$ (or $t_{mh_{req}}$) and then solving out $t_{mh_{req}}$ (or $s_{mh_{req}}$) from the equation (3.6) to forge a valid pair $(s_{mh_{req}}, t_{mh_{req}})$. However, the difficulty is at least equal to that of computing discret logarithm over finite fields.

In view of the above analysis, we conclude that:

THEOREM 4.1. *The difficulty of breaking the signature scheme in the proposed architecture is at least equivalent to the difficulty of computing discrete logarithm over finite fields.*

4.2. Security of the Key Distribution Scheme in the Architecture. We now take the key distribution between the home agent and the mobile host as an example. Except the home agent and the mobile host, the others do not know how to compute $y_{ha}^{-t \cdot x_{mh}}$ (or $y_{mh}^{-t \cdot x_{ha}}$) because at least one of the home agent's and the mobile host's signature secret keys is definitely required in the computation. Therefore, the others can not separate the shared secret key K_2 between the home agent and the mobile host from $y_{mh}^{-t \cdot x_{ha}} \cdot K_2 \ (mod\ p)$ like the process in (3.20). The idea is similar to the scheme proposed by El Gamal [**13**], in which message M is sent to an entity A in the form $(g^r, M \cdot g^{r \cdot x_a})$, where r is a random number, x_a and g^{x_a} are the secret and public key of the entity A respectively. It suffices for the sender to know g^{x_a}, whereas A can recover M by computing first $g^{-r \cdot x_a}$. An eavesdropper faces the problem of computing discrete logarithms.

On basis of the above analysis, we conclude that:

THEOREM 4.2. *The difficulty of breaking the key distribution scheme in the proposed architecture is equivalent to the difficulty of breaking the El Gamal scheme.*

4.3. Protection Against Replaying Attack. The adoption of time stamp at each stage of the architecture makes the replaying attack is insignificant.

Firstly, because the existence of time stamp and signature in the structure of an intelligent agent, an eavesdropper can not use an expired intelligent agent to perform the replaying attack. In addition, the time stamp is considered in key distribution such as (3.10). It increases the difficult of breaking the key distribution scheme and makes replaying attack vainness.

4.4. Mutual Authentication. The mobile host can authenticate its home agent in the following two ways:

1. On basis of the certificate, the verification reply, the signature of its home agent in the return intelligent agent, the mobile host can verify whether the above information has truly come from its home agent.
2. The mobile host can also authenticates its home agent by verifying whether it can elicit the true shared secret key K_2 in the process of (3.20). Actually, the mobile host can not ensure whether K_2 is true in the beginning. However, it will be verified by testing whether it can be used to recover the ciphertext to significant plaintext in the later communication between the mobile host and its home agent. If no problem, the mobile host has authenticated its home agent because only the home agent can generate the shared secret key K_2 which can only be elicited by the mobile host.

The foreign agent also can authenticate the home agent in the above same ways. As a consequence, the architecture provides double secure authentications for the mobile host and the foreign agent.

4.5. Protection of Servers against Malicious Intelligent Agents. An intelligent agent is unique in that its code is executed by a server. Thus an executing intelligent agent has automatic access to some of a server resources. With this level of access intelligent agents can mount attacks by propagating viruses, worms and Trojan horses, impersonating other users and mounting denial of service attack. The standard approach to this problem is to reject all unknown code from entry into servers. It is not a viable solution in a mobile agent environment. Both Telescript [**14**] and Safe-Tcl [**15**] offer approaches to solve the problem.

In the proposed architecture, before a server supplies an agent execution environment for a visiting intelligent agent to run its code, the server will verify the signature. Once any problem occurs when the server runs the code on an agent execution environment, the owner of the intelligent agent is probably malicious and will be accused.

4.6. Protection of Agents against Malicious Servers. In order for an intelligent agent to run, it must expose its code and data to the host environment which supplies the means for this agent to run. The host can always scan the intelligent agent for information, alter the agent state and code, even kill the agent. Thus, the intelligent agent is unprotected from the host. Fortunately, it does not affect our architecture very much from the economic benefit point of view. Both the foreign agent and the home agent are reluctant to attack an intelligent agent dedicating to mobile IP registration and key distribution.

Current consensus is that it is computationally impossible to protect mobile agents from malicious hosts. Instead of tackling the problem from a computational (difficult) point of view, current research [**16**] is looking at sociological means of enforcing good host behavior.

5. Performance Analysis of the Architecture for Securing Mobile IP

The proposed architecture has two significant performance characteristics. One is intelligent characteristic. The missions of mobile IP registration and key distribution are automatically completed by the intelligent agent. The intelligent agent automatically roams the network and executes its code in an agent execution environment supplied by a server. The manual interference is reduced to the least. The another characteristic is its mobility. The intelligent agent migrates itself to the server to fulfil its task so that the network transportation load in the architecture has been great reduced because only a few communications are needed.

6. Conclusion

Many of the most important application of intelligent agents will occur in fairly uncontrolled, heterogeneous environment. In this paper, we have proposed an agent-based architecture for securing mobile IP, which can provide (1) the mutual authentication among mobile host, home agent and foreign agent; (2) key distribution; (3) privacy protection for mobile user and home agent; (4) protection against replaying attack.

The individual properties of the proposed agent-based architecture for securing mobile IP can be concluded as follows:

1. The mission of mobile IP registration and key distribution are completed by intelligent agents;
2. A new signature scheme suitable for mobile hosts is adopted in the architecture; The difficult of breaking the signature scheme is equivalent to that of computing discrete logarithm.
3. A new key distribution is used in the architecture. The difficulty of breaking the key distribution scheme in the proposed architecture is equivalent to the difficulty of breaking the El Gamal scheme.
4. Only a block cipher is used to act as both encryption algorithm and underlying block cipher of the hash function.
5. In view of the existence of signature in each stage of intelligent agent, non-repudiation of origin guarantees a host to falsely deny having sent a message. So any malicious action will be detected and the breeder will be captured.

Our further works will focus on optimal improvement and then implementation of the agent-based architecture for securing mobile IP. The authors will be grateful to receive any suggestion to the architecture.

Acknowledge

We would like to appreciate Dr.Tada for the helpful discussion.

References

[1] C. Perkins (ed.) *IP mobility support*, RFC2002, proposed standard. IFTF mobile IP working Group, Oct., 1996.

[2] R. Atkinson, *Security architecture for the Internet protocol*, request for comments (proposed standard) RFC 1825, IFTF mobile IP working group, August 1995.

[3] R. Atkinson, *IP authentication header*, request for comments (proposed standard) RFC 1826, IFTF mobile IP working group, August 1995.

[4] R. Atkinson, *IP encapsulation security payload (ESP)*, request for comments (proposed standard) RFC 1827, IFTF mobile IP working group, August 1995.

[5] P. Metzger and W. Simpson, *IP authentication using keyed MD5*, request for comments (proposed standard) RFC 1827, IFTF mobile IP working group, August 1995.
[6] P. Metzger, P. Karn and W. Simpson, *The ESP DES-CBC transform*, request for comments (proposed standard) RFC 1827, IFTF mobile IP working group, August 1995.
[7] *The digital signature standard*, Comm. ACM, Vol.35, No.7, pp.36-40, 1992.
[8] ISO/IEC 8696-8 (1993), Information Technology - Open System Interconnection -The Directory: *Authentication Framework*.
[9] X.J.Lai and J.L.Massey, *A proposal for a new block encryption standard*, Advances in Cryptology, Proc. of EUROCRYPT'90, Lecture Notes in Computer Science **473** (1991)389–404.
[10] X.Yi and K.Y.Lam, *Hash function based on block cipher*, IEE Electronics Letters, 33(23), 1997.
[11] D.Chess, C.Harrison and A.Kershenbaum, *Mobile agents: are they a good idea*, Technical Report, March 1995, IBM T.J.Watson Research Center, NY.
[12] D.Chess, B.Grosof, C.Harrison, D.Levine, C.Parris and G.Tsudik, *Itinerant Agents for Mobile Computing*, Technical Report, October 1995, IBM T.J.Watson Research Center, NY.
[13] T.ElGamal, *A public key cryptosystem and a signature scheme based on discrete logarithm*, IEEE Trans. Info. Theory, Vol.IT-31, No.4, pp.468-472, July 1985.
[14] J.White, Telescript Technology: *The foundation of the electronic market place*, General Magic white paper 1995.
[15] N.Borenstein, *EMail with a mind of its own: The Safe-Tcl language for enabled mail*, IFIP WG 65 Conference, Barcelona, May, 1994, North Holland, Arnsterdam, 1994.
[16] L.Rasmusson and S.Janson, *Simulated social control for secure Internet commerce*, In New Security Paradigms'96, ACM Press, September 1996.

JAIST, 1-1 Asahidai, Tatsunokuchi, Nomi, Ishikawa 923-1292, Japan
E-mail address: `xyi, shigeki, sakazaki, okamoto@jaist.ac.jp`

Dept.Computer and Information Science, Fordham University, NY 10023, USA
E-mail address: `hsu@murray.fordham.edu`